Architecting Networked Engineered Systems

Jelena Milisavljevic-Syed · Janet K. Allen ·
Sesh Commuri · Farrokh Mistree

Architecting Networked Engineered Systems

Manufacturing Systems Design for Industry 4.0

Springer

Jelena Milisavljevic-Syed
Systems Realization Laboratory
Division of Industrial Design
University of Liverpool
Liverpool, UK

Sesh Commuri
Department of Electrical
and Biomedical Engineering
University of Nevada
Reno, NV, USA

Janet K. Allen
Systems Realization Laboratory, School
of Industrial and Systems Engineering
University of Oklahoma
Norman, OK, USA

Farrokh Mistree
Systems Realization Laboratory, School
of Aerospace and Mechanical Engineering
University of Oklahoma
Norman, OK, USA

ISBN 978-3-030-38609-2 ISBN 978-3-030-38610-8 (eBook)
https://doi.org/10.1007/978-3-030-38610-8

This Springer imprint is published by the registered company Springer Nature Switzerland AG
The registered company address is: Gewerbestrasse 11, 6330 Cham, Switzerland

Foreword

Global markets, rapidly evolving technologies and changing customer preferences have given rise to a new generation of decentralized distributed manufacturing systems, vertically or horizontally integrated through the industrial Internet of Things (IIoT). They rely on digitization as their backbone and main driver for innovation. Their control systems and underlying processes operate on vast amounts of data, captured by a multitude of sensors to realize new forms of value creation across the global value chain.

Today's businesses need to be able to respond to and fulfil customer requirements, regardless of whether this relates to mass production, mass customization, or even product personalization based on a lot size of 1. Agility is key.

So-called digital twins of manufacturing systems and their underlying control processes are integrated with their physical counterparts, allowing production engineers to analyse vast amounts of data captured by sensors. The results of such big data analytics may be used for process simulation and optimization, and if necessary, in direct response to events, such as progressing tool, predicted machine downtime due to required maintenance, varying energy consumption prizes for production cycles based on the time of day, or capabilities and bottlenecks of the related supply chain networks and its logistics.

In a nutshell, the preceding has become known as Cyber-Physical Production Engineering in the context of Industry 4.0. It is a true paradigm change and a transformative event often referred to as the Fourth Industrial Revolution. Unlike previous Industrial Revolutions, this one is happening at a much more rapid pace, bringing about many challenges.

Perhaps, the biggest of all challenges across the manufacturing sector is the current lack of understanding as to how to systematically and strategically design—or architect—such new cyber-physical production ecosystems in a holistic manner. At present, this is uncharted territory for most, if not all, companies in the industry. Unfortunately, traditional approaches are not suited to the design and analysis of such networked engineered systems. Indeed, we are dealing with a true game changer and it is of utmost importance to get the design right from the outset, and with a holistic Industry 4.0 systems perspective in mind. This is the only way to ensure the benefits of investments made into new technologies, infrastructures, processes and business models, and a new workforce will bear fruit in the long run.

Although many companies may feel the urge to conform Industry 4.0 in a rush, the old saying "You have to go slow in order to go fast." applies. The founder of the Roman Empire, Augustus, would use the Latin phrase "Festina Lente". This translates to: "Make haste, slowly". It served as a reminder for Augustus to perform activities with a proper balance of urgency and diligence.

With this in mind, in their monograph *Architecting Networked Engineered Systems: Manufacturing Systems Design for Industry 4.0*, Dr. Jelena Milisavljevic-Syed and Profs. Janet K. Allen, Sesh Commuri, and Farrokh Mistree address the challenges of architecting networked engineered systems that are anchored in the Industry 4.0 construct.

They present a computational framework, Design for Dynamic Management (DFDM), conforming an architecture for the realization of digitized manufacturing. This framework is designed to facilitate the harmonization of conflicting needs associated with the design of networked engineered systems (NES), such as flexibility, robustness, resilience, operability, connectivity, and the incorporation of real-time analytics for smart sensing.

In the context of Industry 4.0, Design for Dynamic Management takes the design of networked manufacturing systems to a new level. The beauty of their approach is that it is generic in nature and thus not limited to the application domain of manufacturing systems. Rather, it can be applied to the design of networked engineered systems in general, for example the design of chemical plants and processes or the design of smart digital healthcare systems of the future, which all function on the same fundamental principles.

I have thoroughly enjoyed reading the manuscript of this book and can highly recommend it to all those concerned with the conceptualization, design, and realization of next-generation digital/smart manufacturing systems. It is a valuable resource for decision-makers in industry, especially for those tasked with leading their enterprises into the new, digital era.

Likewise, I consider this monograph to be of great value to faculty colleagues and Ph.D. students in the design and manufacturing research and education community.

October 2019

Professor Dirk Schaefer
Chair in Industrial Design
School of Engineering
University of Liverpool
Liverpool, UK

Preface

Manufacturing is driven by the need to meet customers' individual preferences in a timely and cost-effective manner. The integration of smart sensors and networked manufacturing systems has resulted in cyber-physical manufacturing systems that are able to address these preferences on a global scale. The ability to bring together technologies such as Internet of Things (IoT), big data analytics, and machine intelligence with traditional technologies such as smart automation, supply chain, logistics, and cloud computing has resulted in a new wave of advances in manufacturing technology collectively called Industry 4.0. Factories conforming to Industry 4.0 can integrate services across the entire manufacturing process and address dynamic market changes due to variations in customer preferences or disruptions in supply chains/production. The vertical integration of IoT and analytics will make it possible for these factories to optimize supply and logistic networks, implement policies based on predictive behaviour, improve end-to-end throughput, and provide services and products at a lower cost.

Industry 4.0 represents the Fourth Industrial Revolution and provides a framework to address the challenges arising in the integration of cyber-systems and physical resources and cover all aspects of manufacturing including: robust and flexible automation; data collection, analysis, learning, and decision-making; distributed systems; Internet of things (IoT); and supply chain integration. Central to the implementation of Industry 4.0 is the concept of the "digital twin"—a virtual replica of every entity in the manufacturing process—that allows the bridging of the physical and cyber worlds. The digital twin represents a high-fidelity model of the physical processes and can be used to analyse the overall performance of the system, diagnose performance issues, and use data analytics and learning strategies to improve system performance over time.

Design of manufacturing systems in the age of Industry 4.0 requires a new paradigm that considers the distributed and networked aspect of the manufacturing process. In this monograph, we address the challenges in architecting networked engineered systems that are anchored in the Industry 4.0 construct. We present a computational framework, Design for Dynamic Management (DFDM), conforming the architecture for realization of digitized manufacturing. The computational framework, DFDM, is designed to facilitate the harmonization of conflicting requirements in the design of networked engineered systems (NES), such as flexibility, robustness, resilience, operability, connectivity, and the incorporation of real-time analytics and smart sensing—requirements that are necessary for global manufacturing.

The central problem that we address is *designing for dynamic management* that conforms to the Industry 4.0 construct. To address this issue, we propose architecting networked engineered systems, or in other words, creating an architecture for implementing "smart" dynamic management in the realization of digitized manufacturing. We adopt the definition proposed by Perry and Wolf (1992): an architecture consists of "elements, form, and rationale". In the context of an architecture for networked engineered systems, the elements, form, and rationale furnish answers to the *what, how*, and *why* questions, respectively, for dynamic management. Starting with the elements, we identify six different components or methods for dynamic management as shown in Fig. 1, namely *flexibility, robustness, resilience, connectivity, smart sensing, and real-time analytics*. We express the form of our architecture that is compatible with the notion of a *smart platform* that embodies mathematical constructs that capture the weights and relationships between the different elements. The elements and their interrelationships are captured using properties like cost and service level.

We summarize the constituents of the architecture for a *smart platform* and our interpretation in Table 1, which represents a single view catering to multiple users.

This monograph comprises seven chapters (Fig. 2). While we recommend reading the chapters sequentially, each chapter is self-contained, and chapters can be read independently or in any preferred sequence; we suggest reading Chaps. 1 and 7 first. Following this introductory chapter, we discuss overview of the computational framework for the realization of dynamic management in networked engineered systems in Chap. 2. In Chaps. 3–6, we cover elements of the networked manufacturing system. In Chap. 7, the state of the art in implementing Industry 4.0 construct is examined and the technical challenges arising from this rapid evolution of technology are discussed. The digital transformation of the manufacturing industry and its impact on the workforce and society is elucidated. We conclude by

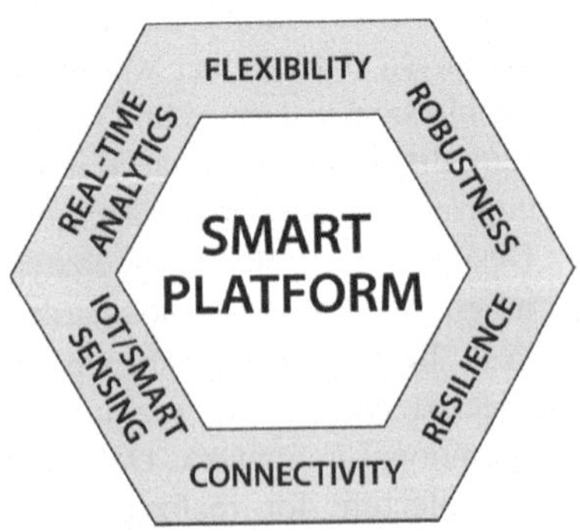

Fig. 1 Elements for realization of networked engineered systems

Table 1 Summary of architecture for networked engineered systems

Elements	Form	Rationale	Chapters
Flexibility	Observe	Identify errors in the system	Chapter 3
	Detect	Identify cause of errors	
	Mitigate	Reduce the effect of errors	
Robustness	Observe	Identify uncertainty	Chapter 3
	Recognize	Recognize type of uncertainty	
	Manage	Manage uncertainty	
Resilience	Withstand and recover	Withstand disruption with minimum loss and quickly recover afterwards	
Operability	Steady-state operability	Analysing functionality of the system in the presence of disruption	Chapter 4
	Dynamic-state operability	Analysing dynamic performance of the system in the presence of disruption	Chapter 5
Connectivity	Connectivity in the production line	Examine tools and operations within the machining station	Chapter 6
	Connectivity of the inspection system	Examine sensors and operations within the inspection station	
	Connectivity of the manufacturing system	Examine connectivity between elements in the system, machining and inspection station connection	
Real-time analytics	Data and information analysis	Analysis of data from low-level process and information from high-level process	Chapter 7
Cyber-security	Protection of data and information	Protection of systems from theft or damage to data, from disruption or misdirection of the services they provide	Chapter 7

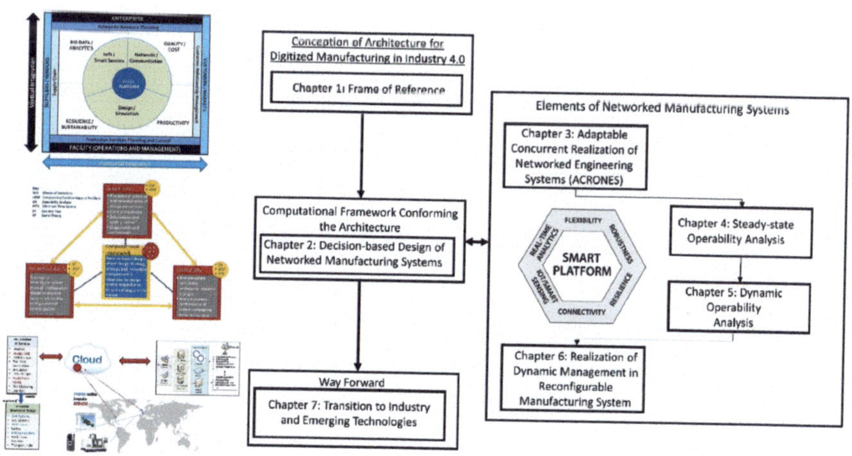

Fig. 2 Organization of the monograph

framing the questions to foster discussion in the larger community on the roles and responsibilities of individuals, companies, and governments in promoting sustainable and equitable growth of manufacturing industry. We suggest how the framework and techniques in this monograph are applicable beyond the manufacturing domain for architecting networked engineered systems in other industries that are being transformed through the adoption of the Industry 4.0 construct.

In closing, we observe that the material in this monograph is anchored in the doctoral dissertation of Jelena Milisavljevic-Syed whom we were privileged to mentor at the University of Oklahoma, Norman, USA. Funds from the LA Comp Chair account, the John and Mary Moore Chair account, and the Tata Consultancy Services, India, were used to support Milisavljevic-Syed through her doctoral studies in the School of Aerospace and Mechanical Engineering at the University of Oklahoma, Norman, USA.

Liverpool, UK
Norman, OK, USA
Reno, NV, USA
Norman, OK, USA
October 2019

Jelena Milisavljevic-Syed
Janet K. Allen
Sesh Commuri
Farrokh Mistree

Reference

Perry, D. E., & Wolf, A. L. (1992). Foundations for the study of software architecture. *ACM SIGSOFT Software engineering notes,* 4(17), 40–52.

Contents

List of Figures

List of Tables

Chapter 1
What is this Monograph About?

The Industry 4.0 construct is characterized by a digital model of end-to-end supply chain, all the manufacturing processes and provides a mechanism to transfer autonomy from the physical realm to the cyber-physical realm. Central to the implementation of the Industry 4.0 construct is the concept of the 'Digital Twin'. The Digital Twin is a virtual replica of every entity in the manufacturing process and allows the bridging of the Physical and Cyber worlds. Such a bridge allows for seamless data exchange in real-time between the physical world and the virtual representations. Since the Digital Twin represents a high-fidelity model of the physical processes, it can be used to analyze the overall performance of the system, diagnose performance issues, and use data analytics and learning strategies to improve system performance over time.

The architecting of networked engineered systems that conforms to the Industry 4.0 construct is the subject of this monograph. In this monograph we describe architecting this network in terms of robustness, flexibility, resilience cybersecurity, and real-time analytics. A Design for Dynamic Management (DFDM) computational framework is proposed to design systems conforming to this architecture. Methods for theoretical and structural validation of the design process are discussed using illustrative examples from the manufacturing industry. The state of the art in implementing Industry 4.0 construct is examined and the technical challenges arising from this rapid evolution of technology are discussed. The digital transformation of the manufacturing industry and its impact on the workforce and society is elucidated. The monograph is concluded by framing the questions to foster discussion in the larger community on the roles and responsibilities of individuals, companies, and governments in promoting sustainable and equitable growth of manufacturing industry.

J. Milisavljevic-Syed et al., *Architecting Networked Engineered Systems*,
https://doi.org/10.1007/978-3-030-38610-8_1

1.1 Architecting Smart Networked Manufacturing Systems

In global markets characterized by dynamic market changes necessitated by changes in customer preferences timely adjustments are required in manufacturing to meet these changing demands. Conventional manufacturing processes are designed for mass manufacturing and are not suited for agile, flexible and highly reconfigurable smart manufacturing. Foundational to the Industry 4.0 construct is digitization that should be harnessed for smart manufacturing. Hence, we propose architecting smart Networked Manufacturing Systems (NMS) that embody Design for Dynamic Management (DFDM). The DFDM Framework is a service-oriented product/system development computational framework that facilitates the harmonization of conflicting needs associated with the design and manufacture of engineered systems.

With a view to providing context on architecting smart networked manufacturing systems, in this section, we present the trend in manufacturing, bridging physical and cyber worlds, benefits of Industry 4.0, an architecture for realization of Industry 4.0 as a digital thread and a digital twin, technical requirements for successful design of networked manufacturing systems, addressing the real industrial problem via DFDM the Framework, and way forward and plans for transition to industry.

What is the trend? Bringing together technologies such as Internet of Things (IoTs), Big Data Analytics, Machine Intelligence with conventional technologies such as Smart Automation, Supply Chain, Logistics, and Cloud Computing has resulted in advances in Manufacturing Technology collectively called Industry 4.0 (Xu et al. 2018). Industry 4.0 represents the Fourth Industrial Revolution and provides a conceptual framework for digitization of manufacturing systems.

How to connect Physical and Cyber worlds? Industry 4.0 is characterized by a digital model of an end-to-end supply network and all the manufacturing processes and provides a mechanism to transfer autonomy from the physical realm to the cyber-physical realm (Monostori 2014). Central to the adoption of Industry 4.0 is the concept of the 'Digital Twin'. The Digital Twin is a virtual replica of every entity in the manufacturing process and allows the bridging of the Physical and Cyber worlds. Such a bridge allows for seamless data exchange in real-time between the physical world and the virtual representations. Since the Digital Twin represents a high-fidelity model of the physical processes, it can be used to analyze the overall performance of the physical system, diagnose performance, and invoke data analytics and learning strategies to improve system performance.

What are the benefits? To appreciate the benefits of Industry 4.0 it is essential to realize the full value chain that includes the supply networks, customer and markets. Digitization of processes across the value chain enable new capabilities such as personalization, real-time alerts and interventions, innovative service models, dynamic product improvement, increased productivity, higher up-time and, ultimately, new business models.

Digitization of manufacturing systems that conforms to the Industry 4.0 construct is not a trivial task and can only be realized through the adoption of an appropriate computational architecture; see Fig. 1.1. The architecture must facilitate the integration of digital threads in two ways. The *vertical integration* of a digital thread facilitates the flow of information between the enterprise and resource planning entities to manufacturing services and planning and control entities within a manufacturing facility, see Fig. 1.1. In the vertical integration of low-level components such as IoT sensors, machine controllers and their digital twins across the manufacturing enterprise are networked with big data analytics to monitor the manufacturing process and improve system performance and product quality. Vertical integration also allows for the creation of interoperable 'systems of systems' that can integrate seemingly diverse set of machines, sensors and controllers to produce a resilient manufacturing process that is capable of withstanding disruptions in the supply network, environment and in the market. *Horizontal integration* deals with data and material flows across the end-to-end value chain, see Fig. 1.1. The horizontal integration is represented by several digital twins from supply chain, manufacturing processes, data flows and IT systems in the product development, to logistics, distribution and ultimately the customer. Elements of the architecture are expanded on in the context of digitized manufacturing processes in steel industry and are discussed next.

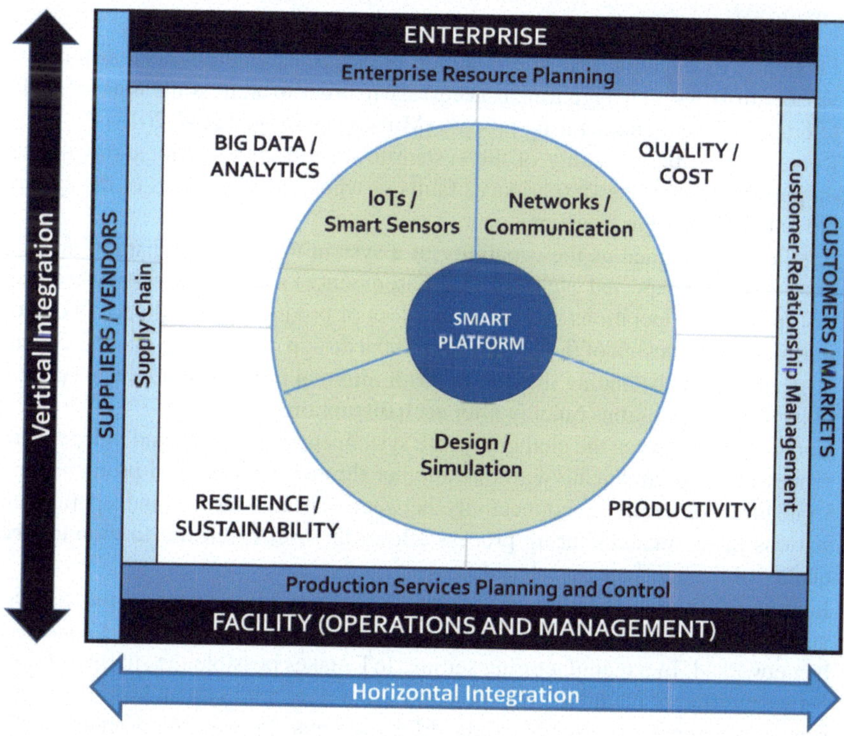

Fig. 1.1 Architecture for realization of digitized manufacturing in industry 4.0 (Milisavljevic-Syed et al. 2019a, b)

Fig. 1.2 Elements for realization of networked engineered systems

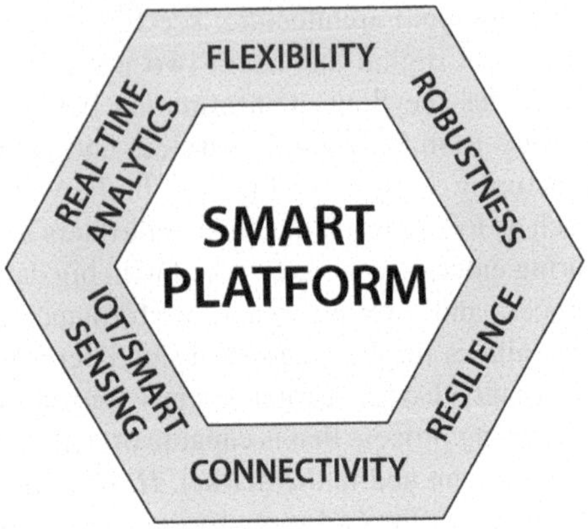

Implementation of systems conforming to the architecture laid out in Fig. 1.1 begins with the specification for the design of smart networked engineered systems, see Fig. 1.2. The technical requirements for networked engineered system are flexibility, robustness, resilience, connectivity, IoT/Smart sensing, and real-time analytics, see Table 1.1.

- Flexibility in selection and determination of design parameters, such as tool and sensor attributes, at design time is necessary in order to address the issue of fault-tolerance in the manufacturing process (Milisavljevic-Syed et al. 2018).
- Robustness is the capability of the system to meet design specifications for the output when the system experiences faults or when the parameters of the system are unknown/change over time.
- Resilience is defined as the capability of a system to withstand disruptions (Gu et al. 2006). Unexpected events in the environment of a manufacturing system can disrupt normal operations and result in a loss of production. Resilience is one of the goals in the design of these systems and the design must ensure that the system is capable of withstanding disruptions with minimal production loss and recover to its normal operations quickly after each disruption.
- Connectivity between the elements in the system enables distinct and independent processes to communicate with one another through well-defined protocols and strategies. For example, connectivity between sensing stations and operational stations in the manufacturing process allows the user/controller to monitor the quality of the part being produced.
- Internet of Things (IoT)/Smart Sensing is a term used to represent the convergence of systems and Internet Technology and allows ‘stand-alone’ systems to be networked. In a manufacturing setting, IoT makes possible the sharing of data between different manufacturing processes as well as with system level processes such as enterprise resource planning (ERP) systems. Networking of manufacturing processes also enables the collection of large volumes of data that can be used to glean information on the performance of the overall manufacturing system and to learn optimal strategies to improve system performance.

Table 1.1 Requirement list for realization of networked engineered systems

Elements	Requirements list	Short definition	Chapters
Flexibility	Observe	Identify errors in the system	Chapter 3
	Detect	Identify cause of errors	
	Mitigate	Reduce the effect of errors	
Robustness	Observe	Identify uncertainty	Chapter 3
	Recognize	Recognize type of uncertainty	
	Manage	Manage uncertainty	
Resilience	Withstand and recover	Withstand disruption with minimum loss and quickly recover afterwards	
Operability	Steady-state operability	Analysing functionality of the system in the presence of disruption	Chapter 4
	Dynamic-state operability	Analysing dynamic performance of the system in the presence of disruption	Chapter 5
Connectivity	Connectivity in the production line	Examine tools and operations within the machining station	Chapter 6
	Connectivity of the inspection system	Examine sensors and operations within the inspection station	
	Connectivity of the manufacturing system	Examine connectivity between elements in the system, machining and inspection stations connection	
Real-time analytics	Data and information analysis	Analysis of data from low-level process and information from high-level process	Chapter 7
Cyber security	Protection of data and information	Protection of systems from theft or damage to data, from disruption or misdirection of the services they provide	Chapter 7

- Big-data Analytics is an evolving area of research that addresses analysis of large volumes of data. Datasets in Big data are characterized by large volumes of data (typically exceeding Terabytes of data), large variety of data in the dataset, and large velocity of incoming data. Data mining and analytical techniques designed for large data can provide statistically better results even when raw source data, with its transactional schema, non-standard data, and poor-quality data is used (Russom 2011). In the context of manufacturing, analysis of low-level data is necessary to detect failures and their causes as well as monitoring the process flow during manufacturing operations. Results from analyzing large volumes of data can be used to learn optimal production strategies.

In the following section we identify key requirements in the context of digitized manufacturing, see Fig. 1.2, for a computational framework for a smart networked manufacturing system, Design for Dynamic Management (DFDM), see Fig. 1.3, conforming the architecture embodied in Fig. 1.4, and accounting for elements of networked engineered system, see Fig. 1.5.

Digitized Manufacturing in the Steel Industry

A steel manufacturing process consists of number of unit operations, such as ladle refining, tundish processing, continuous casting, rolling, heat treatment, etc., where each operation requires proper attention and control (Shukla et al. 2015). A representation of the different operations involved in the production of sheets (product) from slabs (semi-product after casting) is shown in Fig. 1.3. Each of the operations in the process is performed within specified tolerances, in the rows in Fig. 1.3.

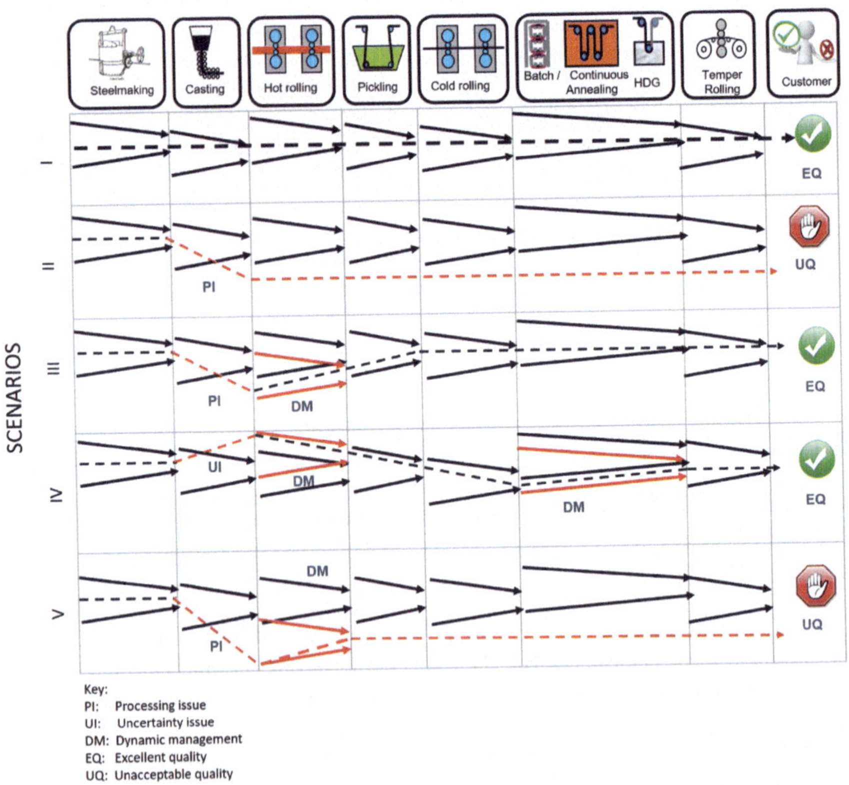

Fig. 1.3 Problem description—design and analysis of steel manufacturing process (Milisavljevic-Syed et al. 2019b)

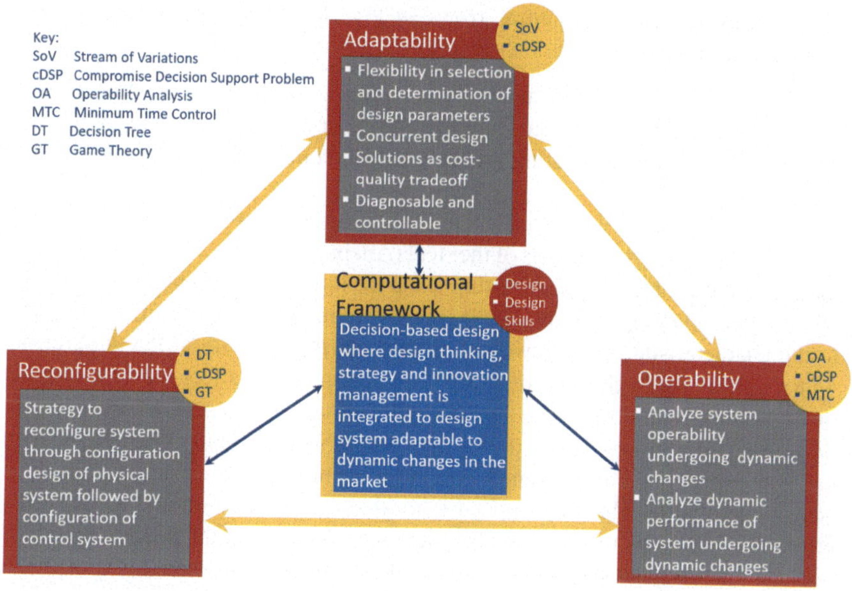

Fig. 1.4 Computational framework, design for dynamic management (Milisavljevic-Syed et al. 2019a)

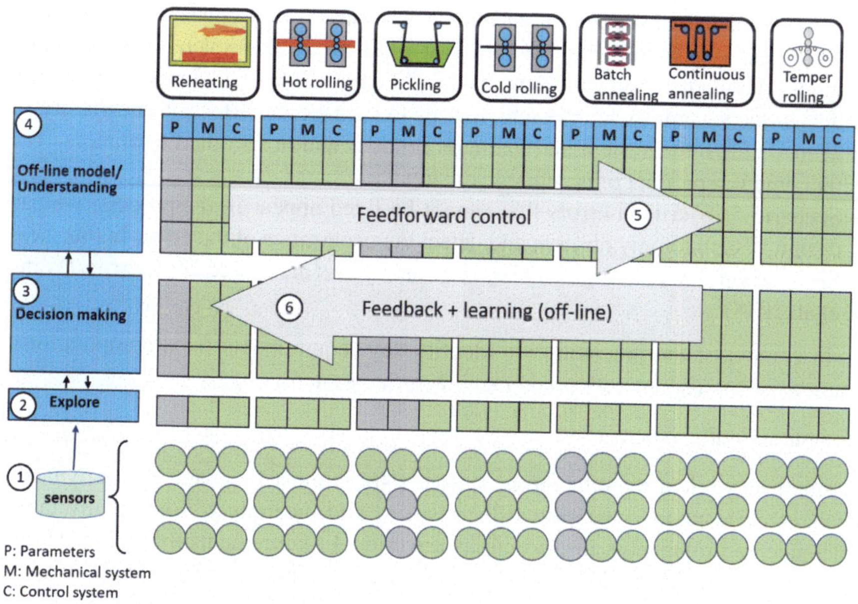

Fig. 1.5 Feedforward dynamic management (Milisavljevic-Syed et al. 2019b)

We describe different scenarios, (rows in Fig. 1.3) that may occur in a steel manufacturing process:

- Scenario 1: If we have an ideal situation, a process without errors where models used to represent the process are complete and accurate, and all data are known then we have an ideal manufacturing process. The outcome is a product of acceptable quality.
- Scenario 2: A lack of thermal and chemical homogeneity results in an error associated with the reduced cleanness of the steel (Anapagaddi et al. 2014). Such an error may appear during continuous casting when the molten metal is in the tundish, the second column in Fig. 1.3, due to the inability to maintain superheat. Further, due to the characteristics of a networked manufacturing system (NMS) these errors will propagate through the remainder of the process and affect the final product quality (Ding et al. 2002). The outcome is a product of unacceptable quality.
- Scenario 3: If a processing error that appears in Stage 1 cannot be fixed then we can sometimes make improvements in subsequent stages to ensure a product of acceptable quality. The way to do it is to identify the location and cause of the error, and take steps to accommodate the error. As a result the process rebounds and continues within specified tolerances. The outcome is a product of acceptable quality.
- Scenario 4: Since computational models that are used to simulate the process are incomplete or inaccurate we must manage the uncertainties embodied as a result in the mathematical control algorithms. Uncertainties may appear due to inherent randomness or unpredictability of a system, or from incomplete information on model parameters and model structure (Allen et al. 2006). Hence, we need to design the system to be operable, that is, to manage the effect of uncertainty so as to stabilize the system and maintain product within specified tolerances. The outcome is a product of acceptable quality.
- Scenario 5: If critical errors that cannot be fixed appear in the process, we can design system to only observe and detect the occurrence of the error. In this case, the decision is to reconfigure the system and bypass the source of the errors (Wang et al. 2018).

In context of the preceding we identify key requirements for a computational framework for a smart networked manufacturing system.

Computational Framework

A smart networked manufacturing system (NMS) must be adaptable to dynamic changes and respond to unexpected disturbances, and uncertainty. Additionally, fault-tolerance and robustness to disruptions are critical requirements that must also be addressed in the design. Maintaining connectivity among elements of the process is one of the key characteristics necessary for these systems to be integrated into the architecture. Ultimately, the long-term viability of the system depends on the capability to glean information from the various data streams in the process and learn in real-time to optimize and manage the process. This information is also necessary for the long-term health and resilience of the system. Current manufacturing systems have been designed for large-scale manufacturing of a product in the most efficient manner. Hence, we advocate the design of smart NMS that is able to adapt to rapid changes in product requirements, system failures that lead to breakdown of interconnectivity between elements of the NMS; events that existing approaches cannot easily handle. Foundational to a smart NMS is a computational framework, Design for Dynamic Management (DFDM), see Fig. 1.4.

DFDM has three critical components: adaptable and concurrent design, operability analysis and reconfiguration strategies (Milisavljevic-Syed et al. 2019a, b). Adaptable and concurrent design methods offer flexibility in selection of design parameters and the concurrent design of the mechanical and control systems (Milisavljevic-Syed et al. 2018). Operability analysis is used to determine the functionality of the system undergoing dynamic change (Milisavljevic et al. 2018). Reconfiguration method allows multiple configurations of elements in the system.

The limitation of the DFDM is that we cannot design processes to mitigate errors when the processing takes place at a station but the effect of the error in the process is only seen in subsequent stations.

Way Forward

We contemplate expanding DFDM to address feedforward dynamic management, Fig. 1.5, where tools and sensors communicate with each other via the Internet of Things (IoT), and sensors data is used to create enriched digital system models (Digital Twin) to achieve process control, advanced measurements, perfect products, and fast process development.

With feedforward dynamic management we are able to use information picked up by sensors (Step ① in Fig. 1.5), to explore the solution space (Step ② in Fig. 1.5), manage the process through smart design and decision-making process (Step ③ in Fig. 1.5), capture and reuse the knowledge with off-line models (Step ④ in Fig. 1.5) for future design or reconfiguration of the process (feed forward control of the process, Step ⑤ in Fig. 1.5). In next iteration we will obtain new information (feedback learning, Step ⑥ in Fig. 1.5) that will be further captured and reused. This is an iterative cycle of exponential learning that is foundational to incorporating intelligence in feedforward dynamic management of NMS.

The resulting outcome is a next generation of smart networked manufacturing systems adaptable to rapidly changing market requirements, resulting in higher quality products and at a lower rework cost. To transition this framework to industry we suggest that DFDM be extended to a Smart Platform, see Fig. 1.1. Some possible extension of DFDM in various industries are illustrated in Fig. 1.6:

- *Cyber-physical design*. Model-based and intelligent decision-based design of systems. We speculate that there are applications in manufacturing and process industries. There are several applications in Industry 4.0 that can benefit from the DFDM procedure. These are discussed in Chap. 7.
- *Cyber-physical-social system design*. Model cyber-social design decision network in the realization of intelligent cyber-physical-social systems. We speculate that there are applications in smart healthcare systems, etc.

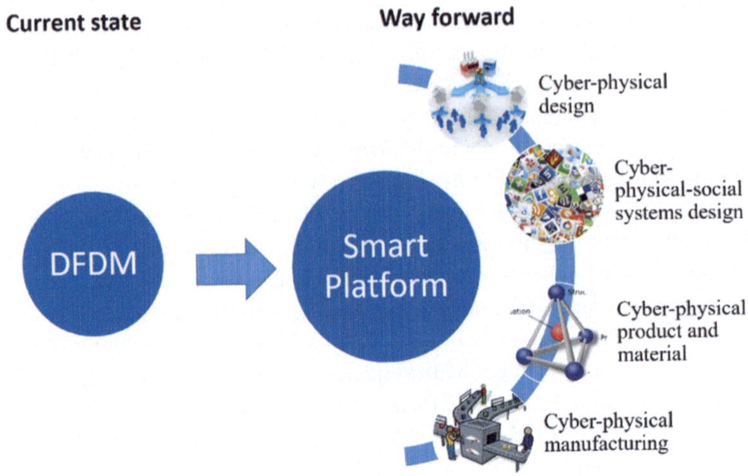

Fig. 1.6 Extension of DFDM (Milisavljevic-Syed et al. 2019a)

- *Cyber-physical product and material design.* Integration of materials, products, and manufacturing processes. We speculate that there are applications in the design of smart sports equipment, etc.
- *Cyber-physical manufacturing*. Design of reconfigurable intelligent manufacturing systems. We speculate that there are applications in additive manufacturing, and semiconductor lithography processes, etc.

In furtherance of Industry 4.0 we advocate the realization of smart Networked Manufacturing Systems. We suggest that Design for Dynamic Manufacturing is foundational to the realization of smart NMS. In the International System Realization Partnership (ISRP) we are creating a Cloud-based Design and Manufacturing (CBDM) platform with potential applications in cyber-physical design, cyber-physical manufacturing, cyber-physical product and material, and cyber-physical-social system design. Further, we speculate that CBDM will be transitioned to industry and be applied in the steel making processes, additive manufacturing, design of smart healthcare systems, etc.

The background and the frame of reference of decision-based design computational framework, Design for Dynamic Management is presented in Sects. 1.2 and 1.3.

1.2 Background: Design for Dynamic Management in Realization of the Networked Engineered Systems

Most new designs are related to innovation as adaptations of the existing systems (Bers et al. 2009; Wagner 1993; Pahl et al. 2013) but rarely to inventions. The reason is designers tend to guide their solution toward existing designs and limit their own creativity by a fixation on particular design (Carryer et al. 2011; Condoor and LaVoie 2007; Linsey et al. 2010). Creating a new architecture is a highly challenging exercise that most designers (Ottino 2004) and engineering organizations are averse to (Collopy and Hollingsworth 2011; Ross and Hastings 2005; Weigel and Hastings 2004). On the other hand, innovation's greatest enemy is an aversion toward challenges (Assink 2006). Novelty brings a potential for performance breakthroughs. However, it increases complexity in product development, uncertainty, cost, and time to market (Bers et al. 2009; Veryzer 1998; Tatikonda and Rosenthal 2000; Buede 2000). New architectures are often designed to overcome these challenges (Bers et al. 2009). Conventional design processes (Veryzer 1998; Benner and Tushman 2003; Vajna et al. 2005) are not conducive to a successful realization of unconventional design. Moreover, a designer tends to adopt new solutions for system design only when they are out of options (Steltzner et al. 2006). We believe new design processes are needed in order to move forward breakthrough innovations (Veryzer 1998; Benner and Tushman 2003; Magnusson et al. 2003; Williams 1999; Allen et al. 2011). Exploration of a new architecture helps designers to escape from conventional design solutions (Chakrabarti et al. 2011).

In this monograph, we describe the model-based realization of networked engineered systems, see Sects. 1.2.1 and 1.2.2.

1.2.1 Model-Based Design of Complex Systems When Models Are Abstractions of Reality

Concept exploration is a process of evaluating different design approaches and is influenced by the way overall design requirements (goals and constraints) are used. Overall design requirement can be used in two ways (Luger and Stubblefield 1990) (1) a priori constraining for generating possible design structures to be consistent with them, (2) a posteriori testing where a possible design satisfy design requirements. A priori use of design requirements involves analysis and transformation of the design requirement that directly influences the generation of solutions. A posteriori use of design requirements includes analysis and evaluation of a possible solution and the degree to which possible solution satisfy these requirements (Chen et al. 1996).

There are two ways to determine top-level design specifications based on the use of design requirements (1) optimization-based approach (a priori constraining), and (2) simulation-based approach (a posteriori testing). However, an ideal concept exploration approach supports both activities a priori constraining and a posteriori testing.

In this monograph, we propose a framework for model-based design of engineered systems when the models are abstractions of reality.

1.2.2 Design Decisions in the Model-Based Design

In design, computational models used to represent networked engineered systems are assumed to be incomplete, inaccurate, without sufficient data. Hence, when sheer size and complexity of design problem is considered attaining optimal solution is impossible. It is expected that there is the best solution for any problem. However, that is not the case and as the best solutions are "satisficing" solutions (Simon 1980), that satisfies the most important goals and constraints. Satisficing solutions go with agenda, heuristics or assumptions made by a designer based on available information at certain point of time.

Acknowledging and managing uncertainty, solution space exploration, and identifying robust solutions of NES is one important element described in this monograph, see Table 1.1. In keeping with the preceding in Sect. 1.3.1, we present the Stream of Variation (SoV) model from Control Theory. In Sect. 1.3.2, we present the compromise Decision Support Problem (cDSP) construct from Decision-Based Design. In Sect. 1.3.3, we present solution space exploration. In Sect. 1.3.4, we present uncertainty in design.

1.3 Design for Dynamic Management in Design of the Networked Engineered Systems

We suggest that designing is a decision-based activity wherein the principal (not only) role of a human designer is to make decisions. The outcome is an improved or new product or process that satisfies a set of requirements, for example, see Table 1.1. We use computational models to simulate the behavior of a product being designed. These models are abstractions of reality necessitating the management of uncertainty. Hence, it is imperative that we focus on finding the design that is relatively insensitive to uncertainties because the computational models are abstractions of reality, see Mistree et al. (1990), Smith et al. (2014).

In the design of NES, the management of uncertainty involves accounting for issues such as imprecise process definition, the lack of knowledge of the process, lack of knowledge in models and information propagation in a chain of models, unpredictability of a physical system, and the necessity to explore alternatives. Furthermore, there are the challenges in capturing the implicit knowledge, extract knowledge from data and scenarios, and developing design methods for decision-based design. In decision-based design, design choices are explored a posteriori due to ramifications of models being abstractions of reality, the emergent properties can only become evident by exercising the Decision Support Problems (DSPs).

In model-based design of complex systems, designers have two options (1) to use the exact equations for predicting the system behavior, or (2) to generate an approximation of the system behavior. We adopt the second option because the first option is highly demanding on computational resources. Further, the method to approximate system behavior must be both accurate and efficient according to Chen et al. (1996).

In model-based design, a designer simulates system performance for different design concepts and assesses the merits of possible designs. The advantages are facilitating the exploration of the whole design space, generating and investigating new ideas, providing insight into dependency between parameters, etc. The most used simulation-based design methods are (1) Grid Search used on ships Georgescu and Boonstra (1990), however, they are impractical in design of complex systems; (2) Random Generation used on ships design (Smith 1993), extensive computational time in design of complex system; (3) Monte Carlo Simulation (Siddall 1984), is accurate but computationally expensive; and (4) Design of Experiment (DOE) Techniques (Mistree et al. 1993) (Fig. 1.7).

We propose a model-based, decision-based approach for designing NES under uncertainty wherein the computational models are abstractions of reality. Further, we address the concurrent design of both mechanical and control system in design of NES. This necessitates establishing connectivity between design parameters of the mechanical (tools on operational stations) and control systems (measuring sensors on sensing stations) through Stream of Variation (SoV) model, and cDSP models, see Sects. 1.3.1 and 1.3.2, and Fig. 1.8.

1.3.1 Control Theory: Stream of Variation (SoV) Model

In a NES such as multistage manufacturing processes (MMPs), products are manufactured through multiple operations or stages where the product quality is typically reflected by the variations of Key Product Characteristics (KPCs) (Zhong 2009). During production, due to part variations and process variations, such as tool error, at each stage, the KPCs of a subassembly will deviate from nominal values. These variations will be carried to the next stage and further interact with the assembly process. Further, these variations are propagated to the downstream stages and accumulated into the final product. In case that the size of final accumulation is large enough the quality of the production process is diminished.

The propagation of variation in MMPs raises a challenge in achieving the quality of the production process. Minimizing the variation of subassembly at the current stage alone may not lead to the best final product quality. In order to achieve effective control in an MMP, three components are necessary (1) a model that captures the variation flow, that is, the Stream of Variation (SoV) model, (2) real-time sensing technologies to measure the variation, and (3) Programmable Tooling (PT) to perform control actions to suppress the variation (Zhong 2009).

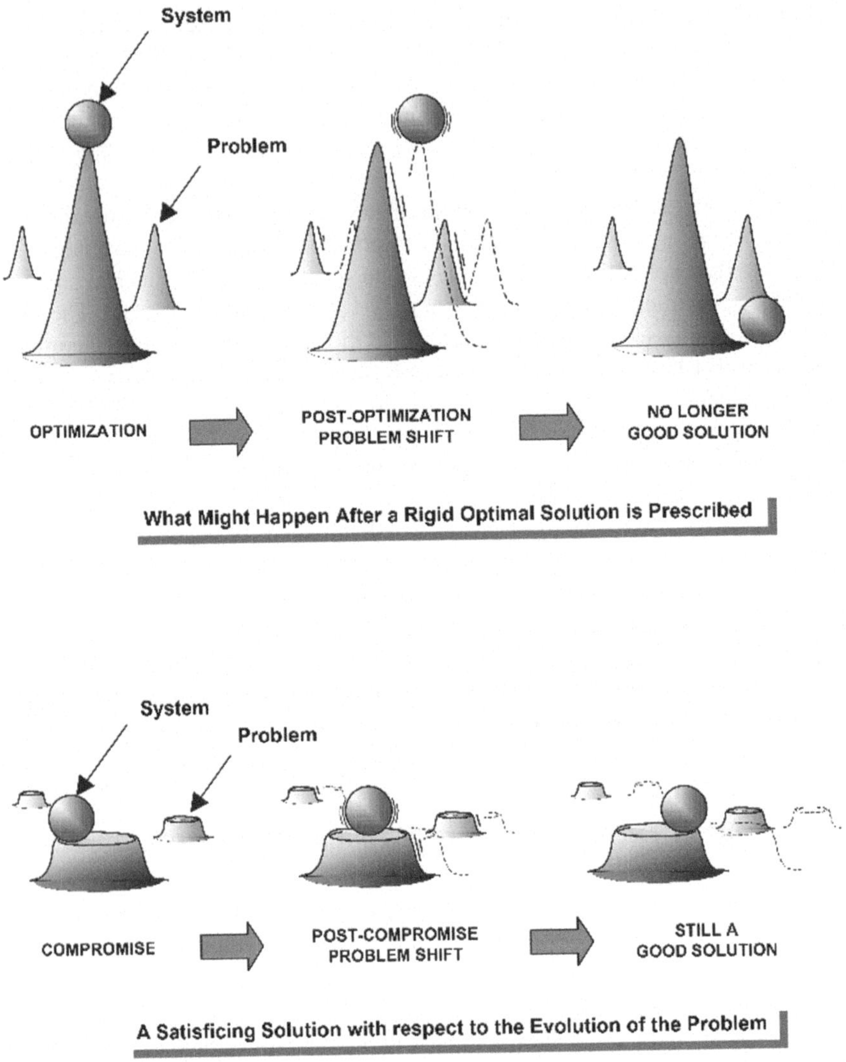

Fig. 1.7 Difference between an optimum solution and a satisficing, robust solution

The propagation of variation in MMPs, is described by the SoV modeling, by exploring the relationship of variation sources and geometric information of each operating station based on design information, especially product and process geometry (Jin and Shi 1999). SoV modeling has been utilized as the mathematical basis in various applications such as process modeling, design evaluation, diagnosis, tolerance synthesis, active control, and other areas (Shi 2006).

The SoV model, used to describe what impact corrections have on the final quality of the product (Jin and Shi 1999; Shi 2006), is illustrated in Fig. 1.9, and

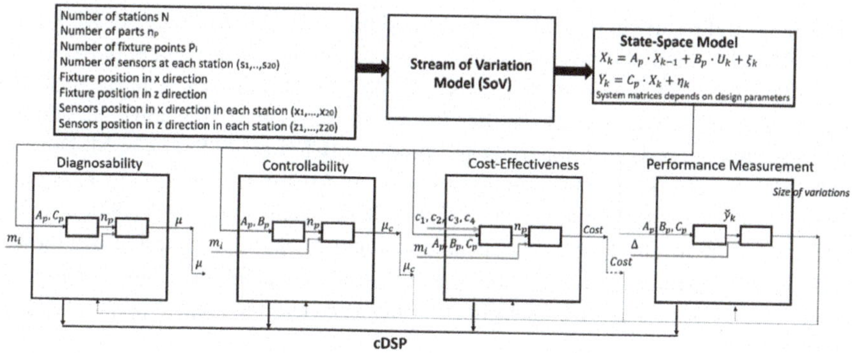

Fig. 1.8 Connection between SoV and cDSPs (Milisavljevic-Syed et al. 2018)

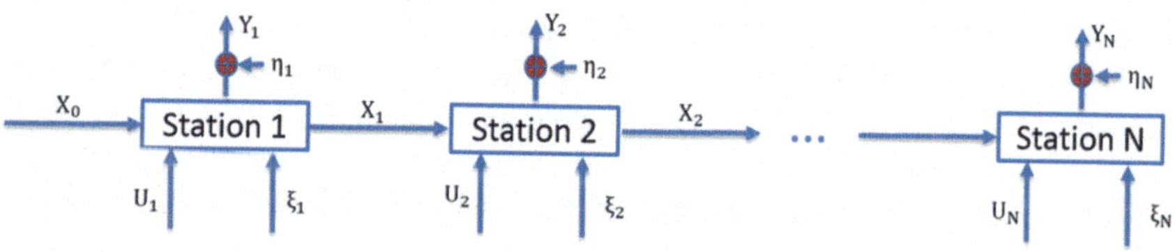

Fig. 1.9 Diagram of multistage assembling process (Ding et al. 2002)

mathematically in Chap. 3, Sect. 3.31.

In a multistage assembling process, Fig. 1.9, parts enter the production line from Station 1 with initial fabrication errors, x_0. At Station 1, the part control action u_1 is first applied through PT, while other unmodeled process errors, ξ_1, add to the variation of the parts. The designed operation at Station 1 then takes place and the state of the subassembly changes to x_1. The subassembly is then transferred to the next station, and variations propagate and accumulate similarly as more parts/subassemblies are joined together, until the finished assembly exits the production line at the final Station N. The KPC is measured at the final Station N as well as intermediate stations, such as Station k. The measurement y_N is obtained with sensor errors η_N (Zhong 2009).

1.3.2 Decision-Based Design: Compromise Decision Support Problem (CDSP)

Many methods have been proposed for implementing Decision-Based Design. Our implementation is anchored in the Decision Support Problem Technique (Marston et al. 2000; Muster and Mistree 1988). A key construct of this technique is the compromise Decision Support Problem (cDSP). Applications include the design of aircraft, mechanisms, thermal energy systems, the design of ships, damage tolerant structural and mechanical systems, composite materials, and the concurrent design of multi-scale and multi-functional materials, and products (Mistree et al. 1990; Smith et al. 2014). Key applications spans inter alia specification development (Chen et al. 1996, 1997; Chen and Lewis 1999; Lewis et al. 1999), robust design (Allen et al. 2006; Seepersad et al. 2006), product families (Simpson et al. 1998; Simpson et al. 2001a, b), the integrated design of materials and products (Choi et al. 2008a, b; McDowell et al. 2009; Panchal et al. 2007; Seepersad et al. 2005), different mechanical systems (Chen et al. 1994; Hernandez and Mistree 2000; Koch et al. 1998; Sinha et al. 2013), and concurrent design of a mechanical and a control system (Milisavljevic et al. 2018). Further, in this monograph, we are adding flexible, functional, and reconfigurable design of NES under uncertainty as requirements to considered at the design stage.

In cases where the decisions relate to complex systems, such as NES, actions taken have high impact, the process is iterative, and certain rationale is required, see Fig. 1.10. It is possible to develop new perspectives through understanding emergent properties and discover new solutions in the process.

Mistree and coauthors offer two types of decisions, namely, selection and compromise, (Mistree et al. 1990, 1991). Mistree and coauthors (Mistree et al. 1995) and Wang and coauthors (Wang et al. 2019) demonstrate that design workflows (networks) can be formulated using selection and compromise Decision Support Problems.

In this monograph, NES is represented through the computational modeling of a network of compromise decisions where the compromise Decision Support Problem is framed as:

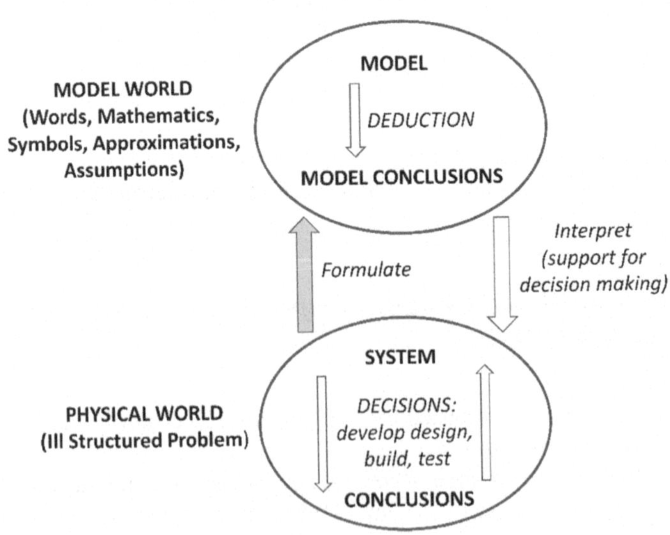

Fig. 1.10 Modelling the physical world

Given

An alternative to be improved through modification,
Assumptions used to model the domain of interest,
The system parameters (fixed variables), and
The constraints and goals for the design;

Find

The independent *system variables* values (they describe the artifact's physical attributes), and
The *deviation variables* values (they indicate the extent to which the goals are achieved); which

Satisfy

The *system constraints* that must be satisfied for the solution to be feasible,
The *system goals* that must achieve, to the extent possible, a specified target value, and
The lower and upper *bounds* on the system variables and *bounds* on the deviation variables, in order to

Minimize

The *deviation function* that is a measure of the deviation of the system performance from that implied by the set of goals and their associated priority levels or relative weights.

The parallel between "demands" and "wishes" (Pahl et al. 2013) can be drawn with the compromise DSP. The demands are constraints and bounds, and the wishes are goals, in the DSP (Smith et al. 2014). Further, the feasible design space is defined by the constraints and bounds, the aspiration space by the goals, and the solution space by the feasible and aspiration space.

1.3.3 Solution Space: Exploring and Understanding the Requirements for Successful Design of Networked Engineered Systems

In model-based design, typically, a designer has to use models of varying accuracy, completeness and fidelity. Hence, the ability to rapidly identify the solution space within which various designs can be explored is important. A strategy for identifying and exploring a possible solution space, a tailored computational environment created to solve the DSPs includes:

1. Discover regions where feasible designs exist based on satisfying the system constraints and bounds or where feasible designs might exist by minimizing the constraint violation (Smith et al. 2014).

2. Frame the feasible design space extremities from the neighborhood of the feasible or near feasible regions using a preemptive (lexicographic minimum), representation of the goals in a higher order search (Smith et al. 2014).
3. Having framed the space and the zones of greatest interest, move between the extremes generating deeper understanding and exploring tradeoffs using an Archimedean (weighted sum) formulation of the goals (Smith et al. 2014).

In this monograph all three steps are included and exercised, Steps 1–3 are presented in Chap. 3. A variety of tools and methods are appropriate in each step and draw on a variety of mathematical foundations. However, obtaining feasible designs for a complex system is not an easy process especially when nonlinearities (dynamic operability model in Chap. 5) and overly tight constraints (reconfiguration model in Chap. 6) can limit design opportunity.

1.3.4 Managing Uncertainty: Robust Design

In the design of NES acknowledging and managing uncertainty is necessary and there are two main approaches (1) reducing uncertainty itself, and (2) designing a system to be relatively insensitive to uncertainty without reducing or eliminating the source of uncertainty, better known as robust design. In this monograph we consider the second approach, designing a system to be insensitive to uncertainty. Another name for robust design is parameter design due to the fact that it is used to make the system response insensitive to uncontrollable system input variations, thus improving the quality of designed product (Choi et al. 2008a, b). However, parameter design alone does not always lead to sufficiently high quality. Further improvements are achieved by controlling the source of variations that is associated with higher cost. Design at lower cost by sacrificing the achievement of optimal performance is the reason why robust design approach is introduced in design.

Typically, design parameters are divided into three categories (1) control factors, (2) noise factors, and (3) responses. Control factors are parameters that a designer adjusts. Noise factors are exogenous uncontrollable parameters that affect the performance of a product/process/methods (Choi et al. 2008a, b). Responses are performance measures for the product or process (Choi et al. 2008a, b). The sources of uncertainty reside in system design models, based on which designers make their decisions. However, there are other sorts of uncertainty in design that cannot be managed by Type I and II robust design approaches, such as uncertainty due to assumptions made while formulating the models, and propagated uncertainty in multiscale simulation chain of models. Further, these uncertainties cannot be directly configured in parameters. It is important for designers to identify where the uncertainty sources reside in a system model in order to employ an appropriate uncertainty management method (Choi et al. 2008a, b).

Type I Robust Design: Managing Uncertainty in Uncontrollable Parameters

Uncertainty in a system model in uncontrollable independent system parameters, known as "noise factors" (Choi et al. 2008a, b). Environmental factors, boundary conditions, operating conditions, materials property variances are "noise factors". Noise factors are usually in parametric form, characterized and quantified as continuous numbers with or without probability information, and cannot be controlled. Natural uncertainty is the most prevailing type of uncertainty, measured in a statistical way.

Taguchi (1993) robust design, that is, robust design Type I addresses design of a system that is insensitive to uncertainty in noise factors.

"***Type I Robust Design**: Identify control factor (design variable) values that satisfy a set of performance requirement targets despite variation in noise factors*" (Taguchi 1993).

Robust design Type I, as in the upper portion of Fig. 1.11, is used to design systems that satisfy a set of performance requirements targets despite variations in

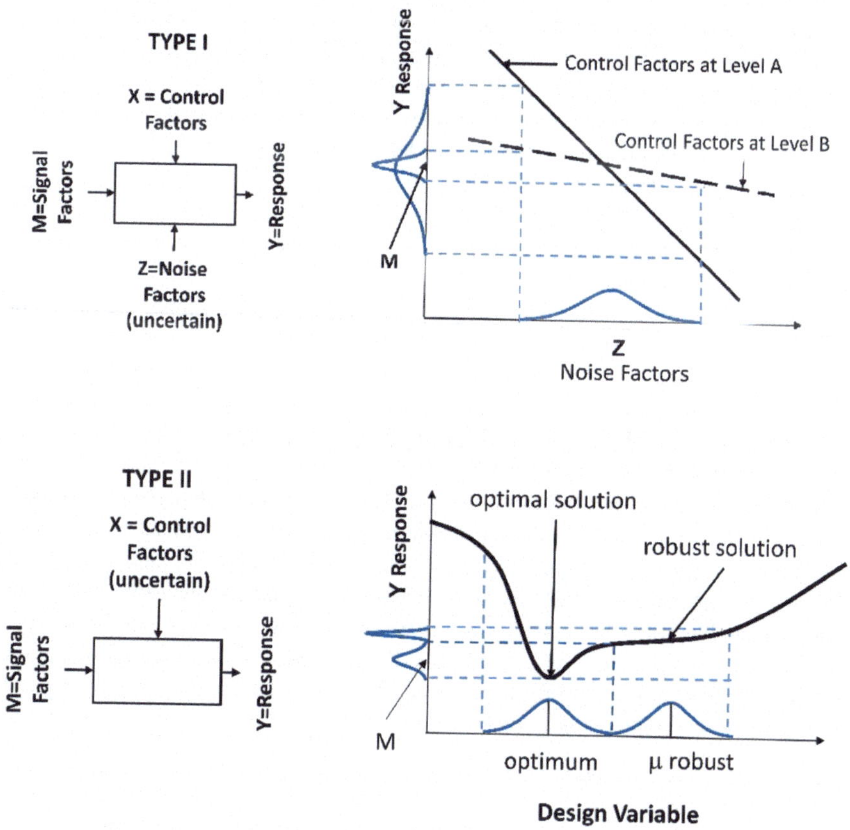

Fig. 1.11 Robust design for uncertainty Type I and Type II (Chen et al. 1996)

noise factors that are uncertain/uncontrollable/independent system parameters (Choi et al. 2008a, b). These noise factors introduce variability into the system response. Despite the fact that Taguchi's robust design principles (Taguchi 1993) are advocated widely in both industrial and academic settings, his statistical techniques have been criticized extensively, and improving the statistical methodology has been an active area of research (Chen et al. 1996; Myers and Montgomery 1995; Nair 1992; Tsui 1992, 1996). Robust design Type II is proposed by Chen and coauthors (Chen et al. 1996) in order to design a system that is insensitive to uncertainty in control factors, see the lower portion of Fig. 1.11. Further, robust design methods have been extended for a variety of applications in engineering design by numerous researchers (Cagan and Williams 1993; Chen et al. 1996; Chen and Lewis 1999; Mavris et al. 1999; Otto and Antonson 1993a, b; Parkinson et al. 1993; Renaud 1997; Yu 1994).

Type II Robust Design: Managing Uncertainty in Controllable Parameters

Addressing the uncertainty in variables representing the controlled system, that is, the "control factors" represents the Type II Robust Design problem (Choi et al. 2008a, b). System parameters such as mass, applied voltage, percentage composition of materials, process control inputs, are some of the control factors. Control factors can be measured and characterized as continuous numbers with or without probability distributions. However, the deviations of control factors may not be controllable and should be treated as noise factors (Choi 2005). Robust design Type II is proposed by Chen and coauthors (Chen et al. 1996) in order to design a system that is insensitive to uncertainty in control factors, see Fig. 1.11.

"***Type II Robust Design***: *Identify control factor (design variable) values that satisfy a set of performance requirement targets despite variation in control and noise factors*" (Chen et al. 1996).

Type II robust design is used to design systems that are invariant to variations in system parameters as the design evolves. In robust design of systems with uncertainties Type II, designers search for means of control factors that satisfy a targets of the set of performances against variation in control factors (Choi et al. 2008a, b). Robust Concept Exploration Method (RCEM) proposed by Chen and coauthors (Chen et al. 1996) is a method that combines two types of robust design to account for the incompleteness of computational models. RCEM is domain-independent approach for generating robust and multidisciplinary design solutions. Robust solutions in multi-objective design problems are tradeoffs between preference and weights of expected performance and sensitivity of performance due to deviations in design or uncontrollable variables (Choi et al. 2008a, b). RCEM has been employed successfully for a simple structural problem and design of a solar powered irrigation system, general aviation aircraft, high speed civil transport, product platforms, and other applications (Choi et al. 2008a, b).

Type III Robust Design: Managing Uncertainty Embedded in System Functions

Uncertainty inserted in system functions is "model structural uncertainty" due to assumptions and idealization in a system (Choi et al. 2008a, b). For example, model structural uncertainty includes employment of uncertain knowledge, errors in computer codes, linearization and discretization errors in finite element analysis, and other assumptions due to limited information.

"***Type III Robust Design***: *Identify adjustable ranges for control factors (design variable), that satisfy a set of performance requirement targets and/or performance requirement ranges and are insensitive to the variability within the model*" (Chen et al. 1996).

The differences among the optimal solution, Type I and II robust solution, see yellow rectangle in Fig. 1.12, and Type I, II and III robust solution, see orange rectangle in Fig. 1.12, is shown for the same objective curve of the function. The system's response, a deviation (or objective) function, is illustrated as a solid curve. The limits of uncertainty are due to the un-configured variability, non-parametric variability, and model parameter uncertainty, see two dotted curves in Fig. 1.12, that are added around the objective function. It can be concluded that Type I, II, and III robust solutions have smaller performance deviations as the optimal and Type I and II robust solution. Type III robust design becomes more important due to increased complexity and stochastic behaviors of systems (Choi et al. 2008a, b).

The most accurate way to incorporate the inserted uncertainty in control and noise factors during design exploration is to perform simulation using statistical techniques (simulation-based design) (Choi 2005). Measure variations of performance by simulating input variations is conducted by Monte Carlo Simulation as most popular method, however, applied on a relatively simple problem by Du and Chen 2000.

Type IV Robust Design: Managing Propagated Uncertainty in a Design and Analysis Process Chain

Uncertainty in a system model used in the design and analysis process chain often arises from the complex design and analysis process chain and not from the system model itself. This type of uncertainty is often seen in multidisciplinary uncertain

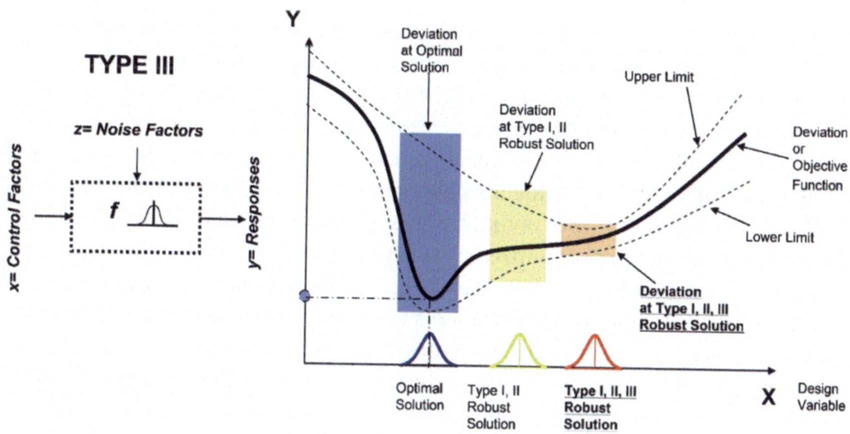

Fig. 1.12 Robust design Type III (McDowell et al. 2010)

system design problems, such as Multidisciplinary Design Optimization (MDO), and includes errors in decisions made by other designers and accumulated errors (propagated uncertainty) by subsequent series of uncertain subsystem models (Choi et al. 2008a, b).

For example, complex multidisciplinary system design requires multiple experts to collaborate to make decisions for designing a system where the outputs of other experts' decisions in a subsystem could be input parameters, constraints, or design spaces of other subsystems or systems design (Choi et al. 2008a, b). In these design activity interactions, a subsystem design error can be propagated to another subsystem or system. Additionally, in complex systems design there is a tendency to employ multiple analyses and simulations in a series to predict system responses.

Type IV Robust Design*: Identify adjustable ranges of control factor (design variable) values under potential uncertainty and uncertainty propagation in a design and analysis process chain; account for uncertainty in downstream activities and uncertainty propagation.*

Type IV robust design is focused on uncertainty associated with the design process chain that emanates from (1) changes in design specifications as a result of downstream or concurrent decisions and design activities, or (2) the propagation and potential amplification of uncertainty due to the combined effect of analysis tasks performed in series or in parallel (Choi et al. 2008a, b). In multidisciplinary design and analysis, including multiscale, multi-physics materials design, with an abundance of shared or coupled variables and analyses performed on multiple length and time scales both sources of design process uncertainty are common and important.

In complex design processes, the output of a given design (or analysis) task could provide input to subsequent tasks, increasing or decreasing the amount of uncertainty in the system via interdependent variables and parameters, see Fig. 1.13 (Choi et al. 2008a, b). Further, due to possible deviation propagation in design process a small deviation in an input variable can produce large variation in the final performance, therefore, it is necessary to represent the size of uncertainty propagating within a design process chain in a graphical manner so that engineers/designers can easily identify which design tasks significantly increases the whole system uncertainty.

Probabilistic and robust design techniques are being infused into a number of multidisciplinary design optimization (MDO) approaches proposed for formulating and concurrently solving decomposed or partitioned complex system design problems, such as single level optimization approaches in which analyses are distributed and independently executed while a single optimizer is used to evaluate the results (Haftka 1985), and multi-level optimization approaches in which analyses are distributed and subsystems are optimized by subsystem-level and system-level optimizers (Sobieszczanski-Sobieski 1988; Wujek et al. 1996) or collaborative optimization (Kroo et al. 1994). Furthermore, few authors consider probabilistic and deterministic variables and parameters in an MDO context (Du and Chen 2000), replacing complex, computationally intensive analysis routines with approximate models (Giunta 1997; Koch et al. 1998; Mavris et al. 1999). On the other hand, in highly nonlinear

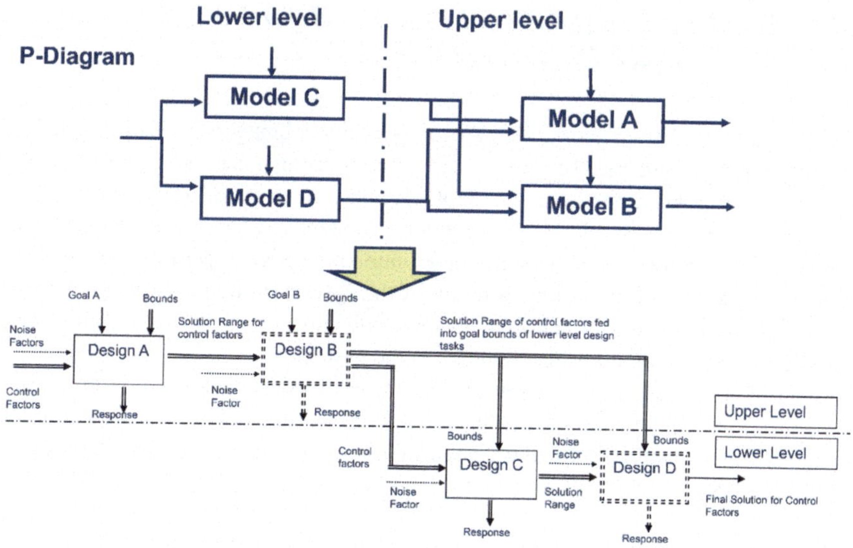

Fig. 1.13 Robust design Type IV for a bi-level system

model or one governed by a large number of parameters, approximate models may ease the computational burdens of MDO but they can be inaccurate or inefficient.

If decision-making is distributed fully among multiple designers who make decisions independently of one another, the difficulty lies in managing interdependencies among the decisions and game theoretic approaches, protocols are needed (e.g., cooperation, non-cooperation, and Stackelberg leader/follower) and mathematical coordination mechanisms, such as rational reaction sets or best reply correspondences.

In Sect. 1.4, we discuss goals and focus in this monograph, and identify the gaps for transition to industry.

1.4 Problem, Goals, Questions and Hypotheses

In this section, we present the problem we address in this monograph, we introduce questions and hypotheses, along with contributions and identified gaps as a way forward. In Sect. 1.4.1, we present the problem we address in the monograph. In Sect. 1.4.2, our focus is on listing the fundamental questions and hypotheses that are addressed in later chapters. The contributions of the monograph are highlighted in Sect. 1.4.3.

1.4.1 Problem: Computational Framework for Design of Networked Engineered Systems

The computational framework, Design for Dynamic Management, is proposed to support adaptable concurrent design, system operability analysis, and reconfiguration of multiple elements in the system. Adaptable and concurrent design is achieved by introducing flexibility in selection of design parameters and concurrent design of the mechanical and control systems, without domain knowledge. Operability analysis is achieved by determining the functionality of the system in the presence of change. Reconfiguration method is achieved by reestablishing connectivity and allowing reconfiguration among multiple elements in the system.

Given that overall design requirements are established and analysis programs exist, the principal motivation in the monograph is to establish a computational framework that is suitable for the design of low-cost and high-quality networked engineered systems adaptable to ambitious market demands in the context of Industry 4.0.

In order to achieve the goal, a method for adaptable design, based on (a) SoV modeling and the compromise DSP (Mistree et al. 1993), see Block 1 in Fig. 1.14; (b) operability analysis of NES, based on operability analysis, cDSP, and minimum time control, see Blocks 2 and 3 in Fig. 1.14; and (c) reconfiguration strategy of NES, based on cDSP and game theory, see Blocks 4 in Fig. 1.14; is proposed.

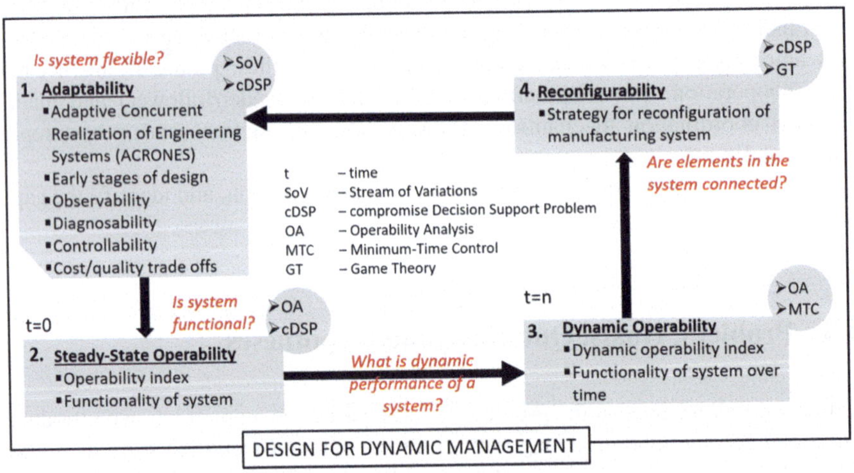

Fig. 1.14 Design for dynamic management computational framework (Milisavljevic-Syed et al. 2018)

1.4.2 Questions and Hypotheses Investigated in the Monograph

Our goal is to propose a computational framework suitable for the design of low-cost and high-quality NES adaptable to ambitious market demands as a support to further digitization (smart manufacturing). As discussed in the previous sections, dynamic changes in the market due to wide variations in customer requirements lead to mass customization where enterprises have to be capable of adjusting the manufacturing process to accommodate a variety of products, each with different features and production quantities. On the other hand, global competition requires enterprises not only to implement cost-effective manufacturing processes but also to improve the quality of the product and shorten time to market. Digitization of manufacturing systems is one technology that is increasingly being adopted to respond to changes in the market. Hence, the need for a computational framework to design a system adaptable to dynamic changes in the market. Design for Dynamic Management is a decision-based computational framework where design thinking, design strategy and innovation management is integrated to design a system adaptable to dynamic changes in the market.

Given these goals, the primary question we address in the monograph is further partitioned on secondary and tertiary question where the hypotheses are further partitioned on secondary and tertiary question; see Table 1.2.

By adopting decision-based design we are able to connect various elements that are essential for the implementation of the Industry 4.0 construct. However, this does not address the deeper issue of design evolution and need to design a system to be adaptable to dynamic changes in the market. The foundational elements of the proposed computational framework is SoV modeling and compromise DSP, operability analysis, and reconfiguration strategy, see Sect. 1.3.

To answer the SQ1, it is necessary to perceive the mechanical and control system concurrently and insert flexibility in selection and determination of values of design parameters in the design of NES. Further, to answer the SQ2, it is necessary to establish operability and disturbance spaces, examine system performance and dynamic performance of the system in the presence of change. Finally, to answer the SQ 3, we recognize the need for exploration and understanding of different reconfiguration strategies of a machine tool, inspection system, and manufacturing system altogether in order to remain competitive on the market with the low-cost process and high-quality product.

Adaptable Concurrent Design of Networked Engineered Systems

In this section, the SQ1 is addressed in two ways (1) determining requirements of adaptable design, and (2) determining main mechanical and control system drivers and their relation in order to achieve the concurrent design.

As presented in Table 1.2, the SQ1 is separated into two supporting questions. To answer TQ1, TH1 is tested and verified. To answer TQ2, TH2 is tested and verified. In Sect. 1.4.1, some background knowledge related to the mechanical and control

Table 1.2 Questions and hypotheses addressed in the monograph

Level	Primary	Secondary	Tertiary	Chapter
Question	PQ: In the context of Industry 4.0 what is the computational framework that facilitates the decision-based design of networked engineered systems adaptable to ambitious market demands as a support to further digitization	SQ1: What is the computational framework in the design method that facilitates adaptable design in the realization of networked engineered systems?	TQ1_1: What is the algorithm that enables identification of the adaptable design of networked engineered systems?	Chapter 3
			TQ1_2: What are the mechanical and control system drivers and is the computational framework in the design method that facilitates concurrent design?	
		SQ2: What is the computational framework in the design method that facilitates dynamic change in the requirements in the realization of functional networked engineered systems?	TQ2_1: What is the algorithm that enables identification of the system functionality in the presence of change?	Chapter 4
			TQ2_2: What is the algorithm that enables identification of the system dynamic performance in the presence of change?	Chapter 5
		SQ3: What is the reconfiguration strategy for reestablishing connection among elements in the system in the realization of networked engineered systems to remain competitive on the market?	TQ3_1: What is the reconfiguration strategy of the machine tool in the networked engineering system?	Chapter 6
			TQ3_2: What is the reconfiguration strategy of the inspection system in the networked engineering system?	

(continued)

Table 1.2 (continued)

Level	Primary	Secondary	Tertiary	Chapter
Hypothesis	PH: We hypothesize, that by applying computational framework, we can support flexible, operable and rapidly configurable design of the manufacturing processes. Furthermore, by introducing flexibility in selection of design parameters, concurrent design of a mechanical and a control system, analyzing dynamic performance of the systems, and applying different reconfiguration strategies further digitization of NES can be ensured	SH1: Design of mechanical and control system concurrently while accounting for different types of uncertainty and extensive (robust) solution space exploration facilitates the adaptable design of networked engineered system	TH1_1: Inserting flexibility in selection and determination of values of design parameters at design time facilitates the adaptable design of NES	Chapter 3
			TH1_2: Determining the mechanical and control system drivers and their mutual relations creates common ground for concurrent design and provides insights into NES	
		SH2: Determining input ranges (operability and disturbance spaces) that give desired solution range of functional system design at steady-state and dynamic state that will allow a system to adjust and stabilize in presence of change	TH2_1: Determining desired output space and available input space will give us information under which conditions system is functional even in the presence of change	Chapter 4
			TH2_2: Determining dynamic available input space gives us a fraction of operating ranges (if a system can transit and stabilize even in the presence of change) that can be achieved within the response time	Chapter 5

(continued)

Table 1.2 (continued)

Level	Primary	Secondary	Tertiary	Chapter
		SH3: Machine tool reconfiguration followed by the inspection system reconfiguration will allow reconfiguration of the manufacturing system and reestablishing connection among elements in the system	TH3_1: Determining characteristics and requirements of operational stations (machine tools and machining operations) in order to reconfigure and accommodate to rapidly changing requirements (product design and product scale)	Chapter 6
			TH3_2: Determining characteristics and requirements of the inspection system (sensors and sensing operations) in order to reconfigure and accommodate to rapidly changing operational stations (machine tools and machining operations)	

system drivers in the design of NES is presented. The SH1 is introduced in Chap. 2 and verified in Chap. 3. The test example is the 2-D panel assembling process in three stations, see Sect. 2.1.3.

Operability Analysis of Networked Engineered Systems

In this section, the SQ2 is addressed in two ways (1) determining system functionality in the presence of change, and (2) determining a dynamic performance of the system in the presence of change.

As presented in Table 1.2, the SQ2 is separated into two supporting questions. To answer TQ2_1, TH2_1 is tested and verified. To answer TQ2_2, TH2_2 is tested and verified. In Sect. 1.4.2, some background knowledge related to operability analysis in the design of NES is presented. The SH2 is introduced in Chap. 2 and verified in Chaps. 4 and 5. The test examples are 2-D panel assembling process in three stations, and continuous stirred tank reactors, see Sects. 2.1.3 and 2.1.4.

Reconfiguration method of Networked Engineered Systems

In this section, the SQ3 is addressed in two ways (1) determining reconfiguration strategy of the machine tool as part of the manufacturing system, and (2) determining reconfiguration method of inspection system as part of the manufacturing system.

As presented in Table 1.2, the SQ3 is separated into two supporting questions. To answer TQ3_1, TH3_1 is tested and verified. To answer TQ3_2, TH3_2 is tested and verified. In Sect. 1.4.3, some background knowledge related to reconfiguration method in design of NES is presented. The SH3 is introduced in Chap. 2 and verified in Chap. 6. The test examples are transmission box in vehicles use, see Sect. 2.1.3.

1.4.3 Contributions in the Monograph

The expected contributions from the monograph are the following:

Adaptable and Concurrent Design of Networked Engineered Systems

1. Identifying mechanical and control system drivers and their relations in concurrent design;
2. Build in flexibility in selection and determination of values of design parameters in both systems;
3. Managing the structure of the high-complexity mathematical problem;
4. Creating effective and efficient decision network structure as a decision support; and
5. Integration of process- and product-related decision model in the comprehensive model.

Operability Analysis of Networked Engineered Systems

6. Framing operability/disturbance space without domain knowledge;
7. Operability analysis of any engineering systems; and
8. Managing system functionality due to change in the requirements.

Reconfiguration method of Networked Engineered Systems

9. Method for Reconfiguration of Manufacturing System (RMS);
10. Method for Reconfiguration of Manufacturing Tool (RMT); and
11. Method for Reconfiguration of Inspection System (RIS).

In Sect. 7.4 we suggest what, in our opinion, is required to transition what is described in this monograph to industry.

The overview of the monograph and the validation strategy is presented in the next section.

1.5 Overview and Validation Strategy of the Monograph

In the design of NES there are subjective elements involved in decision making that reflects the initial design stage of the computational framework implementation, therefore, there is a need to undertake the verification and validation of a computational framework/result due to the fact that a development of a computational framework includes many abstract elements and there is no unique answer.

The "Validation Square" is a method where designers build confidence in the utility of methods and examples that are used to verify the method (Seepersad et al. 2000), see Sect. 1.3.4. Further, the validation square is used to determine whether the method provides correct design solutions regarding structural validity and regarding performance validity (Choi 2005).

The validation square is the process of validation and the validation quadrants are:

- ***Quadrant 1***. *Theoretical Structural Validity.* Examining the structural/ logical validity and overall consistency of the proposed computational framework.
- ***Quadrant 2***. *Empirical Structural Validity.* Includes building the confidence of the example problems chosen to verify a suggested design computational framework.
- ***Quadrant 3***. *Empirical Performance Validity.* Used to build confidence in the applicability of a computational framework for the example problems that are chosen.
- ***Quadrant 4***. *Theoretical Performance Validity*. Building confidence in the general use of the computational framework and determining is it useful for other problems beyond the example problems.

In this monograph, the validation square is adopted to illustrate the layout and the connections between the chapters, see Fig. 1.15.

Quadrat 1 Tasks: Theoretical Structural Validation

- Critically review of the relevant literature and identify opportunities, see Sect. 1.4.
- Justify the hypotheses are logically formulated and appropriately cover the research opportunities, see Sect. 1.4.
- Discuss the decision-based design of NES in the context problem discussion and formulation, mathematical background of proposed models, the computational framework, Design for Dynamic Management, of adaptable and concurrent design, operability analysis, and reconfiguration strategy are constructed to verify the hypotheses in intellectual and methodological aspects, see Sects. 2.1–2.3.
- Identify utility, constraints, application domains for the developed computational framework, design for dynamic management, see Sect. 2.4.

Quadrat 2 Tasks: Empirical Structural Validation

- Discuss the adaptability models challenges, Test Example 1, and testing the Hypotheses 1, see Chap. 3.

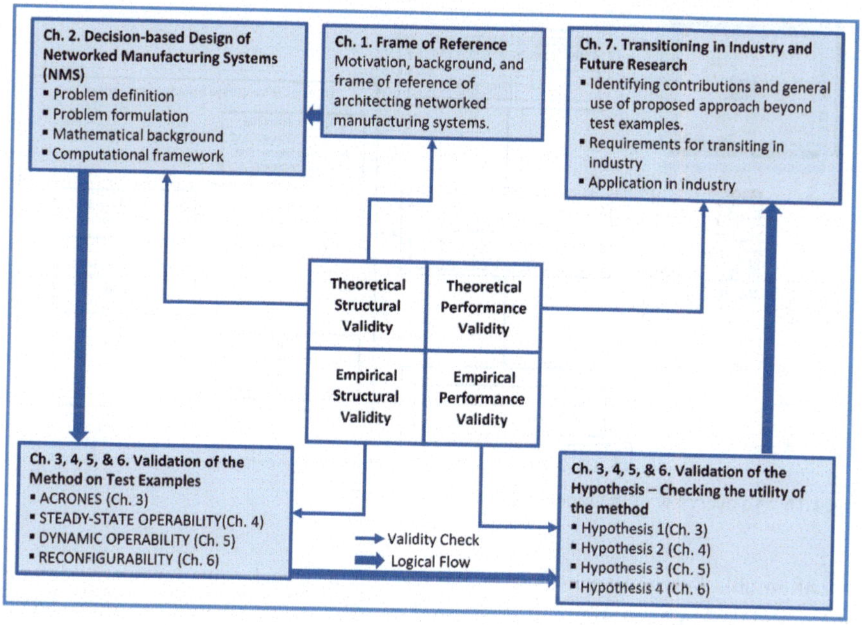

Fig. 1.15 Validation strategy of monograph

- Discuss the operability models challenges, Test Example 2, and testing the Hypotheses 2, see Chaps. 4 and 5.
- Discuss the reconfigurability models, Test Example 3, and testing the Hypotheses 3, see Chap. 6.
- Prove that data is useful for testing the hypotheses; see Sects. 3.5, 4.5, 5.4, and 6.6.

Quadrat 3 Tasks: Empirical Performances Validation

- Validate Hypotheses 1 based on the results in the adaptability model, see Sect. 3.5.
- Validate Hypotheses 2 based on the results in the operability models, see Sects. 4.5 and 5.4.
- Validate Hypotheses 3 based on the results in the reconfigurability models, see Sect. 6.6.

Quadrat 4 Tasks: Theoretical Performances Validation

- Prove the hypotheses are valid but limited for the design of NES, see Sect. 7.3.
- Identify the research gaps necessary to extend the framework presented in this monograph beyond NES to cyber-physical-social systems, see Sect. 7.4.

The organization and a roadmap of the monograph is presented in Fig. 1.16.

Chapter 1. The motivation and foundation are discussed for proposed architecture where architecting NES accounting for adaptable and concurrent, operable, and

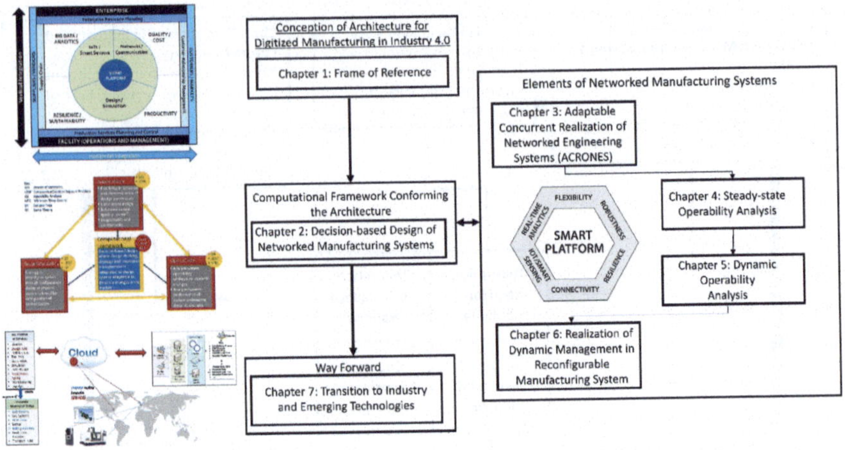

Fig. 1.16 An overview of the monograph

reconfigurable design is considered. The theoretical foundations for adaptable and concurrent design, operability analysis, and reconfiguration methods are introduced and discussed. Relevant literature in each of these research areas is referenced, discussed, and critically evaluated in order to prove theoretical structural validity. The availability, strengths, and limitations of methods and constructs are discussed, which is a foundational design of NES, the stream of variation modeling in engineering design of NES, and to identify research opportunities addressed in the monograph. The primal questions, and hypotheses are introduced. The expected contributions are summarized, research gaps for future work are introduced, and a validation strategy is established in the monograph.

Chapter 2. A problem definition of the design of NES is introduced, the mathematical formulation is considered as a background for later chapters, how to structure and partition mathematical problem of high complexity, and how to create an appropriate decision network structure. Further, desired NES characteristic are postulated and a computational framework, Design for Dynamic Management, is presented.

Chapter 3. The adaptability models are considered, (i) without uncertainty (D), and (ii) with uncertainty where its effect on a designer‘s decision is analyzed through solution space exploration. Further, characteristics of these models, compromise DSP formulation, and a strategy to find appropriate solutions are proposed. A numerical example is used to verify the utility of the decision models that are proposed. Structural and performance validity of adaptability models follows validation square structure. In summary, this chapter is uniting the material from Chaps. 1, 2, and 3 together in order to answer the Questions 1.

Chapters 4 and 5. The operability models are considered and its influence on decision making through observation of operability space. The inclusion of different types of uncertainties is considered and its effect system operability is analyzed through disturbance space exploration. Further, characteristics of these models, compromise DSP formulation, and a strategy to find appropriate solutions are proposed. Numerical examples are used to verify the utility of the performance operability models that are proposed. Structural and performance validity of operability models follows validation square structure. In summary, this chapter unifies the material from Chaps. 1, 2, 3 and 4 together in order to answer the Questions 2.

Chapter 6. The inclusion of different reconfiguration methods are considered and its effect on the reconfiguration of the manufacturing system is analyzed through strategies exploration. Further, characteristics of models, compromise DSP formulation, and a strategy to find appropriate configuration/solutions is proposed. Numerical examples are used to verify the utility of the reconfiguration methods. Designer insights are taken into account in these numerical examples. Analysis of the results of these numerical studies is used to demonstrate the practical applicability of the proposed approach. In summary, this chapter unifies the material from Chaps. 1, 2, 3 and 5 in order to answer the Questions 3.

Chapter 7. In Chaps. 2 through 6 of this monograph, an architecture for networked engineered systems (NES) that conforms to the Industry 4.0 construct is proposed and the use of Design for Dynamic Management (DFDM) computational framework for designing systems conforming to this architecture is discussed. Methods for theoretical and structural validation of the design process are discussed using illustrative examples from the manufacturing industry. In this chapter, the results of the DFDM framework are summarized and the challenges in extending these results to cyber-physical systems encountered in the industry are delineated. The state of the art in implementing Industry 4.0 construct is also examined and the technical challenges arising from this rapid evolution of technology are discussed. The digital transformation of the manufacturing industry and its impact on the workforce and society is elucidated. The chapter concludes by framing the questions to foster discussion in the larger community on the roles and responsibilities of individuals, companies, and governments in promoting sustainable and equitable growth of manufacturing industry.

1.6 Synopsis

In this chapter, the conception of the architecture for design of Networked Engineered Systems (NES), gaps, and the questions worth of investigation are introduced. The motivation for the computational framework, Design for Dynamic Management, is introduced through discussion on topic frontiers in Industry 4.0 and what are the requirements for realization of digitized manufacturing systems. The background anchored in concurrent design, concept exploration, and decision-making when the computational models are approximations of reality is established. The frame of reference built in Stream of Variation (SoV) modeling, the compromise Decision Support Problem (cDSP) construct, solution space exploration, verification and validation, and robust design are presented. Further, in this monograph the foundation is created for a way forward by framing the problem, identifying the gaps and questions worthy of investigation.

New trends in production engineering and design are introduced in Sect. 1.1. In Sect. 1.2, we present the background for Design for Dynamic Management followed by concept exploration and making design decisions in the early design stages. In Sect. 1.3, we present the scope of the monograph, such as Stream of Variation (SoV) modeling, the compromise Decision Support Problem (cDSP) construct, solution space exploration, verification and validation, and robust design are presented. In Sect. 1.4, we establish the questions and corresponding hypotheses. Furthermore, we summarize the contributions of the monograph followed by identified gaps and further addressed as a way forward, transition to industry. In Sect. 1.5, we present the validation of the proposed hypotheses strategy. Evaluation of the structural soundness of the monograph and answer questions are performed by revisiting this chapter.

In Chap. 2 we present a computational framework, Design for Dynamic Management, conforming the architecture, see Fig. 1.17. In Sect. 2.1, we discuss the problem in design of networked engineered systems. In Sects. 2.2 and 2.3, we present the comprehensive problem formulation followed by mathematical background of the computational framework, In Sect. 2.4, we introduce the key elements of the computational framework. In the end, we examine the structural soundness of the computational framework, see Sect. 2.5.

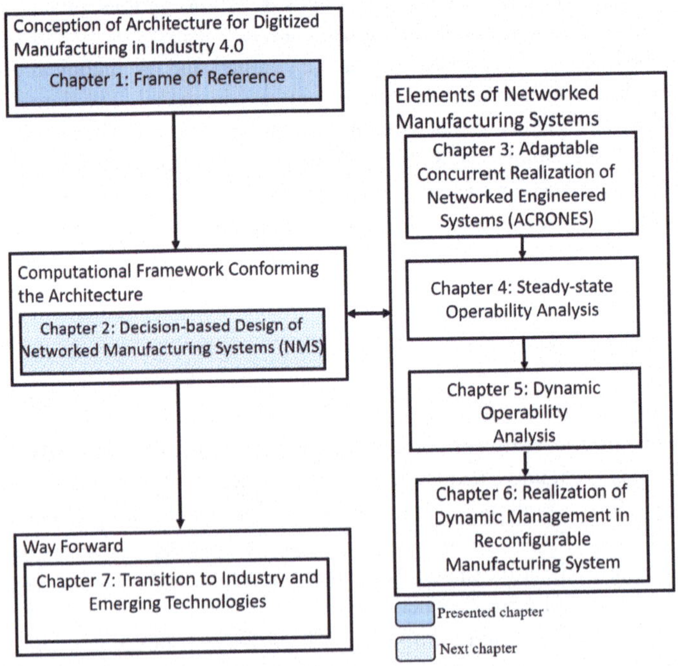

Fig. 1.17 Organization of the monograph—presented and next chapter

Glossary

ACRONES Adaptable Concurrent Realization of Networked Engineering Systems
Big-data Analytics Analysis of large volumes of data
cDSP Compromise Decision Support Problem
Connectivity between the elements in the system Enabling distinct and independent processes to communicate with one another through well-defined protocols and strategies
DFDM Design for Dynamic Management
DSIDES Software for solving Decision Support Problems (DSPs)
DSP Decision Support Problem
Flexibility Selection and determination of design parameters at design time to address the issue of fault-tolerance in the manufacturing process
Internet of Things (IoT)/Smart Sensing Convergence of systems and Internet Technology and allows 'stand-alone' systems to be networked
KPC Key Product Characteristics
MDO Multidisciplinary Design Optimization
MMP Multistage Manufacturing Process
NES Networked Engineered Systems
NMS Networked Manufacturing Systems
PT Programmable tooling
Resilience Ability of a system to withstand disruptions
Robustness Ability of the system to meet design specifications for the output when the system experiences faults or when the parameters of the system are unknown/varying
Smart Manufacturing Digitized Manufacturing
SoV Stream of Variation model
Uncertainty Inherent randomness or unpredictability of a system, model parameters uncertainty, and model structure uncertainty.

References

Allen, J. K., Seepersad, C. C., Choi, H.-J., & Mistree, F. (2006). A survey of robust design with applications to multidisciplinary and multiscale systems. *Journal of Mechanical Design, 128*(4), 832–843.

Allen, J. K., Azarm, S., & Simpson, T. W. (2011). Design complex engineering systems. *Journal of Mechanical Design, 133,* 10.

Anapagaddi, R., Shukla, R., Goyal, S., Singh, A. K., Allen, J. K., Panchal, J. H., & Mistree, F., (2014). Exploration of the design space in continuous casting tundish. In *ASME design Automation Conference*. New York: Buffalo. Paper Number DETC2014-34254.

Assink, M. (2006). Inhibitors of disruptive innovation capability: A conceptual model. *European Journal of Innovation Management, 9*(2), 215–233.

Benner, M. J., & Tushman, M. L. (2003). Exploitation, exploration, and process management: The productivity dilemma revisited. *Academy of Management Review, 28,* 238–256.

Bers, J. A., Dismukes, J. P., Miller, L. K., & Dubrovensky, A. (2009). Accelerated radical innovation: Theory and application. *Technological Forecasting and Social Change, 76*(1), 165–177.

Buede, D. M. (2000). *The engineering design of systems: Models and methods*. New York: Wiley.

Cagan, J., & Williams, B. C. (1993). First order necessary conditions for robust optimality. *ASME advances in design automation* (pp. 539–549). NM: Albuquerque.

Carryer, J. E., Ohline, R. M., & Kenny, T. W. (2011). *Introduction to mechatronic design*. US: Prentice-Hall.

Chakrabarti, A., Shea, K., Stone, R., Cagan, J., Campbell, M., Hernandez, N. V., et al. (2011). Computer-based design synthesis research: An overview. *Journal of Computing and Information Science in Engineering, 11*(2), 021003.

Chen, W., & Lewis, K. (1999). A robust design approach for achieving flexibility in multidisciplinary design. *AIAA Journal, 37*(8), 982–989.

Chen, W., Meher-Homji, C. B., & Mistree, F. (1994). Compromise: An effective approach for condition-based maintenance management of gas turbines. *Engineering Optimization, 22*(3), 185–201.

Chen, W., Allen, J. K., Tsui, K.-L., & Mistree, F. (1996). A procedure for robust design: Minimizing variations caused by noise factors and control factors. *Journal of Mechanical Design, 118*(4), 478–485.

Chen, W., Allen, J. K., & Mistree, F. (1997). The robust concept exploration method for enhancing concurrent systems design. *Journal of Concurrent Engineering: Research and Applications, 5*(3), 203–217.

Choi, H.-J. (2005). *A robust design method for model and propagated uncertainty*. Ph.D. Dissertation. Atlanta, GA: The George W. Woodruff School of Mechanical Engineering, Georgia Institute of Technology.

Choi, H., McDowell, D. L., Allen, J. K., Rosen, D., & Mistree, F. (2008a). An inductive design exploration method for robust multiscale materials design. *Journal of Mechanical Design, 130*(3), 031402.

Choi, H.-J., McDowell, D. L., Allen, J. K., & Mistree, F. (2008b). An inductive design exploration method for hierarchical systems design under uncertainty. *Engineering Optimization, 40*(4), 287–307.

Collopy, P. D., & Hollingsworth, P. M. (2011). Value-driven design. *Journal of Aircraft, 48*(3), 749–759.

Condoor, S., & LaVoie, D. (2007). Design fixation: A cognitive model. In *International Conference on Engineering Design*, no. 507, Paris, France.

Ding, Y., Ceglarek, D., & Shi, J. (2002). Design evaluation of multi-station assembly processes by using state space approach. *ASME Journal of Mechanical Design, 124*(3), 408–418.

Du, X., & Chen, W. (2000). Towards a better understanding of modeling feasibility robustness in engineering design. *Journal of Mechanical Design, 122*(4), 385–394.

Georgescu, C., & Boonstra, V. F. (1990). Concept exploration models for merchant ships. In *TRB, 99th Annual Meeting*, Washington, D.C.

Giunta, A. A. (1997). *Aircraft multidisciplinary design optimization using design of experiments theory and response surface modeling methods*. Doctoral Dissertation. Blacksburg, VA: Virginia Tech.

Gu, X. S., Renaud, J. E., & Penninger, C. L. (2006). Implicit uncertainty propagation for robust collaborative optimization. *Journal of Mechanical Design, 128*(4), 1001–1013.

Haftka, R. T. (1985). Simultaneous analysis and design. *AIAA Journal, 23*(7), 1099–1103.

Hernandez, G., & Mistree, F. (2000). Integrating product design and manufacturing: A game theoretic approach. *Engineering Optimization, 32*(6), 749–775.

Jin, J., & Shi, J. (1999). State space modeling of sheet metal assembly for dimensional control. *Journal of Manufacturing Science and Engineering, 121*(4), 756–762.

Koch, P., Barlow, A., Allen, J., & Mistree, F. (1998). Facilitating concept exploration for configuring turbine propulsion systems. *Journal of Mechanical Design, 120*(4), 702–706.

Kroo, I., Altus, S., Braun, R., Gage, P., & Sobieski, J. (1994). Multidisciplinary optimization methods for aircraft preliminary design. In *5th Symposium on Multidisciplinary Analysis and Optimization Conference* (p. 4325). AIAA.

Lewis, K., Smith, W. F., & Mistree, F. (1999). Ranged set of top-level specifications for complex engineering systems. In U. Roy, J. M. Usher, & H. R. Parsaei (Eds.), *Simultaneous engineering: Methodologies and applications* (pp. 279–303). New York: Gordon and Breach Science Publishers.

Linsey, J. S., Tseng, I., Fu, K., Cagan, J., Wood, K. L., & Schunn, C. (2010). A study of design fixation, its mitigation and perception in engineering design faculty. *Journal of Mechanical Design, 132*(4), 041003.

Luger, G. F., & Stubblefield, W. A. (1990). *Artificial intelligence and the design of expert systems*. Benjamin-Cummings Publishing Co., Inc.

Magnusson, T., Lindström, G., & Berggren, C. (2003). Architectural or modular innovation? Managing discontinuous product development in response to challenging environmental performance targets. *International Journal of Innovation Management, 7*(1), 1–26.

Marston, M., Allen, J. K., & Mistree, F. (2000). The decision support problem technique: Integrating descriptive and normative approaches in decision based design. *Engineering Valuation and Cost Analysis, 3*(2), 107–129.

Mavris, D. N., Bandte, O., & DeLaurentis, D. A. (1999). Robust design simulation: A probabilistic approach to multidisciplinary design. *Journal of Aircraft, 36*(1), 298–307.

McDowell, D. L., Panchal, J., Choi, H.-J., Seepersad, C., Allen, J., & Mistree, F. (2009). *Integrated design of multiscale, multifunctional materials and products*. Butterworth-Heinemann.

McDowell, D. L., Panchal, J. H., Choi, H.-J., Seepersad, C. C., Allen, J. K., & Mistree, F. (2010). *Integrated design of multiscale materials and products*. New York: Elsevier. ISBN-13: 978-1-85617-662-0.

Milisavljevic, J., Commuri, S., Allen, J. K., & Mistree, F. (2018). Steady-state operability in design for dynamic management in realization of networked engineering systems. In *ASME Design for Manufacturing and Assembly Conference*. Quebec City, Quebec, Canada. Paper Number DETC2018–85864.

Milisavljevic-Syed, J., Allen, J. K., Commuri, S., & Mistree, F. (2018). A method for the concurrent design and analysis of networked manufacturing systems under uncertainty. *Engineering Optimization, 51*(4), 699–717.

Milisavljevic-Syed, J., Allen, J. K., Commuri, S., & Mistree, F. (2019a). *Designing smart networked manufacturing systems* (pp. 25–26). Future Steel Forum 2019, Budapest, Hungary.

Milisavljevic-Syed, J., Allen, J. K., Commuri, S., & Mistree, F. (2019b). Design of networked manufacturing systems for industry 4.0. In *52nd CIRP Conference on Manufacturing Systems*. Ljubljana, Slovenia, June 12–14, 2019. Paper Number PROCIR-D-18-00262.

Mistree, F., Smith, W. F., Bras, B., Allen, J. K., & Muster, D. (1990). Decision-based design: A contemporary paradigm for ship design. *Transactions SNAME, 98*, 565–597.

Mistree, F., Smith, W. F., Kamal, S. Z., & Bras, B. A. (1991). Designing *Decisions: Axioms, Models and Marine Applications* (Invited Lecture). In *Fourth International Marine Systems Design Conference (IMSDC91)* (pp. 1–24). Kobe, Japan: Society of Naval Architects of Japan.

Mistree, F., Lautenschlager, U., Erikstad, S. O., & Allen, J. K. (1993). *Simulation reduction using the Taguchi method*. NASA STI/Recon Technical Report N, 94, 17082.

Mistree, F., Bras, B., Smith, W. F., & Allen, J. (1995). Modelling design processes: A conceptual, decision-based perspective. *International Journal of Engineering Design and Automation, 1*(4), 209–221.

Monostori, L. (2014). Cyber-physical manufacturing systems: Roots, expectations and R&D challenges, in variety management in manufacturing. In *47th CIRP Conference on Manufacturing Systems* (pp. 9–13). Procedia CIRP 17, 2014.

Muster, D., & Mistree, F. (1988). The decision support problem technique in engineering design. *The International Journal of Applied Engineering Education, 4*(1), 23–33.

Myers, R. H., & Montgomery, D. C. (1995). *Response surface methodology*. New York: A Wiley Interscience Publication.
Nair, V. N. (1992). Taguchi's parameter design: A panel discussion. *Technometrics, 34,* 127–161.
Ottino, J. M. (2004). Engineering complex systems. *Nature, 427*(6973), 399.
Otto, K. N., & Antonson, E. K. (1993a). Extensions to the Taguchi method for of product design. *ASME Journal of Mechanical Design, 115*(1), 5–13.
Otto, K. N., & Antonson, E. K. (1993b). Design theory and methodology—DTM'93. In T. K. Hight, L. A. Stauffer, & H. R. Parsaei (Eds.), *The method of imprecision compared to utility theory for design selection problems* (Vol. 53, pp. 167–173). New York: ASME.
Pahl, G., Beltz, W., Feldhusen, J., & Grote, K. H. (2013). *Engineering design: A systematic approach*. Berlin: Springer Science and Business Media.
Panchal, J. H., Choi, H.-J., Allen, J. K., McDowell, D. L., & Mistree, F. (2007). A systems-based approach for integrated design of materials, products and design process chains. *Journal of Computer-Aided Materials Design, 14*(1), 265–293.
Parkinson, A., Sorenson, C., & Pourhassen, N. (1993). A general approach for robust optimal design. *ASME Journal of Mechanical Design, 115*(1), 74–80.
Renaud, J. E. (1997). Automatic differentiation in robust optimization. *AIAA Journal, 35*(6), 1072–1079.
Ross, A. M., & Hastings, D. E. (2005). The trade space exploration paradigm. In *INCOSE 2005 International Symposium*.
Russom P. (2011). *Big data analytics*. TDWI Best Practices Report. https://vivomente.com/wp-content/uploads/2016/04/big-data-analytics-white-paper.pdf. Last accessed May 18, 2019.
Seepersad, C. C., Hernandez, G., & Allen, J. K. (2000). A quantitative approach to determining product platform extent. In *ASME Advances in Design Automation Conference*. Baltimore, MD. Paper Number DETC2000/DAC-14288.
Seepersad, C. C., Allen, J. K., McDowell, D. L., & Mistree, F. (2006). Robust design of cellular materials with topological and dimensional imperfections. *Journal of Mechanical Design, 128*(6), 1285–1297.
Shi, J. (2006). *Stream of variation modeling and analysis for multistage manufacturing processes*. Taylor and Francis Group, London, UK: CRC Press.
Shukla, R., Anapagaddi, R., Singh, A. K., Panchal, J. H., Allen, J. K., & Mistree, F. (2015). Exploring the design set points of refining operation in ladle for cost effective desulfurization and inclusion removal. In *ASME Design Automation Conference*, Boston, MA, Paper Number DETC2015-46265.
Siddall, J. (1984). A new approach to probability in engineering design and optimization. *Journal of Mechanisms, Transmissions, and Automation in Design, 106*(1), 5–10.
Simon, H. A. (1980). Cognitive science: The newest science of the artificial. *Cognitive Science, 4*(1), 33–46.
Simpson, T. W., Chen, W., Allen, J. K., & Mistree, F. (1998). Use of the robust concept exploration method to facilitate the design of a family of products, simultaneous engineering: Methodologies and applications. In U. Roy, J. M. Usher, & H. P. Parsaei (Eds.), *Simultaneous engineering: Methodologies and applications*. London, UK: Chapman and Hall.
Simpson, T. W., Maier, J. R., & Mistree, F. (2001a). Product platform design: Method and application. *Research in Engineering Design, 13*(1), 2–22.
Simpson, T. W., Seepersad, C. C., & Mistree, F. (2001b). Balancing commonality and performance within the concurrent design of multiple products in a product family. *Concurrent Engineering, 9*(3), 177–190.
Sinha, A., Bera, N., Allen, J. K., Panchal, J. H., & Mistree, F. (2013). Uncertainty management in the design of multiscale systems. *Journal of Mechanical Design, 135*(1), 011008.
Smith, W. F. (1993). *The modeling and exploration of ship systems in the early stages of decision-based design*. Ph.D. Dissertation. Department of Mechanical Engineering, University of Houston, Houston, TX.

Smith W. F., Milisavljevic, J., Sabeghi, M., Allen, J. K. & Mistree, F. (2014). *Accounting for uncertainty and complexity in the realization of engineered systems* (pp. 195–211). CSDM.

Sobieszczanski-Sobieski, J. (1988). *Optimization by decomposition: A step from hierarchic to non-hierarchic systems.* NASA Technical Report 19810004052.

Steltzner, A., Kipp, D., Chen, A., Burkhart, D., Guernsey, C., Mendeck, G., & Way, D. (2006). Mars science laboratory entry, descent, and landing system. In *Aerospace Conference*, IEEE, U.S.A.

Taguchi, G. (1993). *Taguchi on robust technology development: Bringing quality engineering upstream.* New York: ASME Press.

Tatikonda, M. V., & Rosenthal, S. R. (2000). Technology novelty, project complexity and product development project execution success. *IEEE Transactions on Engineering Management, 47*(1), 74–87.

Tsui, K.-L. (1992). An overview of Taguchi methods and newly developed statistical methods for robust design. *IIE Transactions, 24*(5), 44–57.

Tsui, K.-L. (1996). A critical look at Taguchi's modeling approach for robust design. *Journal of Applied Statistics, 23*(1), 81–95.

Vajna, S., Clement, S., Jordan, A., & Bercsey, T. (2005). The autogenetic design theory: An evolutionary view of the design process. *Journal of Engineering Design, 16*(4), 423–440.

Veryzer, R. W. (1998). Discontinuous innovation and the new product development process. *Journal Product Innovation Management, 15,* 304–321.

Wagner, T. C. (1993). *A general decomposition methodology for optimal system design.* Ph.D. Dissertation. University of Michigan, 9319649.

Wang, G., Shang, X., Yan, Y., Allen, J. K., & Mistree, F. (2018). A tree-based decision method for the configuration design of reconfigurable machine tools. *Journal of Manufacturing Systems, 149,* 143–169.

Wang, R., Wang, G., Yan, Y., Sabeghi, M., Ming, Z., Allen, J. K., et al. (2019). Ontology-based representation of meta-design in designing decision workflows. *Journal of Computing and Information Science in Engineering, 19*(1), 011003.

Weigel, A. L., & Hastings, D. E. (2004). Evaluating the cost and risk impacts of launch choices. *Journal of Spacecraft and Rockets, 41*(1), 103–110.

Williams, T. M. (1999). The need for new paradigms for complex projects. *International Journal of Project Management, 17*(5), 269–273.

Wujek, B. A., Renaud, J. E., Batill, S. M., & Brockman, J. B. (1996). Concurrent subspace optimization using design variable sharing in a distributed computing environment. *Concurrent Engineering, 4*(4), 361–377.

Xu, L. D., Xu, E. L., & Li, L. (2018). Industry 4.0: State of the art and future trends. *International Journal of Manufacturing Research, 56*(8), 2941–2962.

Yu, J. C. (1994). Robust design by matching the design with manufacturing variation pattern. In *Advances in Design Automation-Proceedings of the 20th Annual ASME Design Automation Conference*.

Zhong, J. (2009). *Manufacturing system variation reduction through feed-forward control considering model uncertainties.* Dissertation. University of Michigan.

Chapter 2
Decision-Based Design of Networked Manufacturing Systems (NMS)

Nomenclature

X_k	Part accumulated variation up to Station k including Station k
X_{k-1}	Part accumulated variation up to Station $k-1$ including Station $k-1$
U_k	Control vector at Station k, that is defined as the fixture error vector for both subassembly parts at Station k
Y_k	Measurement obtained on Station k
ξ_k	Noise due to unmodeled effects, independent from other noise
η_k	Sensor noise, independent from other noise
A_{k-1}	Dynamic matrix, characterizes variation change due to part transfer from Station k to/and Station $k+1$
B_k	Input matrix, determines how fixture variation affects part variation at Station k
C_k	Sensor locations information on a station
D_N	Diagnosability matrix
$\rho(\cdot)$	Rank of a matrix
m_k	Number of potential fixture faults at Station k
$\overline{u}_k$	Vector of input parameters at Station k
T_k	Realizability matrix
C_k	Controllability matrix
Y_k	Measurement obtained at Station k
c_1	Monetary cost of total number of sensors
c_2	Monetary cost of sensing station
c_3	Monetary cost for using PT control actions
c_4	Monetary cost for reducing the number of sensing station
s_k	Estimated control actions
Q_k	Weighting coefficient matrix, shows differences in the importance and characteristics of the measured points

J. Milisavljevic-Syed et al., *Architecting Networked Engineered Systems*,
https://doi.org/10.1007/978-3-030-38610-8_2

$x \in \mathbb{R}^{n_x}$	State vector
$u \in \mathbb{R}^{n_u}$	Input/control vector
$d \in \mathbb{R}^{n_d}$	Disturbance vector
$y \in \mathbb{R}^{n_y}$	Output vector of the process
$\dot{x}$	Time derivative of the state vector
$\dot{u}$	Time derivative of the input/control vector
$f : \mathbb{R}^{n_x+n_u+n_d}$	Nonlinear function
$g : \mathbb{R}^{n_x+n_u+n_d} \rightarrow \mathbb{R}^{n_y}$	Nonlinear function
μ	Measure function for calculating the size of the corresponding space
$t_f^*(y_{sp}, d)$	Minimum time necessary to respond to a change in the set-point, y_{sp}, and to a disturbance d, and
$\mathcal{M}$	Include the final-time constraints
$t_f^d(y_{sp}, d)$	Desired dynamic performance, or the maximum allowable response time, in tracking a set-point change, y_{sp}, in DOS and/or recovering from disturbance, d, in EDS
CT	Configuration tree
u	Nodes
v	Edges
U	Node set of the RMT configuration tree
V	Edge set of the RMT configuration tree
u_{ijk}	Element in the node set U
i	Type of a node; see Table 2.1
j	Identifier of the node of the same type
k	Section of the configuration tree to which the node belongs
v	Ordered pair that specifies one edge in the configuration tree
u_{ijk}	Parent node in the edge
$u_{i'j'k'}$	Child node
f_1	Discriminant function of the edge

In the global economy characterized by dynamic markets changes in customer preferences necessitate timely adjustments in manufacturing. Conventional manufacturing processes are designed for mass manufacturing and are not suited for agile, flexible and highly reconfigurable smart manufacturing. Industry 4.0 attempts to address the

Table 2.1 Variable i and corresponding node type

i	u_{ijk}	Node type
1	u_{1jk}	Spindle head
2	u_{2jk}	Fixture
3	u_{3jk}	Slide
4	u_{4jk}	Cross-slide
5	u_{5jk}	Column
6	u_{6jk}	Rotary table
7	u_{7jk}	Base

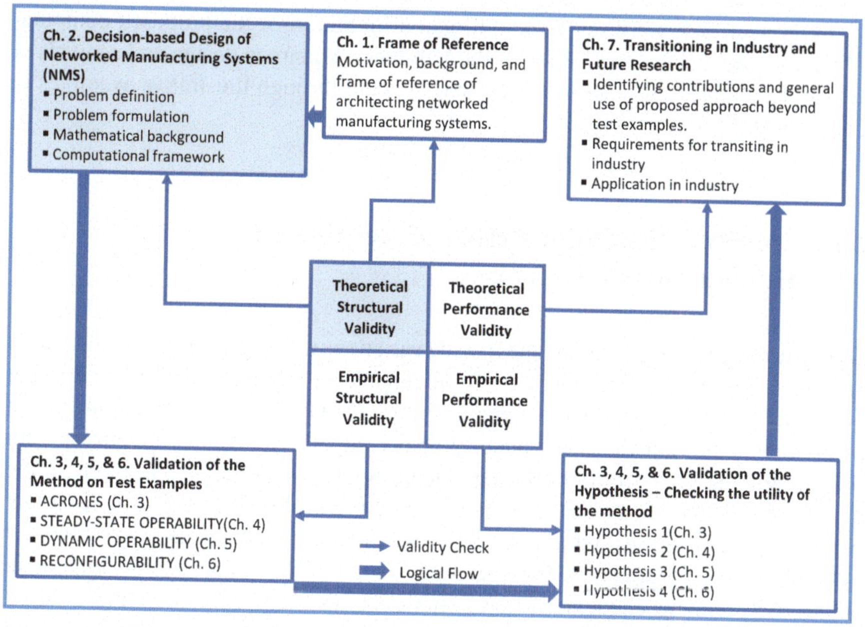

Fig. 2.1 Validation square roadmap

design of smart manufacturing systems by leveraging the advantages of digitization. While digitization helps in connecting erstwhile distributed systems and enables the use of digital models of physical processes in design and analysis, it does not automatically translate into the design of "smart" systems. Therefore, we propose architecting smart Networked Manufacturing Systems that embody the principles of Design for Dynamic Management (DFDM). The DFDM Framework is a service-oriented product/system development computational framework that facilitates the harmonization of conflicting needs associated with the design and manufacture of engineered systems, such as flexibility, robustness, operability, connectivity and resilience, and the incorporation of real-time analytics and smart sensing (see Fig. 1.2). The DFDM framework has three critical components, namely, adaptable and concurrent design, operability analysis and reconfiguration strategies (see Fig. 1.4). These concepts are discussed in detail in Chaps. 3–6 wherein the 'Validation Square' approach in Fig. 2.1 is utilized to illustrate progress through the monograph.

In the design process, we first check the validity of the theoretical structure assumed for the design problem being solved, see Quadrant 1 in Fig. 2.1. We provide a problem formulation, see Sect. 2.2, based on the compromise Decision Support Problem (cDSP) construct further particularized for formulation of the problem, design and analysis of networked manufacturing systems. Furthermore, we build confidence in the soundness of the proposed computational framework by providing a mathematical background of the computational framework, see Sect. 2.3, where we present different components of the computational framework with different models.

Lastly in Sect. 2.4, we discuss the utility and limitation of the proposed computational framework. The application of the Validation Square in design and analysis of networked engineered systems is explored in detail through illustrative examples in Chaps. 3–6.

2.1 Problem Definition: Design of Networked Manufacturing Systems

Dynamic changes in the market due to wide variations in customer needs lead to mass customization where enterprises must be capable of adjusting the manufacturing processes according to the wide variations of product design and substantial change of product scale. On the other hand, global competition requires enterprises not only to provide the cost-effective manufacturing processes but also to improve the quality of product and shorten time to market. Smart manufacturing is increasingly being adopted by companies to respond to these changes in the market. It is required to obtain flexible production, big data analysis in real time, and establish/maintain connectivity among elements in the system. These smart manufacturing systems must be adaptable to dynamic changes and respond to unexpected disturbances, and uncertainty and these are the challenges addressed in this monograph.

In order to maintain low cost and high-quality of the product, the manufacturing system should be capable of producing a wide range of products with high throughput, quality, and low set up and operational costs. It is often difficult to identify one design that meets all the criteria and the design engineer is faced with the task of selecting a design from a set of feasible designs. Further, this choice frequently involves making tradeoffs between requirements and desired performance in the product. In order to facilitate this design approach, we propose a decision-based design computational framework, Design for Dynamic Management (DFDM), to support the design of adaptable, operable and rapidly configurable manufacturing systems, see Fig. 2.2.

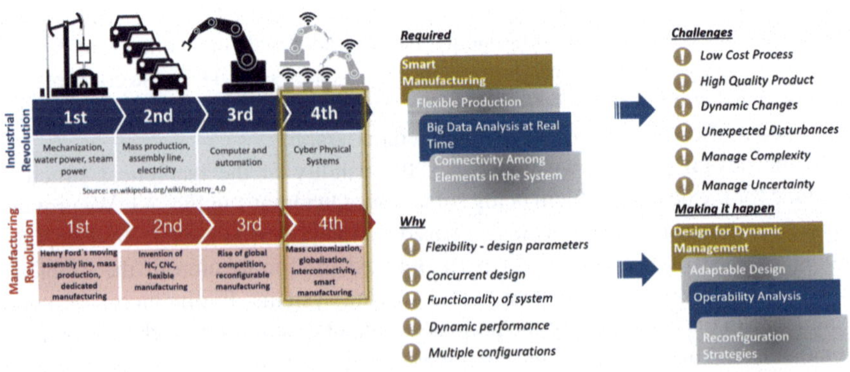

Fig. 2.2 Addressing smart manufacturing features through design for dynamic management

The key components of the computational framework, are (1) adaptability, where flexible production is possible with flexible design; (2) analyzing system functionality through system operability analysis; and (3) reestablishing connectivity among elements in the system through reconfigurability strategies. These components are discussed in detail in Sect. 2.4.

In this section, we state the design problem (Sect. 2.1.1), identify research gaps and provide the motivation for the approach (Sect. 2.1.2), and introduce the illustrative examples that are used for Case Studies in this monograph (Sect. 2.1.3).

2.1.1 Problem Statement: Design of Networked Manufacturing Systems as a Support to Smart Manufacturing

What is the problem? The problem is to design smart NMS in the context of Industry 4.0 that can adapt to changing market demands and produce high quality products at low cost.

What are the challenges? The challenges associated with the problem are (1) designing a flexible, adaptable manufacturing process, (2) ensuring the quality of the product, (3) addressing changes in product requirements in real-time, (4) accommodating unexpected disturbances, and (5) managing system complexity.

What are the requirements? In the design of NMS adaptable to dynamic market demands, foundational requirements are identified in order to address prior mentioned challenges and are then categorized into three groups:

Requirement 1: Flexible Production through Flexible Design

- Flexibility in selection and determination of the values of design parameters with incomplete knowledge of the requirements; and
- Concurrent design of a mechanical and control sub-systems.

Requirement 2: Operability Analysis

- Determining regions in the design space that satisfy performance requirements; and
- Determining the dynamic performance of the system (stability and transient performance) in the presence of change.

Requirement 3: Reconfiguration Strategy

- Strategy for reconfiguration of mechanical and control sub-systems in order to reestablish connection of elements in the manufacturing system.

2.1.2 Identified Gaps

After critical review of the literature (over 250 publications) we identify gaps that need to be bridged to achieve the described functionality; see Sect. 2.1.1.

Gap 1: Adaptability

Gap 1 is associated with Requirement 1. Adaptability is reflected in need for (1) concurrent design of mechanical and control sub-system in order to avoid future rework in later stages of design; (2) managing complexity of mixed variable problems; (3) extensive solution space exploration in order accommodate flexibility in selection of design parameters and determination of the values of design parameters; and (4) include different types of uncertainty in order to explore robust solutions that will result with adaptable NMS.

Gap 2: Operability

Gap 2 is associated with Requirement 2. After critical review of the literature, we suggest that there is a need for operability in the design of NMS; see Chaps. 4 and 5. Operability is reflected in need for (1) analyzing system functionality at steady-state in order to determine system functionality in the presence of change; (2) analyzing system functionality in the dynamic state in order to determine dynamic performance of the system in the presence of change; and (3) overall system functionality in the presence of change.

Gap 3: Reconfigurability

Gap 3 is associated with Requirement 3. After critical review of the literature, we suggest that there is a need for reconfigurability in the design of NMS; see Chap. 6. We identify the need for (1) strategy for reconfiguration of machine tool; (2) strategy for reconfiguration of inspection system; and (3) strategy for reconfiguration of manufacturing system in order to remain competitive in a global market.

Identifying Gaps 1–3 led to the primary question worth investigating "*In the context of Industry 4.0 what is the computational framework that facilitates the decision-based design of Networked Manufacturing Systems adaptable to dynamic market changes as a support to further digitization?*".

2.1.3 Illustrative Examples

In this monograph, four illustrative examples are used to verify different components of the decision-based design computational framework Design for Dynamic Management (DFDM). We use a panel assembling process as the first example to verify adaptability, a continuous stirred tank reactors as the second example to verify operability, a gear box as the third example to verify model-level and device-level reconfigurability, and an engine cylinder block as the fourth example to verify system-level reconfigurability.

Illustrative Example 1: Panel Assembling Process

Panel assembly represents an integral part of majority manufacturing processes, one of those processes is the assembly of automobile bodies. Panel assembling process represents a simplification of an auto body assembling process (Apley et al. 1998) and panel assembling is considered in this monograph. Further, the assumption is that the panels substitute the auto body parts with 100% dimensional accuracy. The dimensional quality of the completed product is highly dependent on the level of accuracy with which the panels are tooled (fixtured). The issue of design and maintenance of accurate fixturing in assembling processes and dimensional quality of the workpieces is important and worth of investigation.

The example used to illustrate adaptability and steady-state operability models in this monograph is a two-dimensional panel assembly process, see Fig. 2.3. There are three stations in the process to assemble four parts (marked as I, II, III, IV in Fig. 2.3) and monitor the assembling process (Ding et al. 2002a, b). Further, parts I and II are assembled at Station 1 (subassembly I + II) is assembled with parts III and IV at Station 2, and the final assembly with four parts are inspected at Station 3 for dimensional defects. Each part is restrained by a set of fixtures consisting of a four-way locator, that controls motion in both x and z-directions, and a two-way locator, that controls motion only in the z-direction. A subassembly with several parts also needs a four-way locator and a two-way locator to completely control its degrees of freedom. The active locating points are marked as Pi, i = 1, …, 8; see Fig. 2.3. In this example, N coordinate sensors are installed on all three stations. Each coordinate sensor measures the position of a part feature, such as a

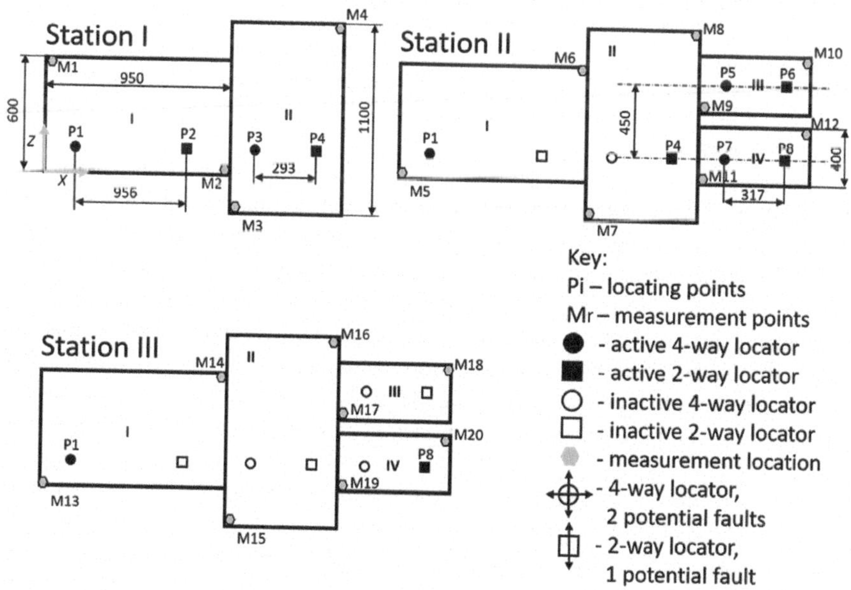

Fig. 2.3 Two-dimensional panel assembling process (Ding et al. 2002a, b)

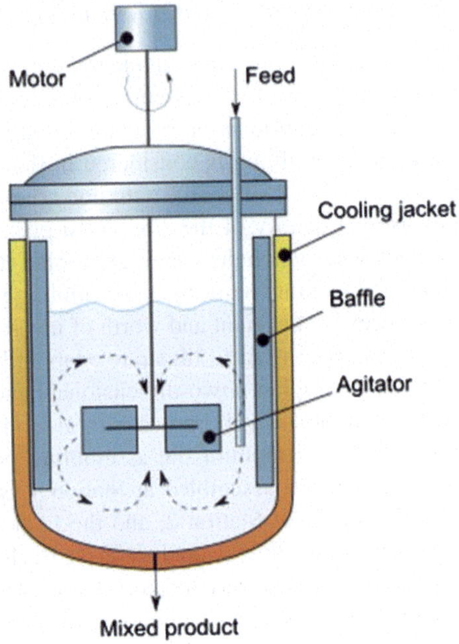

Fig. 2.4 Continuous stirred tank reactors (https://en.wikipedia.org/wiki/Continuous_stirred-tank_reactor)

corner, in two orthogonal directions (x and z). The measurement points are marked as (Mi, i = 1, …, N); see Fig. 2.3.

Illustrative Example 2: Continuously Stirred Tank Reactors

The continuous stirred-tank reactor (cSTR) is a common ideal reactor type in chemical engineering; see Fig. 2.4. A cSTR is a model used to estimate the key unit operation variables when using a continuous agitated-tank reactor to reach a specified output (Schmidt 1998).

The behavior of a cSTR is approximated by a Continuous Ideally Stirred-Tank Reactor (cISTR) wherein perfect mixing is assumed (Schmidt 1998). In a perfectly mixed reactor, the output composition is identical to the composition of the material inside the reactor, which is a function of residence time and rate of reaction. If the residence time is 5–10 times the mixing time, this approximation is valid for engineering purposes. The cSTR model is often used to simplify engineering calculations and can be used to describe research reactors. In practice, it can only be approached, in particular in industrial size reactors. cSTR's are known to be one of the systems which exhibit complex behavior such as steady-state multiplicity, limit cycles, and chaos.

Continuous flow stirred-tank reactors are usually applied in wastewater treatment processes. cSTRs facilitate rapid dilution rates that make them resistant to both high pH and low pH volatile fatty acid wastes. cSTRs are less efficient compared to other types of reactors as they require larger reactor volumes to achieve the same reaction rate as other reactor models such as Plug Flow Reactors (Schmidt 1998).

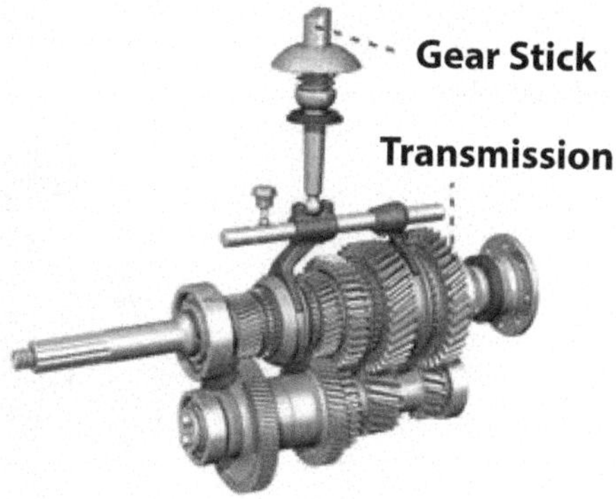

Fig. 2.5 Gear box for vehicles (http://way2science.com/gear-box-in-vehicles/)

Illustrative Example 3: Gear Box

A gear box, see Fig. 2.5, in a vehicle connects the engine to the transmission and is crucial to the transfer of power from the engine to the wheels. The quality of the gear box directly affects the alignment of the engine and drive shafts, gears and affects the life and performance of the vehicle. The gear box is a typical box-type part and the main operations in manufacturing this part includes machining a high-precision plane, drilling bearing holes, screw holes, etc. Due to application demands in the field and an unstable demand/supply relationship, the process of manufacturing multi-gear boxes has to address product variances and fluctuating order demands. A complex market environment forces the manufacturing system to be equipped with Reconfigurable Production Line (RMT) to have the ability to convert functionality and scaling the capability.

In Sect. 6.5.1, we present three design scenarios for gear box production: (1) in the pre-planning phase, the design assumes that configuration of manufacturing system has to conform to a given set of operational requirements; (2) the features of the operations change but the quantity remains the same; and (3) the number of the operations changes but the features remain the same.

Illustrative Example 4: Engine Cylinder Block

Different car manufacturers choose different types of engines based on parameters such as displacement or noise reduction. For example, the Audi S3 uses the I4 engine, the Audi A6 uses the V6 engine, and the Audi A8 uses the V8 cylinder engine. With the premise of maintaining low-cost and high-quality, the enterprises produce more diverse parts to meet the requirements and remain competitive. In the automotive industry reconfiguration of manufacturing systems has unlimited application space. Therefore, the system-level reconfiguration strategy is illustrated via manufacturing the engine cylinder block.

A cylinder block is a unit comprising several cylinders (including their cylinder walls, coolant passages, cylinder sleeves if any, and so forth) (https://en.wikipedia.org/wiki/Cylinder_block). In the early decades of the commercialization of internal combustion engines (1880s to 1920s) the main parts of an engine were distinct

Fig. 2.6 Engine cylinder block (https://en.wikipedia.org/wiki/Cylinder_block)

and made as discrete pieces were bolted together. Further, this is still sometimes used in certain applications where it remains advantageous (especially very large engines, but also some small engines). However, nowadays, there are more efficient ways of designing for manufacture (and also for maintenance and repair) of most petrol engines and diesel engines. These generally involve integrating multiple machine elements into one discrete part, and completing the manufacturing operations (such as casting, stamping, and machining) for multiple elements in one setup with one machine coordinate system (of a machine tool or other piece of manufacturing machinery) (https://en.wikipedia.org/wiki/Cylinder_block). This yields lower unit cost of production (and/or maintenance and repair) (https://en.wikipedia.org/wiki/Cylinder_block).

In Fig. 2.6, we present a modern straight-six engine block for a passenger car integrating the crankcase and all cylinders. The cylinder head bolts to the deck surface at top. Many ribs and bosses can be seen on the side of the casting, as well as the passages for cooling fluid opening into the deck (https://en.wikipedia.org/wiki/Cylinder_block).

2.2 Problem Formulation: Design of Smart Networked Manufacturing Systems

In Sect. 2.2.1, we present the compromise Decision Support Problem (cDSP) construct. In Sect. 2.2.2, we present the cDSP formulation particularized for design of the N-stage NMS.

2.2.1 The Compromise Decision Problem (CDSP) Construct

The Decision Support Problem (DSP) construct is based on the philosophy that design is fundamentally a decision making and model-based process (Smith et al. 2014). Key is the concept of two types of decisions (namely, selection and compromise) and that any complex design can be represented through mathematically modelling a network of compromise and selection decisions (Smith et al. 2014). In this monograph, we use the compromise Decision Support Problem (cDSP) a hybrid formulation that incorporates concepts from both conventional mathematical programming and goal programming, and makes use of some new ones (Mistree et al. 1993).

In this section, we present the word formulation of the cDSP construct supported by mathematical expressions and graphical representation; see Fig. 2.7.

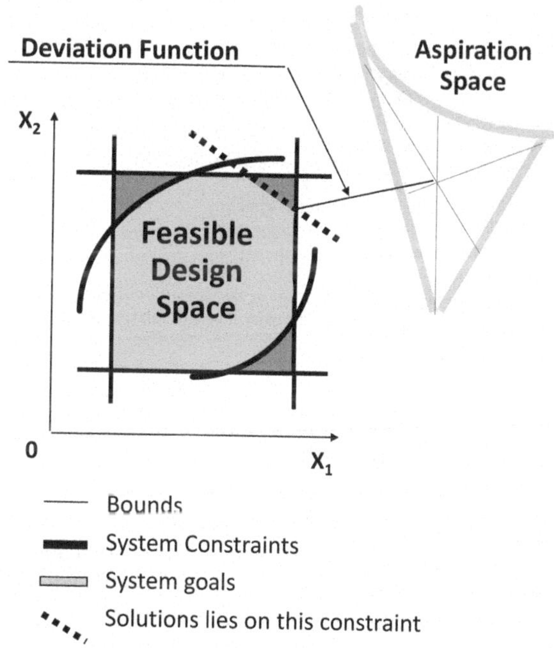

Fig. 2.7 The cDSP model (Milisavljevic-Syed et al. 2018)

Given
An alternative that is to be improved through modification.
Assumptions used to model the domain of interest.
The system parameters.
All other relevant information.

n	number of system variables
$p+q$	number of system constraints
p	equality constraints
q	inequality constraints
m	number of system goals
$g_i(X)$	system constraint function $g_i(X) = C_i(X) - D_i(X)$
$f_k(d_i)$	function of deviation variables to be minimized at priority level k for the preemptive case
W_i	weight for the Archimedean case

Find
The values of the independent *system variables* (they describe the physical attributes of an artifact).

X_j $\quad j = 1,\ldots, n$

The values of the *deviation variables* (they indicate the extent to which the goals are achieved).

d_i^-, d_i^+ $\quad i = 1,\ldots, m$

Satisfy
The *system constraints* that must be satisfied for the solution to be feasible.
There is no restriction placed on linearity or convexity.

$g_i(X) = 0;$ $\quad i = 1,\ldots, p$

$g_i(X) \geq 0;$ $\quad i = p+1,\ldots,p+q$

The *system goals* that must achieve a specified target value as far as possible.
There is no restriction placed on linearity or convexity.

$A_i(X) + d_i^- - d_i^+ = G_i;$ $\quad i = 1,\ldots, m$

The lower and upper *bounds* on the system.

$X_j^{min} \leq X_j \leq X_j^{max};$ $\quad j = 1,\ldots, n$

$d_i^-, d_i \geq 0$ and $d_i^- \cdot d_i^+ = 0$

Minimize
The *deviation function* is a measure of the deviation of the system performance from that implied by the set of goals and their associated priority levels or relative weights:
Archimedean formulation:

$$Z = \sum_{i=1}^{m} W_i(d_i^- + d_i^+); \quad \sum W_i = 1; \ W_i \geq 0.$$

The solution space represents a feasible design space framed by system constraints and bounds; see blue bounded region in Fig. 2.7. The aspiration space framed by a designer's wishes for example, system goals; see red bounded region in Fig. 2.7. The deviation function represents the distance between the aspiration space (red bounded region in Fig. 2.7), and a solution line (dash blue line in Fig. 2.7), which is at the same time system a constraint line. Further, on the solution line we cannot find a single point solution but rather several solutions. The goal is to minimize the deviation function by minimizing the distance between aspiration space and solution space.

2.2.2 *The Compromise Decision Problem (CDSP) Problem Formulation*

The cDSP construct we use to manage the structure and information of decision-making of mixed variable problems. In this monograph, we partition the actual problem on smaller cDSPs and solve in the MATLAB computational environment by the use of different techniques. In the following section we present the cDSP for the problem addressed in this monograph.

Problem Statement: Design of NMS
In Sect. 2.1, we present the problem we address in this monograph. Furthermore, the problem is divided into three parts (1) determine the values of design parameters (tool and sensor attributes) that will give satisficing (robust) solutions regarding the process cost and quality of product; (2) determine input ranges or design configuration that gives functional system design and satisficing dynamic performance of a system in the presence of change; and (3) determine reconfiguration strategy for production line (operational stations and machine tools) and inspection system (sensing stations and sensors) that gives flexible and functional system design while accounting for dynamic changes. The design parameters are: number of working stations is N, number of parts that goes in the manufacturing process is n_p, number of sensors in in the process is M_{Pi}, and number of tools in the process is P_i, and potential fixture failures in the process m_r. Problem requirements are process operability, diagnosability, and –0.8 controllability have to be full, and expected variations, y_k, have to be between –0.8 and 0.8 mm. Problem assumptions that are expected variations parameters follow Gaussian distribution with zero mean and known covariance. The overall objective is to minimize the cost of the process and maximize the quality of the product.

Problem Formulation: Design of NMS
Problem formulation based on the cDSP construct particularized for design of NMS.

Given (parameters)	
Total number of operational stations	N
Number of stamping parts in the process	n_p
Number and position of fixture points	P_i [−]; (x, z) [mm]
Potential number of sensors and position	M_{Pi} [ ]; (x, z) [mm]
Dimensional quality (size of variations) boundary values are set	Y_k [mm]
Find (system variables)	
Total number of sensors and sensing stations	$M_{Pi,M}$; $M_{S,M\,(M=D,C,E)}$
Programmable Tool control actions	PT
Sensing penalties	P
Sensors distribution schemes that are diagnosable, controllable, and cost-effective	$M_{i,k}$
Satisfy (constraints)	
Tooling constraints	Use of programmable tooling; see Eq. 2.6

(continued)

(continued)

Sensing constraints	Number, position and distribution of sensors; see Eq. 2.6
Process diagnosability	100% or partial; see Eq. 2.3
Process controllability	100% or partial; see Eq. 2.5
Satisfy (bounds)	
Lower and upper number of sensors	$M_{Pi,M\,(M=D,C,E)}$; see Sect. 3.2
Lower and upper number of sensing stations	$M_{S,M\,(M=D,C,E)}$; see Sect. 3.2
Use of programmable tooling	PT; see Sect. 3.2
Lower and upper limit of sensing penalties	P; see Sect. 3.2
Deviation variables	$d_i^-, d_i^+ \geq 0$; see Sect. 3.2
Satisfy (goals)	
The mathematical foundations of goals Gi (i = 1, 2, 3) in control theory are described in Sect. 2.3	
Goal 1: Maximize process adaptability	See Sect. 3.2
Goal 2: Maximize process operability	See Sects. 4.4 and 5.4
Goal 3: Maximize process reconfigurability	See Sect. 6.3
Minimize (deviations)	
Minimize deviation function	$Z = min \sum_{i=1}^{3}(W_i d_i^- + W_i d_i^+);\ \sum_{i=1}^{3} W_i = 1;\ W_i \geq 0;$ $d_i^-, d_i^+ \geq 0;\ d_i^+ \cdot d_i^- = 0 (i = 1 - 3);$ See Sect. 2.2.1

2.3 Mathematical Background for the Computational Framework DFDM

In this section, we present the mathematical background of the computational framework, Design for Dynamic Management. The computational framework has three components; see Fig. 2.10. In Sect. 2.3.1, we present the adaptability model under adaptable design (first component of the DFDM). In Sect. 2.3.2, we present the operability model under operability analysis (second component of the DFDM). In Sect. 2.3.3, we present the reconfigurability model under reconfiguration strategy (third component of the DFDM).

2.3.1 Adaptability Model

The adaptability model is a comprehensive computational model based on Stream of Variation (SoV) model. The SoV model we use to simulate NMS. We further partition the adaptability model into process-decision models (diagnosability, controllability, and cost-effectiveness), and product-decision models (performance measurement model) due to its high computational complexity. Both process- and product-decision models are formulated as cDSP's for effective solution space exploration.

Stream of Variation Model

In this monograph, the design of smart NMS that will produce high-quality, low cost products that are produced in a process that is adaptable to ambitious market demands. It is a well-known fact that the better manufacturing process is, the better the quality of the product. Operational tools, such as fixture locators in the assembling process, manage dimensional accuracy of a product are used on each station in the process. For decades process performance was related to the product-inspection-oriented philosophy, where fixtures locators are not directly measured after being installed but the measurements are taken on the finished product (Jin and Shi 1999). The propagation of fixtures variation contributed from each station and its impact on the final product quality are described by the stream-of-variation model (Jin and Shi 1999):

$$X_k = A_{k-1} \cdot X_{k-1} + B_k \cdot U_k + \xi_k \tag{2.1}$$

$$Y_k = C_k \cdot X_k + \eta_k \tag{2.2}$$

In Eq. 2.1, we present the state equation where we imply that part variation on Station k is influenced by two sources (1) accumulated variation up to Station k − 1, and (2) variation at Station k. In Eq. 2.2, we present the observation equation.

The SoV model is primarily developed for multistage assembly and machining processes. The diagram of SoV in NMS for an N-stations assembling process; see Fig. 1.8. Further, the SoV model integrates the process and/or product design and quality information. Diagnosability, controllability, cost-effectiveness, and performance measurement are based on SoV model.

Process Diagnosability

Process diagnosability is the capability of the process to detect faults and identify their causes, according to control theory (Ding et al. 2002a, b). Diagnosability criteria is based on dimensional Stream of Variations (SoV) model; see Eqs. 2.1 and 2.2. The state-space formulation is used to calculate the diagnosability matrix (D_N) (Ding et al. 2002a, b). In Eq. 2.3, we present the diagnosability index in design of NMS.

$$\mu = \frac{\rho(D_N)}{\sum_{k=1}^{N} m_k} \tag{2.3}$$

The diagnosability index lies between [0, 1], where 1 means that the process is completely diagnosable. The process diagnosability index (Ding et al. 2002a, b) is diagnosability criteria and used in the design of NMS.

Process Controllability

Process controllability is the capability of the process to mitigate the errors and drive the system from an arbitrary state to a desired state along specified state trajectories (Mantripragada et al. 1999). In Eq. 2.4, we present the controllability criteria connected with output controllability (Mantripragada and Whitney 1999).

$$U_k = T_k \cdot \overline{u}_k \tag{2.4}$$

The term realizability is a property of the control vector U_k signifying that there are solutions that control the degrees of freedom of the workpiece (Mantripragada and Whitney 1999). Realizability matrix is further transformed into the controllability matrix (Mantripragada and Whitney 1999) and is used in this paper as the process controllability index; see Eq. 2.5.

$$\mu_C = \frac{C_k}{Y_k} \tag{2.5}$$

The controllability index lies between [0, 1], where 1 means that the process is completely controllable.

Cost-effectiveness

Cost-effectiveness is associated with sensing cost. Sensing cost relates to the expense of building sensing stations, using Programmable Tooling (PT) control actions, and penalties for reducing the number of sensing stations. The main assumption is that all parts assembled at any station can be physically measured by sensors at a downstream station for their positional and orientation measurements. There are two ways of measuring product features during production (1) directly where sensors are installed on the assembly station and measurements are taken after the assembly operation is completed, and (2) indirectly where sensors are installed on dedicated stations and workpieces are transferred and measurements are taken. Stations with installed sensors are called sensing stations.

In this monograph, in the cDSP, cost-effectiveness is formulated as a goal function with constraints of the process diagnosability and process controllability; see Eq. 2.6.

$$Cost = \min_{D,C}\left\{ \mathrm{c}_1 \cdot \sum_{\mathrm{i}=1}^{\mathrm{N}} M_{Pi} + \mathrm{c}_2 \cdot \mathrm{M}_s + \mathrm{c}_3 \cdot \mathrm{PT} + \mathrm{c}_4 \cdot \mathrm{P} \right\} \tag{2.6}$$

Performance Measurement

The NMS are modeled as discrete time linear dynamic systems. In this monograph, the main focus is on satisfying the dimensional quality of the workpiece through minimum control actions. Hence, dimensional quality is measured through a performance measurement that is formulated as a minimum effort problem. In order to control the assembling process, we are using a combination of Feedforward Control (FWC) and Programmable Tooling (PT). This combination of FWC and PT is a method for minimizing product variation in NMS. FWC allows minimization of variations on a part-by-part basis using PT. The design of a control system includes variations estimation, modeling and analysis of variation propagation, and process/parts constraints, therefore, a control law is obtained using constrained optimization (Milisavljevic 2015).

The performance measurement model we present as a constrained optimization problem where expected variations are minimized regarding estimated control actions; see Eq. 2.7.

$$PM = \min_{s_k} \sum_{k=1}^{N} \left[\overline{y}_k^T \cdot Q_k \cdot \overline{y}_k \right] \tag{2.7}$$

2.3.2 *Operability Model*

In this section, we introduce steady-state operability in order to determine the input ranges sufficient to achieve the desired output ranges in the presence of expected disturbances. However, for well-designed system steady-state operability is not a sufficient requirement, therefore, dynamic operability is introduced. Dynamic operability is introduced to find the minimum time within the process can respond to a disturbance or move to a new operating point with the available ranges of inputs. Hence, dynamic operability characterizes the inherent operability characteristics of the process. Steady-state operability and dynamic operability gives us information whether the whole system needs to be altered as no other controller will be able to improve the operability. This allows us to address both the servo and regulatory issues over the entire operating space of interest.

Steady-State Operability

The definition of operability is given by Vinson (2001) "A process is operable if the available set of inputs is capable of satisfying the desired steady state and dynamic performance requirements defined at the design stage, in the presence of the set of anticipated disturbances, without violating any process constraints".

Process models are usually given in state-space representation; see Eqs. 2.8 and 2.9.

$$\dot{x} = f(x, u, d) \tag{2.8}$$

$$y = g(x, u, d) \tag{2.9}$$

$$h_1(\dot{x}, x, y, \dot{u}, u, d) = 0 \tag{2.10}$$

$$h_1(\dot{x}, x, y, \dot{u}, u, d) = 0 \tag{2.11}$$

Constraints represent the process, product, and safety specifications, and the bounds on the magnitudes and the rate-of-change of the inputs; see Eqs. 2.10 and 2.11. Further, these constraints are applied to the complete time history of the process and/or a certain time of the process.

In general, based on operational requirements, process outputs are classified into two broad categories (1) set-point controlled—outputs to be controlled to a desired value, and (2) set-interval controlled—outputs to be controlled within the desired range (Georgakis et al. 2003).

Operability Index (OI) is defined by different operating spaces (Vinson and Georgakis 1998, 2000). We will consider next operability spaces:

- Available Input Space (AIS)—inputs of the process able to change over a certain range;
- Achievable Output Space (AOS)—the collection of output points achieved by solving the model for the entire AIS;
- Desired Output Space (DOS)—desired operating window for the process outputs; and
- Desired Input Space (DIS)—the set of input values required to reach the entire DOS.

Achievable Output Space (AOS) is a function of u and d, $AOS_u(d^N)$ where AOS is calculated by considering all the points inside the AIS, denoted by the subscript u, when the disturbances are at their nominal values, d^N.

Desired Input Space (DIS) is a function of y and d, $DIS_y(d^N)$ where DIS is calculated by considering all the points inside the DOS.

Once we have outlined all required spaces we can define the Servo Operability Index in the Output Space (SOIOS); see Eq. 2.12.

$$SOIOS = \frac{\mu[AOS_u(d^N) \cap DOS]}{\mu[DOS]} \tag{2.12}$$

The SOIOS indicates how much of DOS region is achieved with AIS. The value of the SOIS is between 0 and 1, where values bellow 1 implies that designer expectations are greater than designed process can deliver. Further, the SOIOS is useful in analyzing the operability of the existing plant designs (Georgakis et al. 2003).

In nonlinear systems the boundaries of a given input space do not necessarily map to the boundaries of the output region. Behavior of nonlinear systems was exhibited by a vinyl acetate reactor studied by Subramanian and Georgakis (Subramanian and Georgakis 2001).

Servo Operability Index in the Input Space (SOIIS) is an index defined for new plant design or the redesign of an existing plant; see Eq. 2.13.

$$SOIIS = \frac{\mu\left[AIS \cap DIS_y\left(d^N\right)\right]}{\mu\left[DIS_y\left(d^N\right)\right]} \tag{2.13}$$

SOIIS can be used to determine what portion of the DIS is attainable. A value of SOIIS between 0 and 1 indicates that AIS is smaller than DIS. In such a case, it may be necessary to increase the available ranges of some of the inputs or modify the inputs in order to expand AIS (Georgakis et al 2003).

For linear systems the SOIOS and the SOIIS are giving us the same values. However, for nonlinear systems this is not the case.

In order to determine the regulatory operability of the process we need to determine the anticipated ranges of disturbances we need to define the *Expected Disturbance Space (EDS)*. In steady state operability the EDS reflects on uncertainties in model parameters (heat of reaction, heat-transfer coefficients, kinetic constants, pressure, etc.) (Georgakis et al. 2003). In Eq. 2.14, we present the Nominal Operability Index (NOI).

$$NOI = \frac{\mu\left[AIS \cap DIS_d\left(y^N\right)\right]}{\mu\left[DIS_d\left(y^N\right)\right]} \tag{2.14}$$

The NOI can be calculated based on the region of disturbances tolerated with the available inputs, keeping the plant at the nominal operating point (Georgakis et al. 2003). This is defined as *Tolerable Disturbance Space (TDS)*. In Eq. 2.15, we present the NOI for TDS.

$$NOI = \frac{\mu[TDS \cap EDS]}{\mu[EDS]} \tag{2.15}$$

Typically, the other objective is to reject the expected disturbances, and, at the same time, be able to reach all the points in the DOS. Hence, the objective is to design a system that is insensitive to the expected disturbances, and, at the same time, be able to reach all the points in the DOS.

Operability Index (OI) is the overall operability of the process; see Eq. 2.16.

$$OI = \frac{\mu[AIS \cap DIS]}{\mu[DIS]} \tag{2.16}$$

DIS is the total Desired Input Space defined as the union of $DIS_d(y)$ for all y in DOS, or the union of $DIS_y(d)$ for all d in EDS; see Eq. 2.17.

$$DIS = \bigcup_{y \in DOS} DIS_d(y) = \bigcup_{d \in EDS} DIS_y(d) \tag{2.17}$$

The OI values are between 0 and 1, where values bellow 1 implies that designed process is not operable.

Dynamic Operability

Dynamic operability is used to quantify the inherent properties of the process (Uztiirk and Georgakis 2001). Dynamic operability measure is defined as "*The shortest time it would take a system to settle to the desired set point after a set-point change and/or a disturbance occurrence*" (Georgakis et al. 2003).

The operability measure is based on the idea that the time spent away from the desired set point is linked to potential losses due to off-specification products. Different types of feedback controllers can be utilized to evaluate this operability measure. However, a performance measure independent of the feedback controller to be used and capable of assessing the inherent limitations of the process is desirable. Minimum-time optimal controller suits these demands very well (Georgakis et al. 2003).

In Eq. 2.18, we present the minimum-time optimal control problem for continuous systems.

$$\begin{aligned} t_f^*\left(y_{sp}, d\right) &= \min_u \int_0^{t_f} dt \\ \text{s.t.}\ &\mathcal{M}\left(\mathrm{x}_0, \mathrm{u}_0, \mathrm{y}_{\mathrm{sp}}, \ \mathrm{d} \ \ \text{given}\right) \end{aligned} \tag{2.18}$$

Dynamic operating spaces are the extension of the operating spaces used in the steady-state operability to the dynamic problem. We will consider next operability spaces:

- Dynamic Available Input Space (DAIS)—set of input variables (constraints on the magnitudes, and the rate-of-change of the input variables; and
- Dynamic Desired Operating Space (DDOS)—space formed by the combination of the DOS, EDS, and desired response times.

In Eq. 2.19, we present the DDOS.

$$\text{DDOS} = \left\{\left(t_f, y_{sp}, d\right) \middle| t_f \le t_f^d\left(y_{sp}, d\right), \forall y_{sp} \in DOS, \forall d \in EDS\right\} \tag{2.19}$$

- Dynamic Achievable Operating Space (DAOS)—operating space that represents the dynamic performance for a given choice of the DAIS, DOS, and EDS.

In Eq. 2.20, we present the DAOS.

$$\text{DAOS} = \left\{\left(t_f, y_{sp}, d\right) \middle| t_f \le t_f^*\left(y_{sp}, d\right), \forall y_{sp} \in DOS, \forall d \in EDS, u \in dAIS\right\} \tag{2.20}$$

The dynamic Operability index (DOI) is defined as "*The fraction of the operating ranges that can be achieved within the desired response time $t_f^d\left(y_{sp}, d\right)$ given the available input ranges in DAIS*" (Georgakis et al. 2003).

Two additional spaces are introduced in order to mathematically define DOI. First is the operating space, S_1, obtained by the combination of the set points in DOS and disturbances in EDS; see Eq. 2.21.

$$S_1 = \{(y_{sp}, d) | \forall y_{sp} \in DOS, \forall d \in EDS\} \tag{2.21}$$

Second is the operating space, S_2, obtained by projecting the intersection of DDOS and DAOS onto S_1, and it represents the ranges of set points and disturbances within $t_f^d(y_{sp}, d)$; see Eq. 2.22.

$$S_2 = \left\{(y_{sp}, d) \middle| t_f \le t_f^d(y_{sp}, d), \forall y_{sp} \in DOS, \forall d \in EDS\right\} \tag{2.22}$$

DOI values are between 0 and 1, where 0 is the worst performance and 1 is the best performance. In Eq. 2.23, we present the mathematical representation of DOI.

$$DOI = \frac{\mu(S_2)}{\mu(S_1)} \tag{2.23}$$

2.3.3 *Reconfigurability Model*

In this monograph, we advocate reconfiguring both the production line (reconfiguration of the operational station and machine tools) and the inspection system in order to reconfigure the entire manufacturing system (RMS). Hence, the reconfigurability model has two sub-models (1) Reconfiguration of Machine Tool (RMS) and (2) Reconfiguration of the Inspection System (RIS).

Reconfigurable Machine Tool (RMS)

The foundation of the RMS is a configuration tree. In Eq. 2.24, we present the computational model of the configuration tree.

$$CT = \{u, v | u \in U, v \in V\} \tag{2.24}$$

In Eq. 2.25, we present the definition of the node set U.

$$U = \{u_{ijk} | i \in (1, \ldots, 7), j \in N, k \in 0, 1\} \tag{2.25}$$

If $k = 0$, the node belongs to the tool-side branch and if $k = 1$, the node belongs to the workpiece-side branch. In Fig. 2.8, we present the example of a configuration tree to describe RMS configuration.

In addition to the node set, the edge set is important. An edge consists of two nodes in an ordered pair. In Eq. 2.26, we present the edge set, V, Table (2.1).

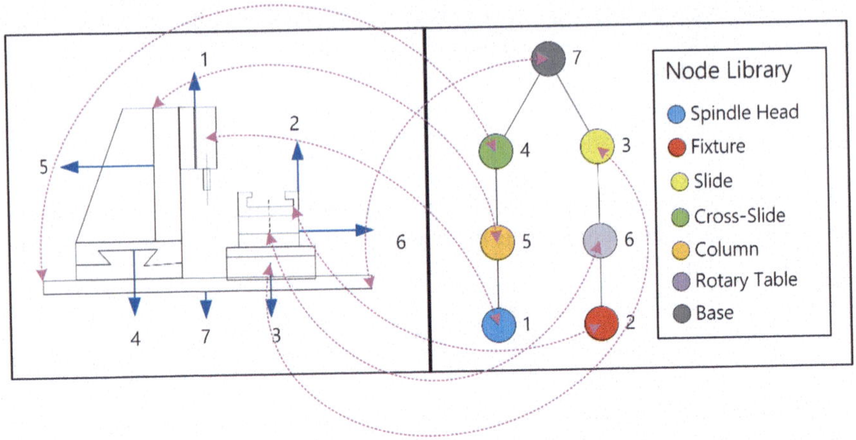

Fig. 2.8 Example of mapping from configuration to configuration tree (Wang et al. 2017)

$$V = \left\{ v = u_{ijk}, u_{i'j'k'} \middle| f_1(v) = 1, \left\langle u_{ijk}, u_{i'j'k'} \right\rangle \in U, i \neq i' \right\} \tag{2.26}$$

$$f_1\left(\left\langle u_{ijk}, u_{i'j'k'} \right\rangle\right) = \begin{cases} 1 \; existing \; edge \; of \; u_{ijk} \; and \; u_{i'j'k'} \\ 0 \; unexisting \; edge \; of \; u_{ijk} \; and \; u_{i'j'k'} \end{cases} \tag{2.27}$$

If the function f_1 equals to 1 the edge between u_{ijk} and $u_{i'j'k'}$ exists in the configuration tree; see Eq. 2.27. On the other hand, if the function f_1 equal to 0 there is no edge between u_{ijk} and $u_{i'j'kp}$. In order to ensure manufacturing stability, the following rules for the assembly relationship among modules are created:

Rule 1: The modules for the same motion are always installed on different branches and there is no assembly relationship among the modules on different branches. In Eq. 2.28, we present the mathematical expression of the Rule 1.

$$\begin{aligned} & u_{ijk}, u_{i'j'k'} = 0 \\ & \sum_{k \neq k'} \sum_{i'} \sum_{j'} u_{ijk}, u_{i'j'k'} = 0 \end{aligned} \tag{2.28}$$

Rule 2: There is at most one assembly relationship between any two modules. The module cannot be assembled with itself. In Eq. 2.29, we present the mathematical expression of the Rule 2.

$$\begin{aligned} & \sum_{j} u_{ijk}, u_{i'j'k'} = 0 \\ & \forall i \neq i' \sum_{j'} u_{ijk}, u_{i'j'k'} \leq 1 \end{aligned} \tag{2.29}$$

Rule 3: The module with the tools and the module for the workpiece are always installed on different branches. In Eq. 2.30, we present the mathematical expression of the Rule 3.

$$\langle u_{1jk}, u_{2j'k}\rangle \, and \, \langle u_{2jk}, u_{1j'k}\rangle = 0 \tag{2.30}$$

Rule 4: A module for rotary motion is always installed on a module for linear motion. In Eq. 2.31, we present the mathematical expression of the Rule 4.

$$\langle u_{6jk}, u_{3j'k}\rangle, \langle u_{6jk}, u_{4j'k}\rangle \, and \, \langle u_{6jk}, u_{5j'k}\rangle = 0 \tag{2.31}$$

Rule 5: For linear motion, a module for up and down motion is always installed on a module for forward and back motion, while the module for forward and back motion is always installed on a module for left and right motion. In Eq. 2.32, we present the mathematical expression of the Rule 5.

$$\langle u_{4jk}, u_{3j'k}\rangle, \langle u_{5jk}, u_{4j'k}\rangle \, and \, \langle u_{5jk}, u_{3j'k}\rangle = 0 \tag{2.32}$$

Reconfigurable Inspection System (RIS)

Typically in the manufacturing system, the inspection station is usually located at the end of the line which is cost-effective and popular in the industry. However, only part of some errors can be detected, while the errors have already caused some quality damage due to delayed detection. On the other hand, in saturated detection, each operational station is followed by an inspection station, which can achieve the complete insurance of product quality but it is very costly and not necessary. Hence, the selection of a minimal but sufficient number of inspection stations to detect errors in the manufacturing process is needed. This can be accomplished by RIS where solutions are located for trade-offs between cost and diagnosability within the RMS manufacturing process. As indicated earlier, diagnosability is based on the SoV model; see Sect. 2.3.1. Accumulation and propagation of errors in the production process are modeled by the SoV model. Therefore, the identified root causes of errors via the SoV model benefits the RIS configuration design in order to achieve the needed error diagnosis. In Fig. 2.9, we present the RMS detection process as a combination of RIS and SoV as a Digital Twin.

We use the phrase Digital Twin to identify the root cause via SoV model and to simulate the manufacturing process. The product quality error refers to the deviation of the actual position and the ideal position, and the main error sources generally include fixturing error caused by the imperfection of locators, datum error caused by the error of datum feature, machine tool error caused by a structural loop of the machine. The fixture error and the machine tool error belong to the current station, while the datum error is influenced by other station. Therefore, the virtual station is simulated by the data of datum measurement. If the virtual station has the same state with the corresponding real station, the real station has good quality, else, the station should be maintained or updated.

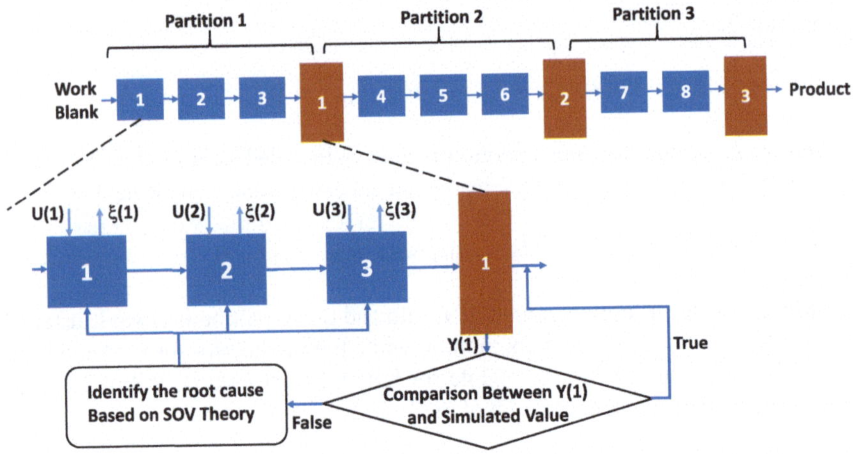

Fig. 2.9 Detection mechanism via integration of RIS and SoV

2.4 Computational Framework: Design for Dynamic Management

In the context of Industry 4.0, we propose a decision-based design computational framework, *Design for Dynamic Management (DFDM)*, as a support to adaptable, operable, and reconfigurable design of NMS. In Fig. 2.10, we present the key components of the computational framework:

i. ***Adaptability in Design of NMS***: Incorporating flexibility in selection and determination of design parameters during (robust) solution space exploration is addressed through the *Adaptable Concurrent Realization of Networked Manufacturing Systems* (ACRONES) method. The ACRONES is a subset of the DFDM, see Adaptability in Fig. 2.10. Two models are developed to support adaptability in the design of NMS. First is the adaptability model for solution space exploration, see Chap. 3. Second is the adaptability model to deal with uncertainty encountered during robust solution space exploration, see Chap. 3.
ii. ***Operability in Design of NMS***: It is based on operability analysis (Fisher et al. 1988; Subramanian et al. 2001), cDSP construct, and minimum-time control (Subramanian et al. 2001). Operability in system design analyzes systems functionality with changes in the requirements; see Operability in Fig. 2.10, and the dynamic performance of the system with a change in requirements. Two models are developed to support operability analysis. First is the Steady-State Operability Model (SSOM) for analyzing system operability at steady-state; see Chap. 4. Second is the Dynamic Operability Model (DOM) for analyzing system operability at dynamic state; see Chap. 5.
iii. ***Reconfigurability in Design of NMS***. It is based on a decision-tree structure (Wang et al. 2017), cDSP construct and a game theoretic approach (Vincent

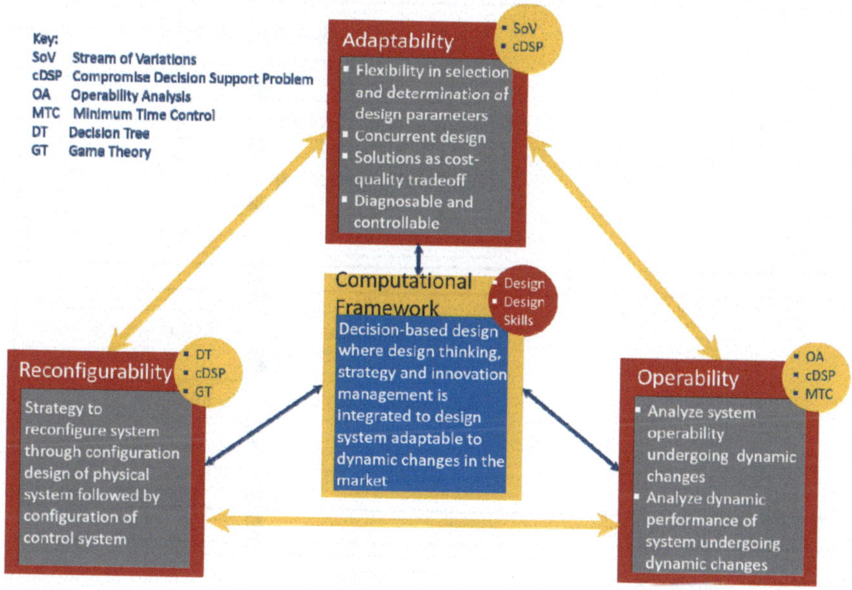

Fig. 2.10 Computational framework, design for dynamic management (Milisavljevic-Syed et al. 2019)

1983). The strategy for reconfiguration supports the repeated systemic reconfiguration; see Reconfigurability in Fig. 2.10. Three models are developed to support reconfirmation strategy; see Chap. 6. First is the RMS strategy for reconfiguration of operational stations and machine tools in the manufacturing process. Second is the RIS strategy for reconfiguration of the inspection system according to needs of the manufacturing processes. Third is the RMS model as a combination of a machine and inspection system.

In Fig. 2.11, we present the steps of the proposed computational framework:

Step A, Fig. 2.11, ***Adaptable Concurrent Realization of Networked Engineering Systems (ACRONES)***: A subset of the Design for Dynamic Management computational framework. The first step, Step A1, is to identify the flexible design parameters, establish their connectivity, and represent the process by a comprehensive state space model. The second step, Step A2, is to determine the interconnections between NMS modules and the mathematical representation of the complete system. Full mathematical representation is presented in Milisavljevic et al. (2017, 2018). The third step, Step A3, is to explore the solution space for appropriate solutions to the design problem. The last step, Step A4, is to identify and manage different types of uncertainty, exploring solution space, and identifying robust solutions to the design problem.

Step B, Fig. 2.11, ***Operability Analysis***: The first step, Step B1, is to connect the output ranges of solutions from ACRONES to input operability spaces of the operability analysis, and output robust ranges of solutions from ACRONES to input operability disturbance spaces in the operability analysis; see Fig. 2.1. The second

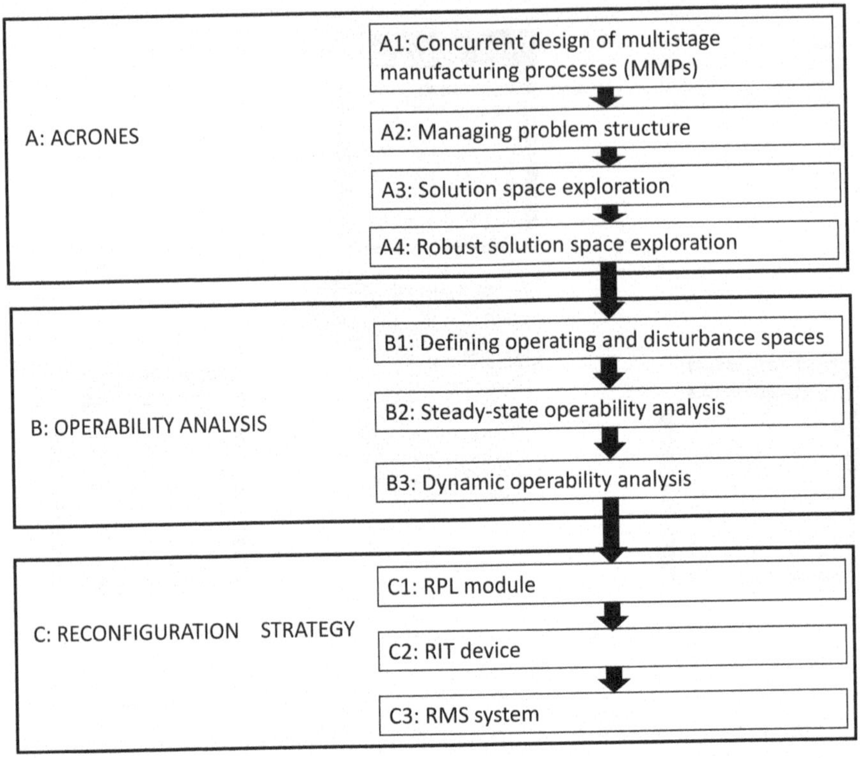

Fig. 2.11 Algorithm, design for dynamic management

step, Step B2, is the steady-state operability analysis. In this step exploration of (a) original system design and obtain input ranges of original system design that gives functional system; and (b) variant system design and obtain a set point for design is taking place. Last step, Step B3, is the dynamic operability analysis in order to explore: (a) functional system due to natural changes in the system over time; and (b) functional system due to change in the requirements.

Step C, Fig. 2.11, ***Reconfiguration Strategy***: The first step, Step C1, of the reconfiguration strategy is to determine the module library and module of RMS in order to obtain the device configuration for the next step. The second step, Step C2, is to use the device configuration in order to determine the sub-system configuration of RIS. The last step, Step C3, is to use sub-system configuration from the previous step in order to determine system configuration of RMS.

2.5 Synopsis

In this chapter we describe the computational framework, Design for Dynamic Management, and address the problem formulation and methods for validating the framework. First, we introduce the problem and formulate it using the compromise Decision Support Problem (cDSP) construct. Second, we describe the mathematics underlying DFDM's key components. The application of this framework is elaborated through illustrative examples, in Chaps. 3–6 we elaborate adaptive and concurrent design where we use a panel assembling process as Illustrative Example 1. In Chap. 4 we elaborate steady-state operability analysis wherein we use a panel assembling process as Illustrative Example 1. In Chap. 5 we elaborate on dynamic operability analysis wherein we use continuous stirred tank reactor as Illustrative Example 2. In Chap. 6 we elaborate different reconfiguration strategies of RMT, RIS, and RMS wherein we use a gear box as Illustrative Example 3 and engine cylinder block as Illustrative Example 4.

In this chapter we present the DFDM framework; see Fig. 2.12. In Chap. 3 we present the adaptable and concurrent design of NMS, in Chaps. 4 and 5 we present the operability analysis of NMS, and in Chap. 6 we present different strategies for reconfiguration of manufacturing systems and elements within the system.

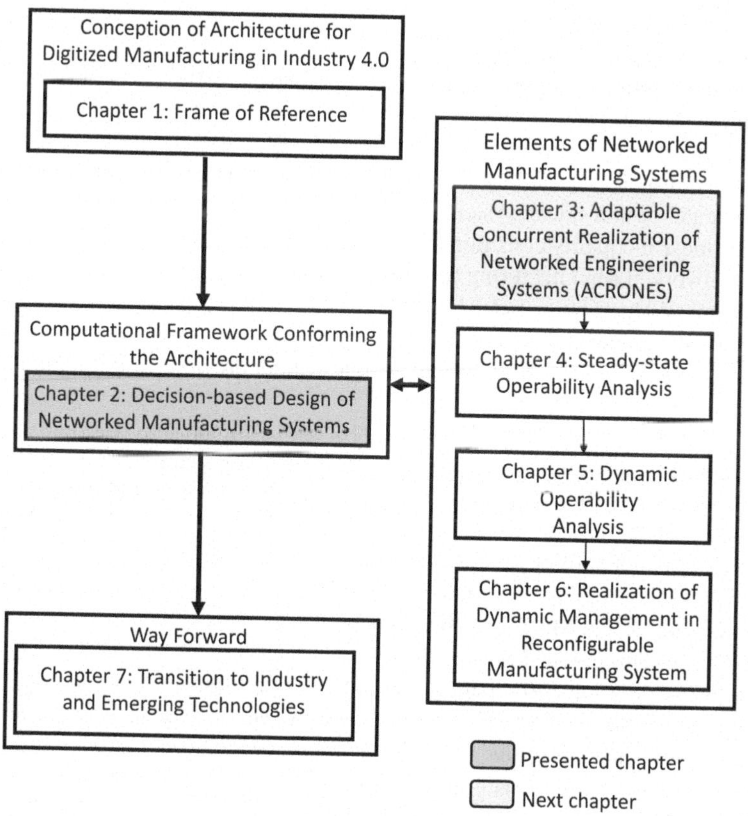

Fig. 2.12 Organization of the monograph—presented and next chapter

Glossary

ACRONES Adaptable Concurrent Realization of Networked Engineering Systems
AIS Available Input Space
AOS Achievable Output Space
Big Data Analysis Ability to analyze large volumes of data
cDSP Compromise Decision Support Problem
Connectivity in the System Enabling distinct and independent processes to communicate with one another through well-defined protocols and strategies
DAIS Dynamic Available Input Space
DAOS Dynamic Achievable Operating Space
DDOS Dynamic Desired Operating Space
DFDM Design for Dynamic Management
DIS Desired Input Space
DOI Dynamic Operability Index
DOM Dynamic Operability Model
DOS Desired Output Space
EDS Expected Disturbance Space
Flexible Production Ability to adapt to changes in the product being manufactured, both in type and quantity
FWC Feedforward Control
Mixed Variable Problems Problems with multiple goals where goal functions are linear and/or non-linear, system variables are continuous, Boolean, linear and/or non-linear inequality constraints, equality constraints, and system boundaries
NMS Networked Manufacturing Systems
NOI Nominal Operability Index
OI Operability Index
Original System Design New system design without previous information to build the model
Process Controllability Capability of the process to mitigate the errors and drive the system from an arbitrary state to a desired state along specified state trajectories
Process Diagnosability Capability of the process to detect faults and identify their cause
PT Programmable tooling
RIS Reconfiguration of Inspection System
RMS Reconfigure of Manufacturing System
RMT Reconfiguration of Machine Tool
Sensing Cost Sensing cost relates to the expense of building sensing stations, using Programmable Tooling (PT) control actions, and penalties for reducing the number of sensing stations
Sensing Stations Stations with installed sensors
Smart Manufacturing Digitized Manufacturing
SOIIS Servo Operability Index in the Input Space
SOIOS Servo Operability Index in the Output Space

SoV Stream of Variation model
SSOM Steady-State Operability Model
TDS Tolerable Disturbance Space
Uncertainty Inherent randomness or unpredictability of a system, model parameters uncertainty, and model structure uncertainty
Variant System Design System design based on the existing information

References

Apley, D., & Shi, J. (1998). Diagnosis of multiple fixture faults in panel assembly. *Journal of Manufacturing Science and Engineering, 120*(4), 793–801.

Continuous stirred-tank reactor. https://en.wikipedia.org/wiki/Continuous_stirred-tank_reactor. Last visited 19 June, 2019.

Ding, Y., Ceglarek, D., & Shi, J. (2002a). Design evaluation of multi-station assembly processes by using state space approach. *Journal of Mechanical Design, 124*(3), 408–418.

Ding, Y., Ceglarek, D., & Shi, J. (2002b). Fault diagnosis of multistage manufacturing processes by using state space approach. *Journal of Manufacturing Science and Engineering, 124*(2), 313–322.

Engine block. https://en.wikipedia.org/wiki/Cylinder_block. Last visited 19 June, 2019.

Fisher, W. R., Doherty, M. F., & Douglas, J. M. (1988). The interface between design and control. 1. process controllability. *Industrial and Engineering Chemistry Research, 27*(4), 597–605.

Gear box in vehicles. http://way2science.com/gear-box-in-vehicles/. Last visited 19 June, 2019.

Georgakis, C., Uztürk, D., Subramanian, S., & Vinson, D. R. (2003). On the operability of continuous processes. *Control Engineering Practice, 11*(8), 859–869.

Jin, J., & Shi, J. (1999). State space modeling of sheet metal assembly for dimensional control. *Journal of Manufacturing Science and Engineering, 121*(4), 756–762.

Mantripragada, R., & Whitney, D. E. (1999). Modeling and controlling variation propagation in mechanical assemblies using state transition models. *IEEE Transactions on Robotics and Automation, 15*(1), 124–140.

Milisavljevic, J. (2015). *Accounting for uncertainty in the realization of multistage manufacturing processes*. School of Aerospace and Mechanical Engineering, Norman, Oklahoma, The University of Oklahoma, MS thesis.

Milisavljevic, J., Commuri, S., Allen, J.K., & Mistree, F. (2017). Concurrent design exploration method (CDEM) of networked engineering systems. In *ASME International Design Engineering Technical Conferences*. Cleveland, Ohio, ASME.

Milisavljevic-Syed, J., Commuri, S., Allen, J. K., & Mistree, F. (2018). A method for the concurrent design and analysis of networked manufacturing systems. *Engineering Optimization, 51*(4), 1–19.

Milisavljevic-Syed, J., Commuri, S, Allen, J. K., & Mistree, F. (2019). Concurrent design exploration method for realizing networked manufacturing systems for Industry 4.0. In *CIRP, 52nd Annual Conference on Manufacturing Systems*. PROC-D-18-00262.

Mistree, F., Smith, W., & Bras, B. (1993). A decision-based approach to concurrent design. In *Concurrent engineering* (pp. 127–158). Springer.

Schmidt, L. D. (1998). *The engineering of chemical reactions (topics in chemical engineering)*. USA: Oxford University Press.

Shang, X., Milisavljevic-Syed, J., Wang, G., Allen, J. K., & Mistree, F. (2019). A key feature-based method for configuration design of reconfigurable inspection system. *International Journal of Production Research, Under Review*. TPRS-2019-IJPR-1260.R1.

Smith, W. F., Milisavljevic, J., Sabeghi, M., Allen, J. K., & Mistree, F. (2014). Accounting for uncertainty and complexity in the realization of engineered systems. CSDM (Posters), Citeseer.

Subramanian, S., & Georgakis, C. (2001). Steady-state operability characteristics of idealized reactors. *Chemical Engineering Science, 56*(17), 5111–5130.

Subramanian, S., Uztürk, D., & Georgakis, C. (2001). An optimization-based approach for the operability analysis of continuously stirred tank reactors. *Industrial and Engineering Chemistry Research, 40*(20), 4238–4252.

Uztiirk, D., & Georgakis, C. (2001). Inherent dynamic operability of processes I: definitions and analysis in the siso case. *Industrial and Engineering Chemistry Research.*

Vincent, T. L. (1983). Game theory as a design tool. *Journal of Mechanisms, Transmissions, and Automation in Design, 105*(2), 165–170.

Vinson, D. R. (2001). A new measure of process operability for improved steady-state design of chemical processes.

Vinson, D. R., & Georgakis, C. (1998). A new measure of process output controllability. *IFAC Proceedings Volumes, 31*(11), 663–672.

Vinson, D. R., & Georgakis, C. (2000) A new measure of process output controllability. *Journal of Process Control*, 10(2–3), 185–194.

Wang, G., Shang, X., Yan, Y., Allen, J. K., & Mistree, F. (2017). A decision tree based method for the configuration design of reconfigurable machine tools. *Journal of Manufacturing Systems, 49,* 143–162.

Chapter 3
Adaptable Concurrent Realization of Networked Engineering Systems (ACRONES)

Nomenclature

X_i	Part accumulated deviation up to station i including Station i
X_{i-1}	Part accumulated deviation up to Station i−1 including Station i−1
U_i	Control vector at station i, is defined as the fixture error vector for both subassembly parts at station i
Y_i	Measurement obtained on station i
V_i	Noise due to unmolded effects
W_i	Sensor noise
A_{i-1}	Dynamic matrix, characterizes deviation change due to part transfer between station i and station i + 1, depends on the change of locating schemes in a production stream
B_i	Input matrix that determines how fixture deviation affects part deviation on station i, based on the geometry of a fixture locating layout
C_i	Information about sensor locations on a station
$L_X(P_i, P_{i+1})$	Distance between coordinate points P_i and P_{i+1} in the body coordinate in the *x*-direction
$L_Z(P_i, P_{i+1})$	Distance between coordinate points P_i and P_{i+1} in the body coordinate in the *z*-direction
x_{P_i}	Coordinate point P_i in the x-direction
$x_{P_{i+1}}$	Coordinate point P_{i+1} in the x-direction
z_{P_i}	Coordinate point P_i in the z-direction
$z_{P_{i+1}}$	Coordinate point P_{i+1} in the z-direction
$L_X(R_{m_r,r}, A_r)$	Distance between first part point A_r in the part and measurement point m_r in x- direction
$L_Z(R_{m_r,r}, A_r)$	Distance between first part point A_r in the part and measurement point m_r in *z*- direction
x_{A_r}	Coordinate first part point A_r in the x-direction
$x_{R_{m_r,r}}$	Coordinate measurement point $R_{m_r,r}$ in the x-direction

J. Milisavljevic-Syed et al., *Architecting Networked Engineered Systems*,
https://doi.org/10.1007/978-3-030-38610-8_3

z_{A_r}	Coordinate first part point A_r in the z-direction
$z_{R_{m_r,r}}$	Coordinate measurement point $R_{m_r,r}$ in the z-direction
M_i	Total number of sensors
M_S	Total number of sensing stations
C	Cost of the process
$M_{i,k}$	Sensor distribution scheme
M_{pi}	Potential number of sensors at station i
P_i	Fixture points

In Chap. 2, we introduced Design for Dynamic Management (DFDM), a computational framework for designing flexible, operable and rapidly reconfigurable manufacturing processes. DFDM has three critical components, namely, adaptable and concurrent design, operability analysis, and reconfiguration method (see Fig. 3.1). *Adaptable Concurrent Realization of Networked Engineering Systems (ACRONES)* (highlighted quadrant in Fig. 3.1) is a method that is used to implement the adaptable and concurrent design component in the DFDM framework. In this chapter, we describe the use of ACRONES in the selection of design parameters and the concurrent design of the mechanical and control systems.

In Sect. 3.1 we present the state of the art in methods for designing adaptable and concurrent NMS. In Sect. 3.2 we identify the gaps that need to be bridged to make possible the design of smart networked manufacturing systems. The gaps are associated with (1) introducing flexibility in the selection of tools and sensors at

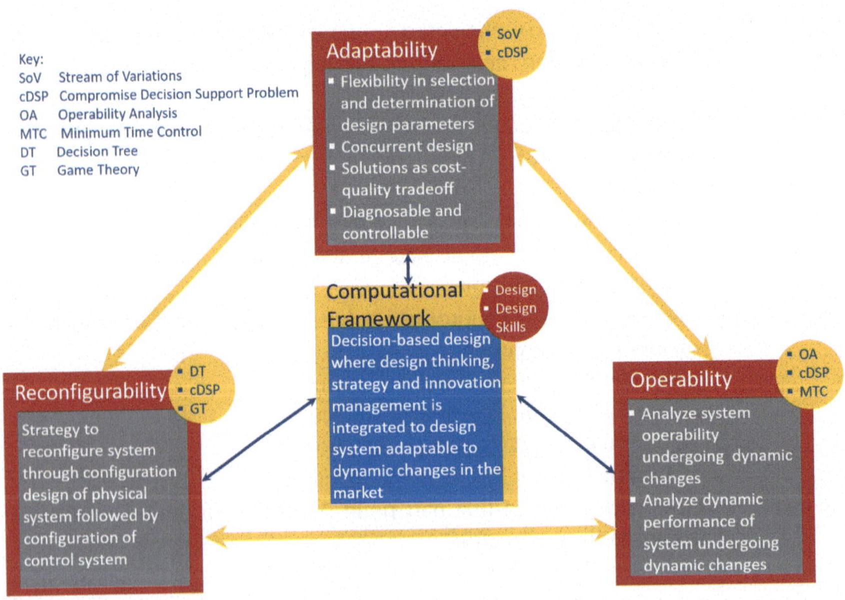

Fig. 3.1 Computational framework, design for dynamic management

design time, (2) managing complexity in the creation of a digital twin of the manufacturing process, (3) exploring the solution space to determine feasible solutions, and (4) managing uncertainty in design of NMS are all addressed. In Sect. 3.3, we present the ACRONES method. In Sect. 3.4, we use Illustrative Example 1 (2-D panel stamping process) to demonstrate the efficacy of the method. In Sect. 3.5, the results are analyzed to develop insights into the factors affecting process cost and dimensional quality of the product, and the tradeoffs between cost and quality during the design of NMS.

3.1 Adaptable and Concurrent Design: A Critical Review

A Networked Manufacturing System (NMS) refers to an interconnected system of operational stations and components required to manufacture a product (Shi and Zhou 2009). Production in an NMS is a continuous process where the product is manufactured stage by stage. Local variations at each stage and interactions amongst multiple stages affect the quality of the manufactured product. Several authors have focused on material planning and control strategies. These strategies are categorized as push, pull, or hybrid depending on the method of releasing the production orders to individual stations. Materials requirement planning (MRP) and Kanban control systems are well-known implementations of push and pull strategies respectively. Hybrid strategies incorporate features of both push and pull systems (Krishnamurthy et al. 2004). Several authors have reported on the performance of these strategies in relation to the production volume and product variability (Spearman and Zazanis 1992; Buzacott and Shantikumar 1993; Womack and Jones 2003; Suri 1998; Spearman and Hopp 1996). Implementations of MRP and Kanban strategies use fixed representation of the manufacturing process and cannot easily address the flexibility that is desired in the design process. Altiok, incorporated stochastic modeling of manufacturing systems wherein the production function and inventory control are both emphasized (Altiok 1997). Gershwin (1994), modeled assembly lines as 'transfer lines' and studied the effect in process time variability and buffer size among stages on the overall production rate and average in-process inventory. While these methods are well known, they are not suited for use in the design of NMS to achieve a specified product quality.

Part manufacture and assembly are one of the largest applications of NMS. In these applications, the tools and sensors at each manufacturing stage ensure that the component or subassembly at that stage meets pre-specified design criteria. However, errors arising from tool wear, incorrect part fixturing, component failure, process uncertainty etc. these propagate from state to stage and can cause degradation of overall product quality (Jiao and Djurdjanovic 2010). One approach is to incorporate the cause of accumulated errors in NMS and reduce their effects on product quality (Jin and Shi 1999). Another approach is to model the NMS as a dynamic system and consider parameters such as process diagnosability and process controllability to study the effect of sensor placement and tooling on the NMS (Ding et al. 2003).

Optimality can also be considered to determine appropriate system parameters to minimize a process cost metric for an NMS (Ding et al. 2003). Regardless of the approach, an understanding of the functional attributes of the mechanical and control systems that comprise the NMS and their effects on the properties of the NMS is necessary. In this monograph, the focus is on the desired dimensional quality of the product and the design of the NMS to meet this requirement. In order to design the NMS, the parameters of the mechanical system, namely, the type, number, and the position of tools; and the parameters of the control system, namely, the type, number, and the position of sensors and sensing stations are treated as parameters to be determined. The design question is how to select these parameters to guarantee that the cause of dimensional variations is diagnosable and to ensure that the overall system is controllable, for example, that the effect of variations on the dimensional quality is eliminated.

The use of Stream of Variation (SoV) model in analyzing process diagnosability and process controllability of an NMS was reported by several authors (Jin and Shi 1999; Ding et al. 2000, 2002a, b, c). However, the analytical methods used in these reports assume that the models representing the NMS are fixed and known, an assumption that is untrue in the design of flexible NMS. The requirement that the NMS be diagnosable and controllable affects the choice of several design variables; see Table 3.1.

An extensive literature survey is documented by Milisavljevic (2015), wherein the following key unresolved difficulties in design are identified (see Table 3.2):

1. the appropriate selection of design parameters (Ding et al. 2002a, b),
2. the concurrent design of the components of NMS (Liu et al. 2006),
3. integrating flexibility during the design process (Mistree et al. 1993),
4. satisfying conflicting requirements such as process diagnosability and process controllability simultaneously (Ding et al. 2002c),
5. overcoming computational complexity (Xiao 2003), and
6. developing a general method applicable to any type of NMS (Milisavljevic 2015; Milisavljevic et al. 2017).

Individual difficulties in NMS design are addressed by several authors; see Table 3.2. In this Chapter we put forward a systematic method to address all the above mentioned difficulties in the concurrent design and analysis of NMS.

3.2 Adaptable and Concurrent Design of Networked Manufacturing Systems

Dynamic changes in customer demands require enterprises to not only provide cost-effective manufacturing processes and at a minimum maintain the quality of product and shorten time to market. Smart manufacturing is increasingly being adopted by companies to respond to these changes in the market. Distributed manufacturing wherein individual components are networked together is one of the characteristics

Table 3.1 Requirements list for design of the NMS

Requirements	Type of requirements	Type of design variables	Mechanical/control system parameters
Type of fixture locator	Flexible	Integer	M
Number of fixture locators	Flexible	Integer	M
Position of fixture locator	Flexible	Continuous	M
Type of sensors	Flexible	Integer	C
Number of sensors	Flexible	Integer	C
Position of sensors	Flexible	Continuous	C
Distribution of sensors	Flexible	Boolean	C
Type of sensing stations	Flexible	Integer	C
Number of sensing stations	Flexible	Integer	C
Programmable tooling control actions	Flexible	Continuous	C
Process diagnosability(%)	Fixed	Continuous	C
Process controllability(%)	Fixed	Continuous	C
Cost?	Flexible	Integer	
Dimensional Quality?	Flexible	Continuous	

of these systems. These systems must be adaptable to dynamic changes in product type and quantity arising from changing customer demands; respond to unexpected disturbances arising out of disruptions in supply chain or manufacturing process; and deal with uncertainty resulting from faulty measurements or network failures. In order to accomplish this, the design of such systems should be flexible enough to investigate all options and select those that meet requirements such as flexibility, robustness, cost and quality. Keeping in mind the challenges associated with such a design, in this chapter we explore a method for adaptable and concurrent design of NMS.

Contribution 1: Identifying Mechanical and Control System Drivers and Their Relations in Concurrent Design

Table 3.2 Overview foundational papers for identifying key unresolved difficulties in the adaptable and concurrent design of NMS (Milisavljevic-Syed et al. 2018)

Paper	Aspects				
	Concurrent design	Solving multi-objective problems	Solution space exploration	Stream of variation Modeling	Design of control system
Xiao (2003)	*				
Smith et al. (2014)		*			
Liu et al. (2006)		*			
Mistree et al. (1993)			*		
Jin and Shi (1999)				*	
Ding et al. (2000)				*	
Ding et al. (2002a, b, c)					*
Mantripragada and Whitney (1999)					*
Ding et al. (2003)					*
Shi and Zhou (2009)					*
Jiao (2012)				*	*
Izquierdo et al. (2007)					*
Mistree et al. (1993)			*		
Apley and Shi (1998)					*
Smith et al. (2015)		*			
Marston et al. (2000)		*			
Milisavljevic (2015)	*	*	*	*	

(continued)

Table 3.2 (continued)

Paper	Methods							
	Collaborative multidisciplinary decision-making methodology	Model-based exploration through decision making	Objectives-orientated sensor allocation by multivariate analysis	Compromise decision support problem construct	State-space model variation propagation in NMS	Diagnosability analysis of NMS	Controllability analysis of NMS	Cost-orientated sensors distribution in design of NMS
Xiao (2003)	*							
Smith et al. (2014)		*						
Liu et al. (2006)			*					
Mistree et al. (1993)				*				
Jin and Shi (1999)					*			
Ding et al. (2000)						*		
Ding et al. (2002a, b, c)						*		
Mantripragada and Whitney (1999)							*	
Ding et al. (2003)								*
Shi and Zhou (2009	*					*		
Jiao (2012)			*				*	
Izquierdo et al. (2007)					*		*	
Mistree et al. (1992)				*				
Apley and Shi (1998)						*		
Smith et al. (2015)		*						
Marston et al. (2000)		*						
Milisavljevic (2015)	*	*		*	*	*	*	*

(continued)

Table 3.2 (continued)

Paper	Literature evaluation				
	Game theoretical approach for coupling	Solution space exploration in design of complex systems solution space exploration in design of complex systems	Model-based approach implemented in DSIDES	Objective-Orientated coupling or interactions in design of NMS	Locating ranges of solutions instead of single-point solutions
Xiao (2003)	*				
Smith et al. (2014)		*	*		
Liu et al. (2006)				*	
Mistree et al. (1993)					*
Jin and Shi (1999)					
Ding et al. (2000)					
Ding et al. (2002a, b, c)					
Mantripragada and Whitney (1999)					
Ding et al. (2003)					
Shi and Zhou (2009)					
Jiao (2012)					
Izquierdo et al. (2007)					
Mistree et al. (1992)					*
Apley and Shi (1998)					
Smith et al. (2015)		*	*		
Marston et al. (2000)		*	*		
Milisavljevic (2015)		*	*	*	*

(continued)

Table 3.2 (continued)

Paper	Literature evaluation						Gaps
	SoV for Systematic Analysis and variation Propagation Control in NMS	SoV for fault diagnosis in design of NMS	SoV for Generic NMS	Design of Diagnosable NMS by use of SoV	Design of Controllable NMS by use of SoV	Design of cost-optimal NMS by use of SoV	
Xiao (2003)							Concurrent design of a mechanical and a control system in design of NMS
Smith et al. (2014)							Solving complex mathematical problems
Liu et al. (2006)							Finding range of solutions in design of NMS
Mistree et al. (1993)							Cost-quality tradeoff in solution space exploration
Jin and Shi (1999)	*						Generalized method that can fit any NMS, requirements, etc.
Ding et al. (2000)		*	*				Generalized method that can fit any NMS, requirements, etc.
Ding et al. (2002a, b, c)		*		*		*	Observe diagnosability and controllability concurrently in design of NMS
Mantripragada and Whitney (1999)				*			Observe diagnosability and controllability concurrently in design of NMS
Ding et al. (2003)					*		Observe cost and process diagnosability and controllability in design of NMS

(continued)

Table 3.2 (continued)

Paper	Literature evaluation						Gaps
	SoV for Systematic Analysis and variation Propagation Control in NMS	SoV for fault diagnosis in design of NMS	SoV for Generic NMS	Design of Diagnosable NMS by use of SoV	Design of Controllable NMS by use of SoV	Design of cost-optimal NMS by use of SoV	
Shi and Zhou (2009)	*						Design of NMS requires fusion of theories, tools and techniques from multiple disciplines to achieve utilization of information
Jiao (2012)	*			*			Generalized method that can fit any NMS, requirements, etc.
Izquierdo et al. (2007)				*			Observe process controllability and product quality concurrently in design of NMS
Mistree et al. (1992)							Cost-quality tradeoff in solution space exploration
Apley and Shi (1998)	*		*				Generalized method that can fit any NMS, requirements, etc.
Smith et al. (2015)							Solving complex mathematical problems
Marston et al. (2000)							Solving complex mathematical problems
Milisavljevic (2015)		*	*	*	*	*	Forward mentioned research gaps

In the design of NMS, it is necessary to design both mechanical and control systems concurrently in order to improve the performance during operation (Milisavljevic-Syed et al. 2018; Liu et al. 2006). Further, the design must satisfy system properties such as process diagnosability, and process controllability (Milisavljevic-Syed et al. 2018). Hence we propose a method that can address the resolution of these requirements at design time; see Sect. 3.3.

Contribution 2: Flexibility in Design of Mechanical and Control Systems

In the design of NMS it is necessary to maintain design freedom throughout the design process to accommodate unexpected downstream design changes to reduce the design timeline and cost in the longer term (Milisavljevic-Syed et al. 2018). Further, we suggest that selection of design parameters affects not only dimensional quality of the final product, but also the cost of the process (Ding et al. 2002a, b), and properties such as diagnosability and controllability (Ding et al. 2002c; Mantripragada and Whitney 1999). Hence, we propose a method that allows us to retain flexibility in selection of design parameters while ensuring that all the resulting choices meet system requirements; see Sect. 3.3.

Contribution 3: Solution Space Exploration of Highly Complex Mathematical Problem

In a design of NMS we are using computational models to represent the process. However, the computational models are complex due to multi-variable problem characteristics and the number of variables we use to simulate the process. Since several different implementations are possible to manufacture the product, it is necessary to explore the solution space of these models to determine the appropriate design for the NMS (Milisavljevic-Syed et al. 2018). Hence, in order to manage the structure of highly complex mathematical problems we propose a method; see Sect. 3.3.

Contribution 4: Decision Network Structure as a Decision Support in Design of NMS

One major challenge associated with design of the NMS is the complex relationships that exist between the various design parameters involved in the process, system properties, and process cost and product quality; see Fig. 3.2. Attributes of tools and sensors, process diagnosability, process controllability, process cost, and product quality are necessary to model the process (Mistree et al. 1993; Xiao 2003).

The relationship between design parameters, system properties, process cost, product quality, and accurate measurement of each of them is required to fully describe the behavior of the NMS; see Fig. 3.2. For example, the number of sensors influences system properties and process diagnosability, and the number of tools influences system properties such as process controllability. Also, the number of tools and sensors influence the process cost and quality of the product. All these complex relationships need to be precisely modeled and are a huge challenge from a design perspective. Hence, we propose a method to integrate effective and efficient decision network structure for supporting design making during the design process; see Sect. 3.3.

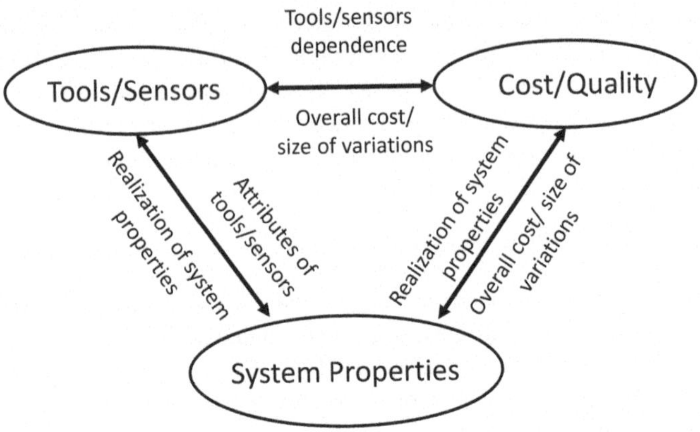

Fig. 3.2 Schematic illustration of the relationship between design parameters, system properties and process cost and quality

3.3 Adaptable Concurrent Realization of Networked Engineering Systems (ACRONES)

The proposed method for adaptable concurrent design and analysis of networked manufacturing systems is carried out in four steps; see Fig. 3.3.

- ***Step A***: Identifying flexible design parameters, establishing their connectivity, and representing the process with a comprehensive state space model; see Sect. 3.3.1.
- ***Step B***: Determining interconnections between NMS modules and the mathematical representation of the complete system; see Sect. 3.3.2.
- ***Step C***: Exploring the solution space for appropriate solutions; see Sect. 3.3.3.
- ***Step D***: Identifying and managing different types of uncertainty, exploring solution space, and identifying robust solutions to the design problem; see Sect. 3.3.4.

3.3.1 Step A, Figure 3.3: Concurrent Design of Networked Manufacturing Systems

The concurrent design of mechanical and control systems of the NMS includes three steps:

- ***Step A1***: Determine the system variables of the mechanical and control systems. Design parameters of the mechanical system are related to tool attributes and operational station attributes. Design parameters of the control system are related to sensor attributes and sensing station attributes.
- ***Step A2***: Establish connectivity among design parameters. For example, there are both tools and sensors on operational stations. The relationship between the design

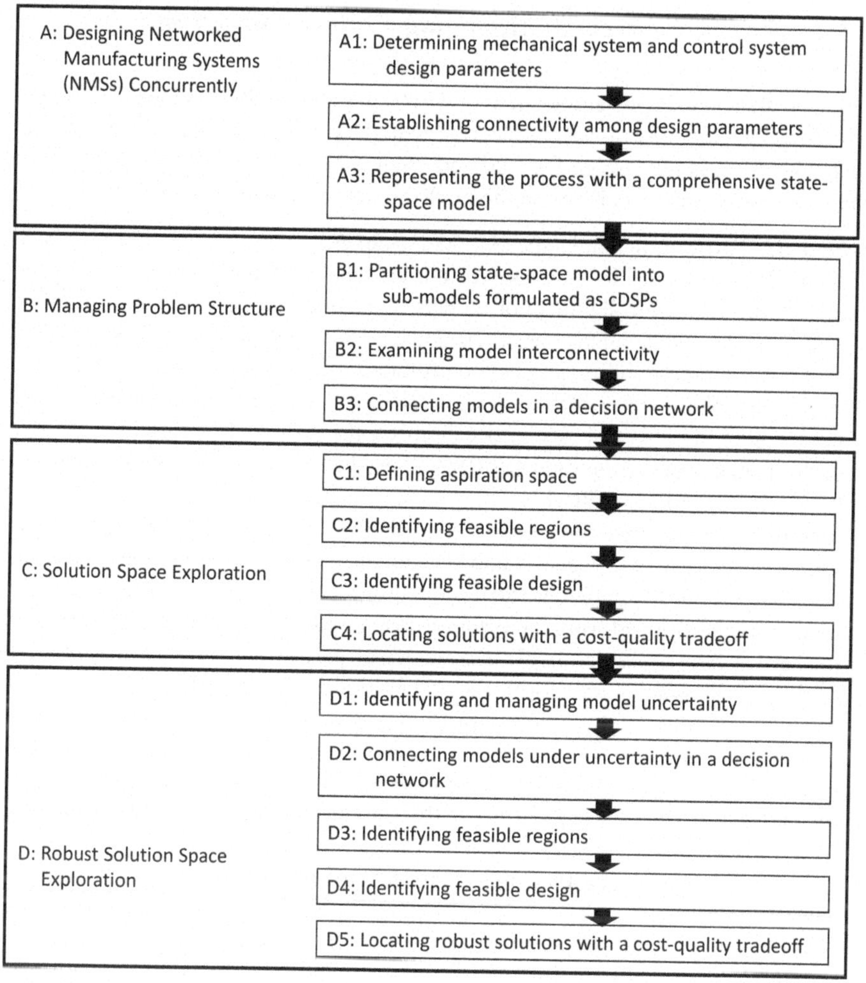

Fig. 3.3 Adaptive concurrent realization of networked engineering systems (ACRONES) framework (Milisavljevic et al. 2017)

parameters and the dynamic behavior of the NMS during runtime is expressed in the state-space form; see Fig. 1.7.

- ***Step A3***: Represent the process with a comprehensive state-space model. The relationships in Step A2 result in models that are often distinct from one another. These models have to be unified to obtain a description of the complete NMS. In this monograph, the propagation of variations in the output of each process is considered using the 'Stream of Variation (SoV)' approach (Jin and Shi 1999). An overall system description in the state-space form is determined using this approach; see Sect. 3.4.

3.3.2 Step B, Figure 3.3: Managing Problem Structure

The state-space model developed in Step A3 is computationally complex and usually requires a large number of variables to represent the process; see Sect. 3.3.1. Solving such a problem is not straightforward and often is computationally expensive. The steps to manage the complexity of the mathematical representation are:

- ***Step B1***: Partition the state-space model, Step A3, into sub-models formulated as cDSPs. These may be process decision models (for example, process diagnosability, process controllability) or models to estimate process cost (for example, cost-effectiveness Izquierdo et al. 2007); see lower part in Fig. 1.7. The cDSPs for process diagnosability, process controllability, and process cost (models D, C and E) are described in Milisavljevic (2015). Process decision models are used to represent the effect of design decisions on the cost of the process. Further, a performance observation model (cDSP) is used to estimate the output quality of the product; see PM in Fig. 1.7.
- ***Step B2***: Establish interconnectivity between the SoV and the cDSP models. The foundation of process decision and performance observation models is a comprehensive state-space model (SoV); see upper part in Fig. 1.7. However, these models are partitioned from the comprehensive state-space model; see upper part in Fig. 1.7, and represent as cDSPs; see lower part in Fig. 1.7. Further, process decision models and the performance observation models have the same mechanical and a control system characteristics and system matrices. For instance, process decision models; see D, C, E in Fig. 3.4, have the same inputs, the same numbers of parts assembled, n_p, and can have the same errors introduced in the process m_k and different outputs (μ, μ_c, etc.) that can be interfaced with a performance observation model to measure the size of variations; see Fig. 3.5.
- ***Step B3***: Connect all cDSPs with a decision network; see Fig. 3.4. To accomplish Step B3, a decision network is developed to identify a possible solution for the NMS design criteria and determine its effect on the process cost. For instance, decisions such as sensing characteristics directly influence the cost of the process;

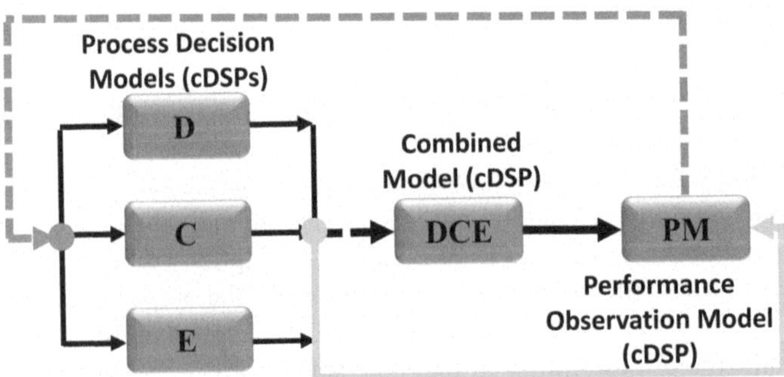

Fig. 3.4 Connecting process decision and performance observation models with decision network structure (Milisavljevic-Syed et al. 2018)

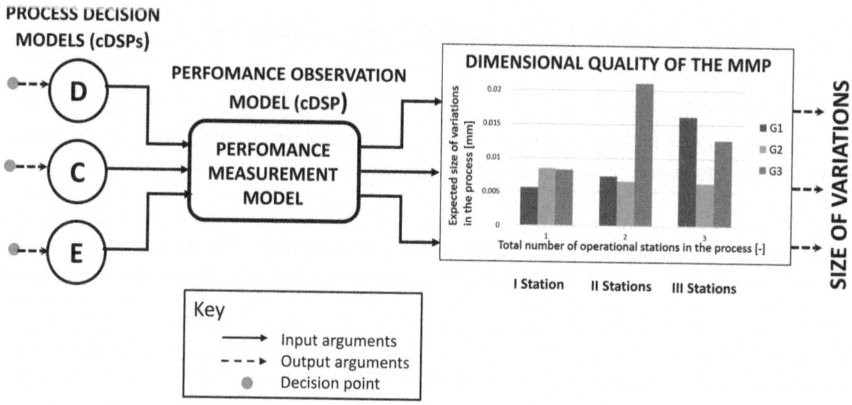

Fig. 3.5 Measuring size of variations with performance observation model (Milisavljevic-Syed et al. 2018)

see light gray link in Fig. 3.4, and different numbers of sensors or types of sensors entails different process costs. If design constraints such as the number of tools and the position of tools, process diagnosability and process controllability are satisfied, the process decision models are integrated; see dashed black link in Fig. 3.4, and the process of searching for additional solutions that are diagnosable, controllable and cost-effective continues. The combined model is used to integrate individual cDSPs such as process diagnosability, process controllability, and cost-effectiveness; see DCE in Fig. 3.4. If there is no feasible solution to the overall NMS design problem, the next step is to return to the process decision models knowing which process decision models must be reconfigured; see dashed gray link in Fig. 3.4.

The next step is to measure the size of variations by connecting the combined model with a performance observation model; see continuous black link in Fig. 3.4. Since the process decision models are connected with the performance observation model; see Fig. 3.5, design decisions have a direct influence on the process cost; see Fig. 3.4, and indirect on the dimensional quality of product. The output of process decision models, such as the number and position of sensors, sensor distribution scheme and the number of sensing stations, is the input to the performance observation model; see continuous black link in Fig. 3.5, where the size of variations at each station is measured; see the right side in Fig. 3.5. Further, the output of the performance observation model is the size of variations associated with the product quality; see dashed black link in Fig. 3.5. If the size of variations lies within the prescribed limits, the decision point is reached; see the gray point in Fig. 3.5, and process decision models are integrated into the combined model; see DCE in Fig. 3.4. However, if the size of variations is above the prescribed limits, then appropriate process decision models have to be reconfigured. Finally, all the feasible solutions are consolidated and a solution space exploration procedure is employed to choose the best solution, Sect. 3.5.3.

3.3.3 Step C, Figure 3.3: Solution Space Exploration

In ACRONES solutions are a tradeoff between the process cost and product quality, for example, sizes of variations in products. A strategy for identifying and exploring a possible solution space in the design of the NMS is:

- ***Step C1***: Define an aspiration space by setting up the goals for a particular case. For instance, a designer may wish to minimize variations in product dimensions or at least keep them within the range $-0.8 \leftrightarrow 0.8$ [mm] while minimizing process cost.
- ***Step C2***: Identify model interconnectivity by determining regions where feasible designs exist based on satisfying constraints and bounds or where they might exist by minimizing constraint violation.
- ***Step C3***: Identify feasible designs from the neighborhood of feasible or near feasible regions, frame the boundary of the feasible design space using a preemptive representation of the goals (Smith et al. 2015).
- ***Step C4***: Locate solutions as a cost-quality tradeoff, having refined an understanding of the cDSPs, process decision models' feasible design space and the regions of greatest interest in Step C3, move along the extreme values generating deeper understanding by exploring tradeoffs by using an Archimedean (weighted sum) formulation of the goals as indicated by Smith et al. (2015). Regions of great interest are guided by what is most important to a design engineer, such as the process cost, quality, etc. The goal is to minimize the deviation function, for example, the distance between the aspiration space and feasible design space. The proposed method is iterative and in each iteration, the deviation function is minimized and satisfying solutions are located.

3.3.4 Step D, Figure 3.3: Robust Solution Space Exploration

Analysis models are abstractions of reality; they are typically incomplete, inaccurate, and with inequal fidelity resulting in uncertainties. Hence, we seek solutions that are relatively insensitive to the impact of these models being abstractions of reality. However, in the design of NMS it is difficult to identify and manage different types of uncertainty in different models, therefore, exploring a solution space and locating robust solutions is a challenging task. A strategy for identifying that are relatively insensitive to uncertainties and exploring this solution space is as follows:

- ***Step D1***: Solution space exploration, Step C, leads to gaining insight into the process characteristics; see Sect. 3.3. In NMS, uncertainty management, raises concerns such as unpredictability of both mechanical and control systems, whether parameters of a given model are subject to variations associated with variations of attributes of the tools and sensors of systems, imprecise processes, the lack of knowledge about some processes, the lack of knowledge about models and

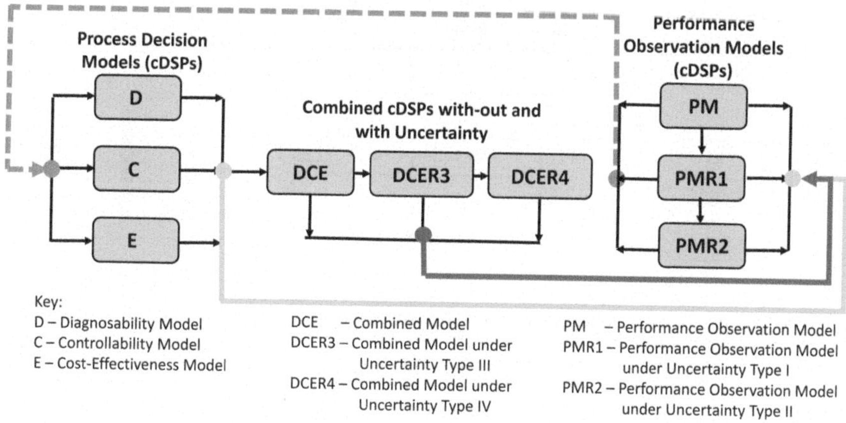

Fig. 3.6 Connecting cDSPs with and without uncertainty with a decision network structure (Milisavljevic et al. 2017)

information propagation through a chain of models, and the necessity of exploring alternatives.

- ***Step D2***: In Step D1 different types of uncertainty are identified and managed. In Step D2 all cDSPs with uncertainty are connected in a decision network; see Fig. 3.6. A decision network is proposed in Step B3 to identify a possible solution for the design criteria for the NMS under uncertainty and its effect on the process cost.

The next step is to connect the combined model without uncertainty (DCE), and the combined model with uncertainty Type III (DCER3) and uncertainty Type IV (DCER4), with the performance observation model without uncertainty (PM), and the performance observation model with uncertainty Type I (PMR1) and uncertainty Type II (PMR2); see Fig. 3.6. Since the combined models relate to the performance observation model, design decisions such as attributes of the tools and sensors have a direct influence on the cost of the process, and an indirect effect on the quality of the product; see Figs. 3.6 and 3.7. The output of the combined models, such as the number and the position of sensors, distribution of sensors in the process, and the number of sensing stations is the input to the performance observation models where the size of variation in the N-stage process for each station is measured; see continuous black link in Fig. 3.7. Further, the outputs from the performance observation models are the sizes of variations which determine the quality of the product; see dashed black link in Fig. 3.7. If the size of variations is within prescribed limits reached and the search for robust solutions continues; see light gray point in Fig. 3.7. Finally, all the feasible solutions are consolidated and a solution space exploration procedure (Steps D3–D5) is employed to pick the preferred solution.

- ***Step D3***: Identify model interconnectivity by determining regions where feasible designs exist based by satisfying the constraints and bounds or where they might exist by minimizing constraint violation.

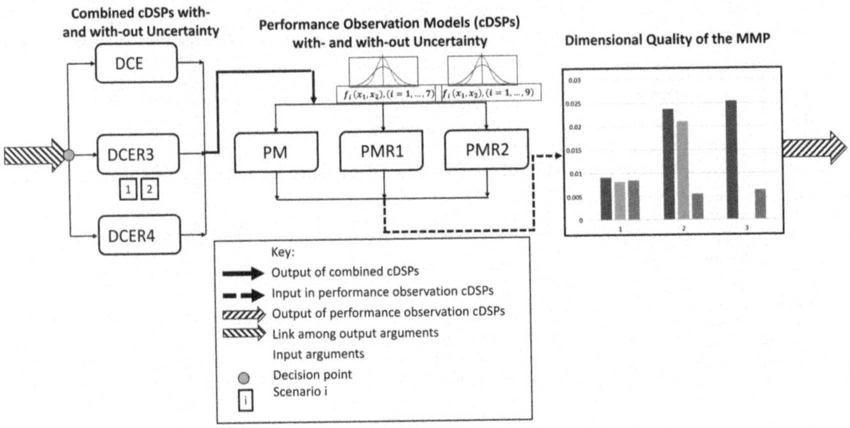

Fig. 3.7 Measuring the size of variations with performance observation models with and without uncertainty

- ***Step D4***: Identify feasible designs in the neighborhood of feasible or near feasible regions, framing the boundary of the feasible design space using a preemptive goal formulation (Ding et al. 2003).
- ***Step D5***: Locate solutions as a cost-quality tradeoff, as previously explained in Step C4. The proposed method is iterative and in each run, the deviation function is minimized, and good robust solutions are located.

NMS are not inherently diagnosable and controllable but rather need to be designed to be so and there is more than one way of accomplishing this. If the model of the system is known and accurate then it is easy to design both diagnosable and controllable system. However, if the model of the system is unknown or inaccurate then the solution space will change. Further, if there is uncertainty in the process how it is possible to identify the right solution? The problem addressed and resolved with the proposed method is how to determine tooling and sensing arguments to design a system that is cost-effective, diagnosable and controllable, and that has a solution which is robust even in the presence of uncertainty.

In Sect. 3.4, we present the compromise Decision Support Problem (cDSP) formulation of the adaptable concurrent model.

3.4 Illustrative Example: Panel Stamping Process

The design method outlined in the previous section is illustrated using an example of an automotive panel stamping process involving N stations; see Fig. 3.8. In the example, four parts are stamped in three stations. Each part is restrained by a set of fixtures (P_i) and each position of a part feature is measured by a coordinate sensor (M_i); see Fig. 3.8. The problem is to determine the number of sensors (T_{sen}) and

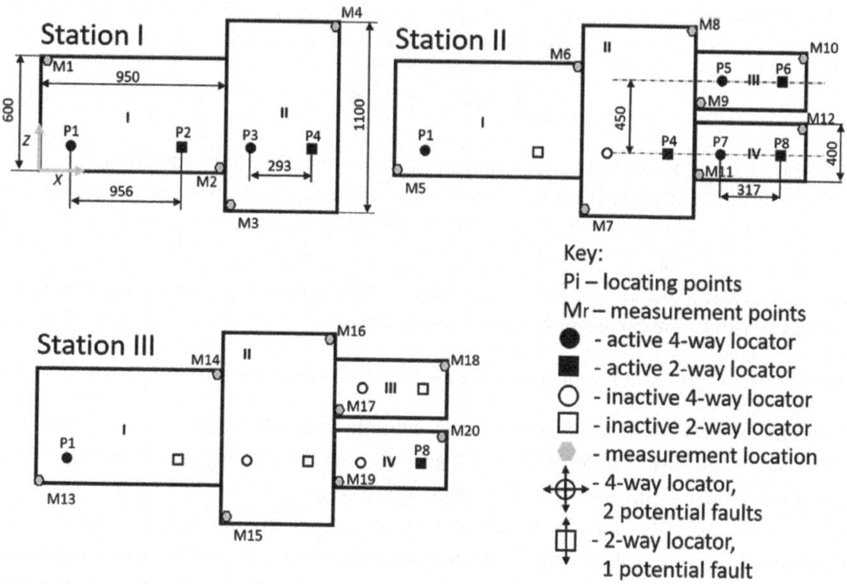

Fig. 3.8 Two-dimensional panel assembling process (Ding et al. 2002a, b, c)

their distribution in the process ($M_{i,k}$), the number of sensing stations (T_{nss}), use of PT's control actions (T_{pt}), and sensing penalties (T_{ptnss}) in order to minimize the overall cost (see Goals 1–3 forward) of the NMS while ensuring the process is diagnosable and controllable (100%). Further, it is desirable to identify the source of dimensioning errors and trace them back to their source and use the method's built in redundancy to make the system robust to these errors. In the discussion that follows, it is assumed that a 3-2-1 fixture is used and all parts are rigid.

3.4.1 Panel Stamping Process Model Equations and Design Specifications

In this section, we present the Stream of Variation (SoV) model and state matrices for 2-D Panel Assembling Process in N—Stations is presented. Furthermore, we present the system variable selection and specifications.

Stream of Variation (SoV) Model

A dimensional SoV model is primarily developed for multistage assembly and machining processes. Fixture locators are tools in the assembling process used at each stage of the process that manages dimensional quality of a product. The propagation of fixtures variation contributed from each station and its impact on the final product quality are described by the stream-of-variation model (Jin and Shi 1999); see Eqs. 3.1 and 3.2:

$$X_i = A_{i-1} \cdot X_{i-1} + B_i \cdot U_i + V_i \tag{3.1}$$

$$Y_i = C_i \cdot X_i + W_i \tag{3.2}$$

In Eq. 3.1, we present the state equation that implies that part variation at Station i, X_i, is influenced by two sources: (1) the accumulated variation up to Station $i - 1$, and (2) the variation at Station i. The observation equation (Eq. 3.2) shows how the variation at Station i of the manufacturing process affects the output, Y_i, of the station. In Eqs. 3.1 and 3.2, U_i is the input at Station i, and V_i and W_i represent the process noise and measurement noise, respectively.

State Matrices for 2-D Panel Assembling Process in N—Stations

Dynamic Matrix: If the fixture locating scheme is unchanged in the consecutive stations, for example, several features are machined by using the same datum in a multi-station machining operation then the dynamic matrix is equal to the unit matrix. However, if a part is positioned by a new set of fixtures, the part will be reoriented on a new fixture set. The dynamic matrix can now be developed using the new fixture set. For the development of a dynamic matrix see Jin and Shi (1999). The difference between three stage and four stage panel stamping processes is the number of dynamic matrices A.

In Eq. 3.3, we present the transfer matrix that reflects the changes in fixture positions,

$$T(i-1) = \begin{pmatrix} H(i) & \Theta \\ \Theta & \Theta \end{pmatrix}. \tag{3.3}$$

In Eq. 3.4, we present the assembly transfer matrix if fixture locators P_1 and P_2 are on the different Parts i and j.

$$H(i) = \left[\, H_{ri}(i) \;\Theta\; H_{ji}(i) \,\right] \tag{3.4}$$

In Eq. 3.5, we present the assembly transfer matrix if fixture locators P_1 and P_2 are on the same Part i.

$$H(i) = \left[\, \Theta \; -M_{i,i}(i) \;\Theta \,\right] \tag{3.5}$$

In Eq. 3.6, we present the subassembly transfer matrix, between Parts r and i.

$$H_{ri}(i) = M_{Ar,P1}(i) \cdot D(i) \cdot M_{P1,Aj}(i) \tag{3.6}$$

In Eq. 3.7, we present the subassembly transfer matrix, between Parts j and i.

$$H_{rj}(i) = M_{Ar,P1}(i) \cdot G(i) \cdot M_{P2,Aj}(i) \tag{3.7}$$

Transformation matrix gives the deviation relationship between the two points of part point A_r and locator point P_1; see Eq. 3.8.

$$M_{Ar,P1}(i) = \begin{pmatrix} 1 & 0 & -L_Z(P_1, A_r) \\ 0 & 1 & L_x(P_1, A_r) \\ 0 & 0 & 1 \end{pmatrix} \tag{3.8}$$

Deviation matrix gives the deviation between fixture points P_1 and P_2; see Eq. 3.9.

$$D(i) = \begin{pmatrix} -1 & 0 & 0 \\ 0 & -1 & 0 \\ 0 & \frac{1}{L_X(P_1, P_2)} & 0 \end{pmatrix} \tag{3.9}$$

Transformation matrix gives the deviation relationship between the two points of part point A_j and locator point P_1; see Eq. 3.10.

$$M_{P1,Aj}(i) = \begin{pmatrix} 1 & 0 & -L_Z(A_j, P_1) \\ 0 & 1 & L_x(A_j, P_1) \\ 0 & 0 & 1 \end{pmatrix} \tag{3.10}$$

Deviation matrix gives deviation between fixture points P_1 and P_2; see Eq. 3.11.

$$G(i) = \begin{pmatrix} 0 & 0 & 0 \\ 0 & 0 & 0 \\ 0 & -\frac{1}{L_X(P_1, P_2)} & 0 \end{pmatrix} \tag{3.11}$$

Transformation matrix gives the deviation relationship between the two points of part point A_j and locator point P_2; see Eq. 3.12.

$$M_{P2,Aj}(i) = \begin{pmatrix} 1 & 0 & -L_Z(A_j, P_2) \\ 0 & 1 & L_x(A_j, P_2) \\ 0 & 0 & 1 \end{pmatrix} \tag{3.12}$$

Dynamic matrix $A_1(2)$ represent assembly of Parts 1 and 2 that are coming from Station 1 to Station 2; see Eq. 3.13.

$$A_1(2) = \begin{bmatrix} A_{11} & \cdots & A_{14} \\ \vdots & \ddots & \vdots \\ A_{41} & \cdots & A_{44} \end{bmatrix}_{12\times 12} \tag{3.13}$$

Dynamic matrix $A_2(3)$ represent subassembly Parts 1 and 2, and Part 3 that are coming from Station 2 to Station 3; see Eq. 3.14.

$$A_2(3) = \begin{bmatrix} A_{11} & \cdots & A_{14} \\ \vdots & \ddots & \vdots \\ A_{41} & \cdots & A_{44} \end{bmatrix}_{12\times 12} \quad (3.14)$$

Dynamic matrix $A_3(4)$ represent subassembly Parts 1, 2 and 3, and Part 4 that are coming from Station 3 to Station 4; see Eq. 3.15. Dynamic matrix A_3 can be found in four stage processes.

$$A_3(4) = \begin{bmatrix} A_{11} & \cdots & A_{14} \\ \vdots & \ddots & \vdots \\ A_{41} & \cdots & A_{44} \end{bmatrix}_{12\times 12} \quad (3.15)$$

Input Matrix B: Matrix B_i is the input matrix that determines how fixture deviation affects part deviation on Station *i*, based on the geometry of a fixture locating layout (Jin and Shi 1999). The rank of B_i equals to the number of degrees of freedom d.o.f. of the supported workpieces restrained by the fixture set. The difference between three stage and four stage panel stamping processes is the number and size of input matrices B.

In Eq. 3.16, we present the input matrix for Station *i*.

$$B(i) = \begin{pmatrix} Q_{A_1,P_1} & \Theta \\ \vdots & \vdots \\ Q_{A_i,P_1} & \Theta \\ \Theta & Q_{A_{i+1},P_1} \\ \Theta & \Theta \end{pmatrix} \quad (3.16)$$

where Θ is an appropriately sized null matrix.

In Eq. 3.17, we present the coordinate transformation matrix from the fixture error to the part locating error represented by the part point A_i at Station i.

$$Q_{A_i,P_1} = \begin{pmatrix} 1 & \frac{L_Z(A,P_1)}{L_X(P_1,P_2)} & -\frac{L_Z(A,P_1)}{L_X(P_1,P_2)} \\ 0 & 1-\frac{L_Z(A,P_1)}{L_X(P_1,P_2)} & \frac{L_Z(A,P_1)}{L_X(P_1,P_2)} \\ 0 & -\frac{1}{L_X(P_1,P_2)} & \frac{1}{L_X(P_1,P_2)} \end{pmatrix} \quad (3.17)$$

In Eq. 3.18, we present the input matrix $B_1(1)$ for Station 1.

$$B_1(1) = \begin{bmatrix} B_{11} & \cdots & B_{12} \\ \vdots & \ddots & \vdots \\ B_{41} & \cdots & B_{42} \end{bmatrix}_{12\times 6} \quad (3.18)$$

In Eq. 3.19, we present the Input matrix $B_2(2)$ for Station 2.

$$B_2(2) = \begin{bmatrix} B_{11} & \cdots & B_{12} \\ \vdots & \ddots & \vdots \\ B_{41} & \cdots & B_{42} \end{bmatrix}_{12\times i} \quad (3.19)$$

where i = 9 in case of 3-stage process, and i = 6 in case of 4-stage process.
In Eq. 3.20, we present the input matrix $B_3(3)$ for Station 3.

$$B_3(3) = \begin{bmatrix} B_{11} & \cdots & B_{12} \\ \vdots & \ddots & \vdots \\ B_{41} & \cdots & B_{42} \end{bmatrix}_{12\times i} \quad (3.20)$$

where i = 3 in case of 3-stage process, and i = 6 in case of 4-stage process.
In Eq. 3.21, we present the input matrix $B_4(4)$ for Station 4.

$$B_4(4) = \begin{bmatrix} B_{11} \\ \vdots \\ B_{41} \end{bmatrix}_{12x3} \quad (3.21)$$

Control Matrix C: Matrix C_i contains the information about sensor locations on a station. When sensors are installed on one or more stations in a production line, the index for the observation Equation 2 is actually a subset of 1,2,..., N, whereas the index for the state Equation 1 is the complete set (Jin and Shi 1999). Similarly, the rank of C_i corresponds to the number of measured degrees of freedom of a part or a subassembly on Station *i*.

In Eq. 3.21, we present the control matrix for Station *i*.

$$C(i) = \begin{pmatrix} C_1(i) & \cdots & \Theta \\ \vdots & \ddots & \vdots \\ \Theta & \cdots & C_n(i) \end{pmatrix} \quad (3.21)$$

In Eq. 3.22, we present the sub-matrix for Station *i*.

$$C_r(i) = \begin{pmatrix} 1\ 0 & -L_Z(R_{1,r}, A_r) \\ 0\ 1 & L_X(R_{1,r}, A_r) \\ \vdots\ \vdots & \vdots \\ 1\ 0 & -L_Z(R_{m_r,r}, A_r) \\ 0\ 1 & L_X(R_{m_r,r}, A_r) \end{pmatrix}_{(2\cdot m_r)\times 3} \quad (3.22)$$

In Eq. 3.23, we present the control matrix $C(3)$ for end-of-line sensing distribution scheme in three stage process in the measurement station.

$$C(3) = \begin{bmatrix} C_{11} & \cdots & C_{14} \\ \vdots & \ddots & \vdots \\ C_{41} & \cdots & C_{44} \end{bmatrix}_{16\times 12} \quad (3.23)$$

In Eq. 3.24, we present the control matrix $C(4)$ for end-of-line sensing distribution scheme in four stage process in the measurement station.

$$C(4) = \begin{bmatrix} C_{11} & \cdots & C_{14} \\ \vdots & \ddots & \vdots \\ C_{41} & \cdots & C_{44} \end{bmatrix}_{16\times 12} \quad (3.24)$$

In Eq. 3.25, we present the control matrix $C_1(1)$ for distributed sensing distribution scheme in three stage process in the first station.

$$C_1(1) = \begin{bmatrix} C_{11} & \cdots & C_{14} \\ \vdots & \ddots & \vdots \\ C_{21} & \cdots & C_{24} \end{bmatrix}_{8\times 12} \quad (3.25)$$

In Eq. 3.26, we present the control matrix $C_2(2)$ for distributed sensing distribution scheme in n- stage process in the second station.

$$C_2(2) = \begin{bmatrix} C_{11} & \cdots & C_{14} \\ \vdots & \ddots & \vdots \\ C_{41} & \cdots & C_{44} \end{bmatrix}_{i\times 12} \quad (3.26)$$

where i = 16 in case of 3-stage process, and i = 12 in the case of a 4-stage process.

In Eq. 3.27, we present the control matrix $C_3(3)$ for distributed sensing distribution scheme in n-stage process in the third station.

$$C_3(3) = \begin{bmatrix} C_{11} & \cdots & C_{14} \\ \vdots & \ddots & \vdots \\ C_{41} & \cdots & C_{44} \end{bmatrix}_{16\times 12} \quad (3.27)$$

In Eq. 3.28, we present the control matrix $C_4(4)$ for distributed sensing distribution scheme in four stage process in the fourth station.

$$C_4(4) = \begin{bmatrix} C_{11} & \cdots & C_{14} \\ \vdots & \ddots & \vdots \\ C_{41} & \cdots & C_{44} \end{bmatrix}_{16\times 12} \tag{3.28}$$

Selection and Specification of System Variables

In Eq. 3.29, we present the cost of the process as a function of total number of sensors M_i and sensing stations M_S in the process.

$$C = M_i + M_S \tag{3.29}$$

Total number of sensing stations in the process is the sum of the actual number of sensors over the potential number of sensors at Station k. If there are no sensors at Station k then there is no sensing station $M_{S,k}$ at a particular Station k. If there is at least one sensor then there is sensing station $M_{S,k}$ at station k. In Eq. 3.30, we present the mathematical formulation.

$$M_S = \sum_{k=1}^{n}\sum_{i=1}^{m}\frac{M_i}{M_{pi}} \text{ given that } \frac{M_i}{M_{pi}}\begin{cases} = 0 \\ > 0 \end{cases} \text{ then } M_{S,k}\begin{cases} 0 \\ 1 \end{cases} \tag{3.30}$$

Total number of sensing stations depends on sensor distribution scheme $M_{i,k}$. In case of *end-of-line* sensing the total number of sensing stations is $0 \le M_S \le 1$, and for *distributed* sensing it is $0 \le M_S \le N$ (N is the number of operational stations).

Sensors distribution scheme present how actual number of sensors M_i is distributed throughout the stations in the process. It is different for *end-of-line* and *distributed* sensing; see Eq. 3.31 (Table 3.3).

$$\begin{array}{cc} \text{Distributed Sensing} & \text{End-of-Line Sensing} \\ M_1 = M_{1,1} + M_{2,1} + \cdots + M_{i,1} & M_1 = 0 \\ \vdots & \vdots \\ M_m = M_{i+1,m} + M_{i+2,m} + \cdots + M_{n,m} & M_m = M_{1,m} + M_{2,m} + \cdots + M_{n,m} \end{array} \tag{3.31}$$

3.4.2 The Compromise Decision Support Problem (CDSP) Formulation

The cDSP construct is used to describe NMS in state-space form; see Eqs. 3.1 and 3.2. The NMS cDSP is a superset of process diagnosability, process controllability, and cost-effectiveness smaller cDSP models and it is obtained by combing these cDSPs; see Fig. 1.7. The comprehensive state-space model for the illustrative example has

Table 3.3 Panel stamping process design parameters

	Coordinate in x- direction [mm]	Coordinate in z- direction [mm]
Fixture points		
P_1	63.0	50.0
P_2	800.0	50.0
P_3	1019.0	50.0
P_4	1312.0	50.0
P_5	1688.0	470.0
P_6	2005.0	470.0
P_7	1688.0	50.0
P_8	2005.0	50.0
Sensor points		
M1	0.0	600.0
M2	950.0	0.0
M3	950.0	−300.0
M4	1630.0	800.0
M5	0.0	0.0
M6	950.0	600.0
M7	950.0	−300.0
M8	1630.0	800.0
M9	1630.0	300.0
M10	2310.0	700.0
M11	1630.0	−150.0
M12	2310.0	200.0
M13	0.0	0.0
M14	950.0	600.0
M15	950.0	−300.0
M16	1630.0	800.0
M17	1630.0	300.0
M18	2310.0	700.0
M19	1630.0	−150.0
M20	2310.0	200.0

8 system variables and 16 constraints; see Sect. 2.1.3. The design problem is to determine the minimum number of sensing stations and sensors, and an adequate sensor distribution scheme, use of PT control actions, and sensing penalties in order to satisfy constraints (4), (6), (14), and (15) and to minimize overall process cost and maximize product quality.

Given

Known	Number of operational stations in the process	N	[–]
	Number of parts in the stamping	n_p	[–]
	Number, and the position of fixture points in the process	P_i	[–; x, z]
	Potential number, and the position of sensors in the process	M_{Pi}	[–; x, z]
	Dimensional quality (size of variations) boundary values are set	y_k	[mm]
Assume	A 3-2-1 fixture is used		
	All parts used in the process are rigid		
	Sum of system goals weight coefficients are equal		

Find

Total number of sensors	M_{Pi}	[–]
Sensors distribution in the process per station	$M_{i,k}$	[–]
Total number of sensing stations	M_S	[–]
Use of PT's control actions in the process regarding cost-effectiveness	PT	[–]
Total sensing penalties in the process	P	[–]
Deviation variables	$d_i^-, d_i^+ (\mathrm{i} = 1, \ldots, 4)$	[–]

Satisfy

Constraints		
C1 Desired process diagnosability index	$\mu_D = 1$	[%]
C2 Desired process controllability index	$\mu_C = 1$	[%]
C3 System variables weight coefficients has to be equal to 1, regarding the cost-effectiveness, that ties in Eq. 9	$c_1 + c_2 + c_3 + c_4 = 1$	[–]
C4 Sum of system goals weight coefficients has to be equal to 1	$w_1 + w_2 + w_3 = 1$	[–]
C5 Fixture points P_i and P_j cannot have the same position (i, j = 1,..., 8)	$P_{i,x} \neq P_{j,x}, P_{i,z} \neq P_{j,z}$	[mm]
C6 Sensors points M_i and M_j cannot be the same points (i, j = 1,..., 20)	$M_{i,x} \neq M_{j,x}, M_{i,z} \neq M_{j,z}$	[mm]
C7 No three sensors can be collinear in x- and z- direction	$M_{i,x} \neq M_{j,x} \neq M_{r,x},$ $M_{i,z} \neq M_{j,z} \neq M_{r,z}$	[mm]
C8 Fixture points $P_1, P_2, P_3, P_4, P_7, P_8$ must be collinear in z- direction	$P_{1,z} = P_{2,z} = P_{3,z} = P_{4,z} = P_{7,z} = P_{8,z}$	[mm]

(continued)

(continued)

C9 Fixture points P_5, P_6 must be collinear in z- direction	$P_{5,z} = P_{6,z}$	[mm]
C10 Product of deviation variables equal 0	$d_i^- \times d_i^+ = 0$	[–]
Goals		
G1 Minimize cost of the process	$(c_1 \cdot M_{Pi} + c_2 \cdot M_S + c_3 \cdot PT + c_4 \cdot P) - d_1^- + d_1^+ = 1$	[–]
G2 Maximize quality of the product	$\left(-\sum_{k=1}^{N}[\bar{y}_k^T \cdot Q_k \cdot \bar{y}_k]\right) + d_2^- - d_2^+ = 1$	[–]

Goal G_1 is normalized and its value is between $0 \leq G_1 \leq 100$. Coefficients $c_j (j = 1, \ldots, 4)$ are monetary costs where $\sum_{j=1}^{4} c_j = 1$.

Bounds		
B1 The total number of sensors has to be between 0 and 20	$0 \leq M_{Pi} \leq 20$	[–]
B2 Distribution of sensors per station	$0 \leq M_{i,k} \leq 8$	[–]
B3 Total number of sensing stations has to be between 0 and 4	$0 \leq M_S \leq 4$	[–]
B4 Use of control actions has to be between 0 and 1	$0 \leq PT \leq 1$	[–]
B5 Sensing penalties has to be between 0 and 2	$0 \leq P \leq 2$	[–]
B6 Dimensional variations at station	$-0.8 \leq y_k \leq 0.8$	[–]
B7 Cost	$0 \leq \text{Cost} \leq n$	[$]
B8 Deviation variables have to be greater than or equal to 0	$d_i^-, d_i^+ \geq 0$	[–]

Minimize

The deviation function (Z): Archimedean formulation

$$Z = \min \sum_{i=1}^{3} \left(W_i d_i^- + W_i d_i^+\right); \sum_{i=1}^{3} W_i = 1; \quad W_i \geq 0$$

In Sect. 3.5, we present the results and discussion of the adaptable concurrent model, where smaller process decision models and combined models with and without uncertainty.

3.5 Results and Discussion

As Milisavljevic-Syed et al. (2018) discuss the specification of system variables and their effect on the process cost is difficult to ascertain prior to implementation. For the sake of illustration, we assumed five sensors and up to three sensing stations are available for use in the NMS. We explore the solution space by undertaking a cost-quality tradeoff and we present results associated with process cost, product quality, and cost-quality analysis; see Sects. 3.5.1, 3.5.2 and 3.5.3. All results are obtained through simulation in MATLAB.

3.5.1 *Cost of the Process*

The data obtained by exercising ACRONES is used by a designer to frame a design space based on feasible bounds. This gives a design engineer the insight into the selection of design parameters (number and distribution of sensors, and number of sensing stations) resulting with diagnosable, controllable and cost-effective system design, and how this influences the process cost even under uncertainty. The cost of the process is related to the number of sensors and sensing stations, use of PT control actions in the process and sensing penalties due to minimized number of sensing stations.

In this illustrative example, we consider the cost of the process of the process decision models; see D, C, E in Fig. 3.4. Further, we consider combined comprehensive models without and with uncertainty; see DCE, and DCER3 in Fig. 3.6.

We consider three different process decision models: (1) G1—diagnosability model where we are minimizing sensing cost regarding process diagnosability; (2) G2—controllability model where we are minimizing sensing cost regarding process controllability; and (3) G3—cost-effectiveness model where we are minimizing sensing and tooling cost regarding both process diagnosability and process controllability; see Fig. 3.9. If one sensing station, M_S, is implemented design G1 is achieved and the process cost is \$16. If two sensing stations, M_S, are implemented all three goals are achieved where process cost for G1 is \$17, G2 is \$24, and G3 is \$36. If three sensing stations, M_S, are implemented in design all three goals are achieved where the process cost for G1 is \$18, G2 is \$26, and G3 is \$39. In Table 3.4, we present the process cost with respect to goals, G1–G3, where it can be seen that the process cost is increasing with the increase of sensing stations in the process and the highest process cost occurs when three sensing stations are implemented in design of NMS.

In Tables 3.4, 3.5, 3.6 and 3.7, we summarize the cost of the process for the combined model without uncertainty; see DCE in Fig. 3.6, and the combined model with uncertainty; see DCER3 in Fig. 3.6. In this section, we compare the cost of the process for feasible in DCE (Milisavljevic-Syed et al. 2018), and DCER3 for two scenarios, DCER3_Q and DCER3_C (Milisavljevic et al. 2017).

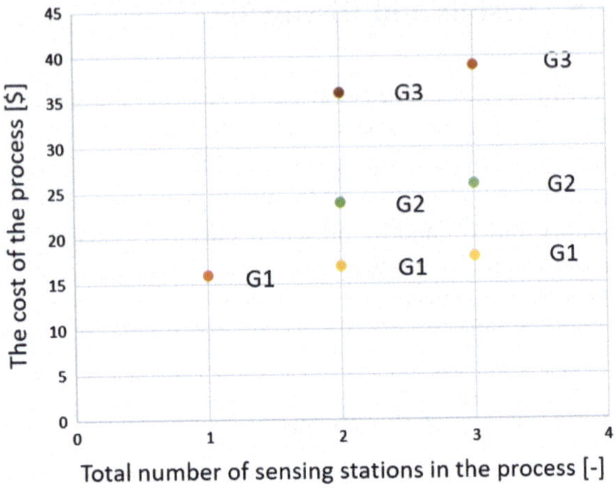

Fig. 3.9 Cost of the feasible designs according to process decision models (Milisavljevic-Syed et al. 2018)

Table 3.4 Solution space of the combined cDSP (Milisavljevic-Syed et al. 2018)

Goal function	Number of sensors distribution schemes $M_{i,k}$ [–]	Total number of sensing Stations M_S [–]	Total number of sensors M_{Pi} [–]	Cost of the process C [$]	End-of-process variations y_k [mm]
G1	38	1	5	16	8.77E−06
	710	2		17	1.20E−02
	832	3		18	0.016088
Total	1580				
G2	8	2	5	24	0.006343
	8	3		26	0.006343
Total	16				
G3	8	2	5	36	8.77E−06
	8	3		39	0.012677
Total	16				

In Table 3.5, we present the solutions of DCE. Further, with the use of two sensing stations, the total number of sensors is between 4 and 12, and 256 solutions are available. On the other hand, with the use of three sensing stations in the process, the total number of sensors is between 5 and 20, and 65,280 solutions are available.

In Table 3.6, we present the solution space of DCER3_Q. Further, with the use of two sensing stations, the total number of sensors is between 5 and 11, and 64 solutions are available. On the other hand, with the use of three sensing stations in the process, the total number of sensors is between 6 and 19, and 16,320 solutions are available.

Table 3.5 Solution space of DCE (Milisavljevic et al. 2017)

Number of sensors distribution schemes $M_{i,k}$ [–]	Total number of sensing stations M_S [–]	Total number of sensors M_{Pi} [–]	Cost of the process C [$]	Range of end-of-process variations y_k [mm]
1	2	4	26	2.56E−06
8		5	31	2.56E−06—0.011195
28		6	36	2.59E−06—0.011261
56		7	41	3.00E−06—0.011263
70		8	46	3.05E−06—0.011264
56		9	51	3.50E−06—0.011264
28		10	56	5.18E−06—0.011264
8		11	61	7.19E−06—0.011264
1		12	66	0.011264
Total 256				
8	3	5	34	2.56E−06
92		6	39	2.56E−06—0.011195
504		7	44	2.56E−06—0.011261
1750		8	49	2.56E−06—0.011263
4312		9	54	2.56E−06—0.011264
7980		10	59	2.56E−06—0.011264
11,432		11	64	2.56E−06—0.011264
12,869		12	69	2.56E−06—0.011264
11,440		13	74	2.57E−06—0.011264
8008		14	79	2.59E−06—0.011264
4368		15	84	3.01E−06—0.011264
1820		16	89	3.06E−06—0.011264
560		17	94	3.50E−06—0.011264
120		18	99	5.19E−06—0.011264
16		19	104	7.19E−05—0.011264
1		20	109	0.011264
Total 65,280				

In Table 3.7, we present the solution space of DCER3_C. Further, with the use of two sensing stations, the total number of sensors required is between 4 and 5, and 8 solutions are available. On the other hand, with the use of three sensing stations in the process, the total number of sensors is limited to 5, and 8 solutions are available.

The cost of the process is lower for DCER3_Q and DCER3_C than for the combined DCE. Further, with use of two sensing stations and four sensors in the process, the cost of the process for the combined DCER3_C is $19; see Table 3.7, where for DCE is $26; see Table 3.5. With the use of two sensing stations and five sensors

Table 3.6 Solution space of DCER3_Q (Milisavljevic et al. 2017)

Number of sensors distribution schemes $M_{i,k}$ [–]	Total number of sensing stations M_S [–]	Total number of sensors M_{Pi} [–]	Cost of the process C [$]	Range of end-of-process variations y_k [mm]
1	2	5	30	2.56E−06
6		6	35	2.59E−06—0.011195
15		7	40	3.02E−06—0.011261
20		8	45	3.49E−06—0.011263
15		9	50	5.17E−06—0.011264
6		10	55	7.19E−05—0.011264
1		11	60	0.011264
Total 64				
8	3	6	37	2.56E−06—2.57E−06
76		7	42	2.56E−06—0.011195
344		8	47	2.56E−06—0.011261
986		9	52	2.56E−06—0.011263
1996		10	57	2.56E−06—0.011264
3002		11	62	2.56E−06—0.011264
3432		12	67	2.56E−06—0.011264
3003		13	72	2.57E−06—0.011264
2002		14	77	2.60E−06—0.011264
1001		15	82	3.03E−06—0.011264
364		16	87	3.50E−06—0.011264
91		17	92	5.19E−06—0.011264
14		18	97	7.19E−05—0.011264
1		19	102	0.011264
Total 16,320				

Table 3.7 Solution space of DCER3_C (Milisavljevic et al. 2017)

Number of sensors distribution schemes $M_{i,k}$ [–]	Total number of sensing stations M_S [–]	Total number of sensors M_{Pi} [–]	Cost of the process C [$]	Range of end-of-process variations y_k [mm]
1	2	4	19	2.57E−06
7		5	23	2.60E−06—0.011195
Total 8				
8	3	5	23	2.59E−06—2.60E−06
Total 8				

in the process cost of the process for DCE is $31; see Table 3.4, for DCER3_Q is $30; see Table 3.6, and for DCER3_C is $23; see Table 3.7. With the use of three sensing stations and five sensors in the process cost of the process for DCE is $34; see Table 3.5, and for DCER3_C is $23; see Table 3.7.

In summary, the feasible design space is wider for DCE than it is for DCER3_Q and DCER3_C. However, the cost of the process is lower once uncertainty is inserted in DCE as previously discussed. The lowest process cost is achieved with DCER3_C when two sensing stations and four sensors are used in the design of NMS. It can be concluded that the cost of the process depends on sensors distribution and process cost can be reduced with adequate distribution of sensors in the process.

3.5.2 *Quality of the Product*

The dimensional accuracy of manufactured components is an important measure of product quality. Therefore, to incorporate this into the NMS design, the cDSP is augmented with the performance observation model (Milisavljevic 2015), to determine the expected size of process variations. Two questions need to be answered (1) "Will dimensional quality increase by increasing the number of sensing stations in the process?" and (2) "Will dimensional quality increase once uncertainty is accounted for in the cDSP?".

In order to answer the first question process decision models; see D, C, E in Fig. 3.5, are connected with performance measurement model; see PM in Fig. 3.5. If one sensing station, M_S, is used than the expected size of variations, y_k, is measured only for G1; see Fig. 3.10. However, dimensional quality is maintained within prescribed boundaries since the size of variations is minimized and close to 0 [mm] in the end-of-line of the process. Further, it can be concluded that if a system is designed to be only diagnosable then this is an adequate solution with the lowest process cost; see Table 3.4, and with satisfactory dimensional quality; see Fig. 3.10.

If two sensing stations, M_S, are used in the process the expected size of variations, y_k, is measured for all 3 goals; see Fig. 3.11. The expected size of variations for G1; see the orange rectangle in Fig. 3.11, at Station 3 is 0.015 [mm], for G2; see the yellow

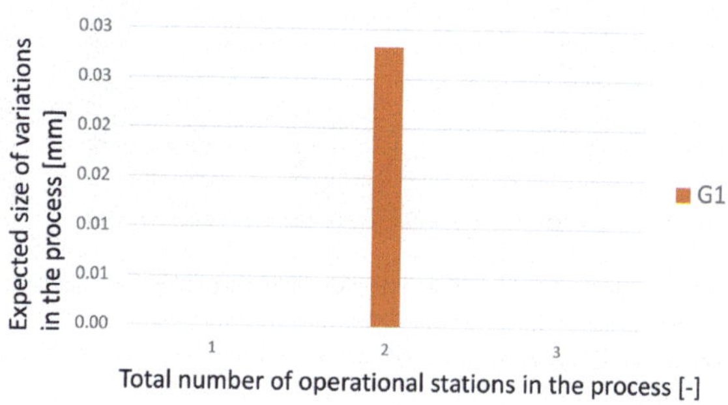

Fig. 3.10 Expected size of variations in the process with one sensing station (Milisavljevic-Syed et al. 2018)

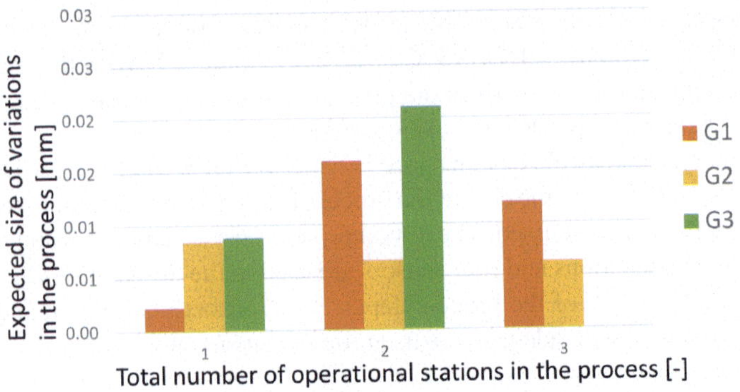

Fig. 3.11 Expected size of variations in the process with two sensing stations (Milisavljevic-Syed et al. 2018)

rectangle in Fig. 3.11, at Station 3 is 0.01 [mm], and for G3; see green rectangle in Fig. 3.11, at Station 3 is 0 [mm].

If three sensing stations, M_S, are used in the process it can be seen that the expected size of variations is measured, y_k, for all 3 goals; see Fig. 3.12. The expected size of variations for G1; see orange rectangle in Fig. 3.12, at Station 3 is 0.02 [mm], for G2; see yellow rectangle in Fig. 3.12, at Station 3 is 0.01 [mm], and for G3; see green rectangle in Fig. 3.12, at Station 3 is 0.015 [mm].

The size of variations is much lower when two sensing stations are used in the process, regarding process diagnosability, G1, and cost-effectiveness, G3; see Fig. 3.11. However, there is no difference in process controllability if the size of variations when two or three sensing stations are used, G2; see Figs. 3.11 and 3.12. Furthermore, the dimensional quality is not improved by increasing the number of sensing stations but rather by the adequate selection of design parameters and sensor distributions.

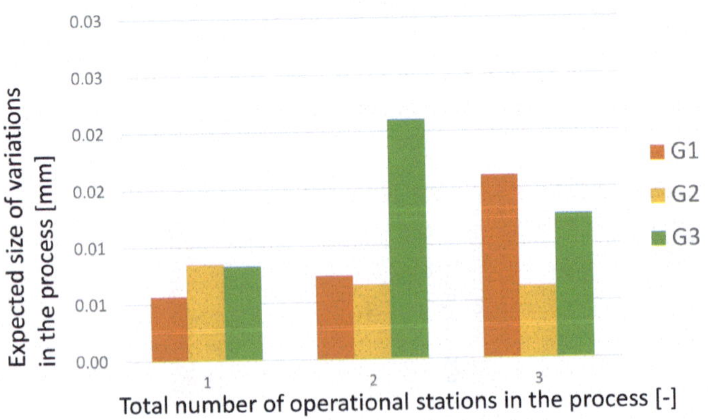

Fig. 3.12 Expected size of variations in the process with three sensing stations (Milisavljevic-Syed et al. 2018)

To answer the second question, the combined model without uncertainty; see DCE in Fig. 3.7, and with uncertainty; see DCER3 in Fig. 3.7, are connected with the performance observation model; see Fig. 3.7. If two sensing stations and four sensors are used in the process, the size of variations is higher when uncertainty is present in DCE. Further, the size of variations is 2.56E−06 [mm] of DCE; see Table 3.5, and 2.57E−06 [mm] of DCER3_C; see Table 3.7. On the other hand, if two sensing stations and five sensors are used in the process, the size of variations is lower when uncertainty is present in DCER3_Q. Further, the size of variations is between 2.56E−06 and 0.011195 [mm] of DCE; see Table 3.5 and Fig. 3.11, 2.56E−06 [mm] of DCER3_Q; see Table 3.6 and Fig. 3.12, and between 2.60E−06 and 0.011195 [mm] of DCER3_C; see Table 3.7. However, if three sensing stations and five sensors are used in the process the size of variations is higher when uncertainty is applied in DCE. Further, the size of variations is 2.56E−06 [mm] of DCE; see Table 3.5 and Fig. 3.11, and between 2.59E−06 and 2.60E−06 [mm] of DCER2_C; see Table 3.7.

In summary, dimensional quality is not improved by increasing the number of sensing stations but rather by the adequate selection of the design parameters and adequate sensor distributions. Further, the size of end-of-line variations, are much lower when two sensing stations are used in the process regarding goals Gi, $(i = 1 - 3)$. The amount of additional complexity associated with adding stages to the manufacturing process grows linearly and does not depend on the level of process parallelization. Further, if the size of variations of the DCE model is compared with the DCER2_Q model it can be concluded that the size of variations is lower once model simplifications are made in a DCE model that is quality orientated. Therefore, the answer to the question is that dimensional quality will increase once uncertainty is present in DCE as long the simplification made in DCE is quality orientated.

3.5.3 The Cost-Quality Relationship Analysis

In this section, the relationship between the process cost and the quality of the product is addressed, solutions are located and analyzed with use of 5 sensors and 2–3 sensing stations in the process; see Step D5 in Fig. 1.7. Further, if one sensing station is used in the process then there are no feasible designs according to DCE and DCER3. However, if two or three sensing stations are used in the process there are feasible designs for DCE and DCER3; see Fig. 3.13.

The cost of the process is lower for DCER3 than for DCE; see Fig. 3.13. The lowest cost of the process is for DCER3_C where simplifications made in DCE are cost orientated; see dots in Fig. 3.13.

If two sensing stations are used in the process, then the size of variations are increased up to 0.025 [mm] in Station 3 for DCE; see orange rectangles in Fig. 3.14, decreased to 0 [mm] in Station 3 for DCER3_Q; see yellow rectangles in Fig. 3.14, and decreased to 0.006 [mm] in Station 3 for DCER3_C; see green rectangles in Fig. 3.14.

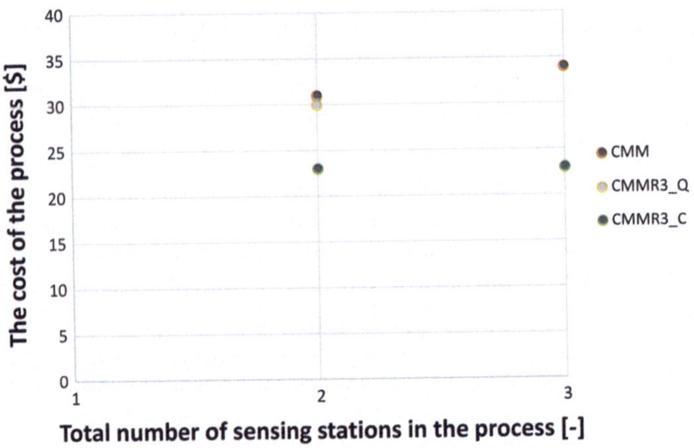

Fig. 3.13 Cost of the feasible designs according to DCE and DCER3 (Milisavljevic et al. 2017)

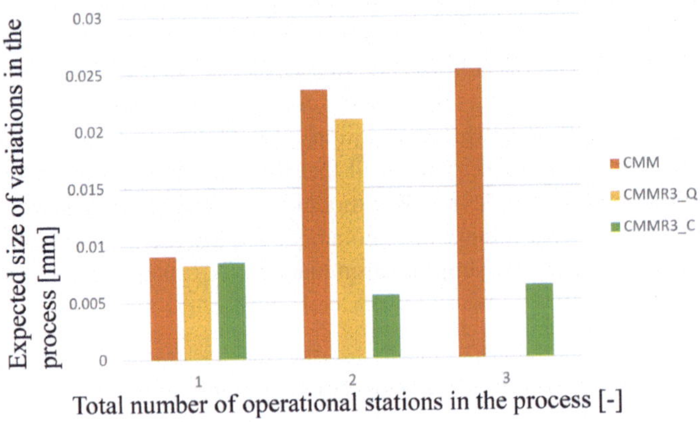

Fig. 3.14 Expected size of variations in the process with two sensing stations (Milisavljevic et al. 2017)

If three sensing stations are used in the process, then the size of variations increased to 0.052 [mm] in Station 3 for DCE; see orange rectangles in Fig. 3.15, and decreased to 0.006 [mm] in Station 3 for DCER3_C; see green rectangles in Fig. 3.15.

In summary, the cost-quality analysis regarding DCE and DCER3 leads to the following observations (1) the cost of the process is higher with the use of three sensing stations than with two sensing stations with the same number of sensors in the process, (2) the feasible design space is larger for the DCE model than it is for the DCER3_Q and DCER3_C models, (3) the cost of the process depends on the distribution of sensors and process cost can be reduced with the adequate distribution of sensors in the process, (4) the cost of the process is reduced once

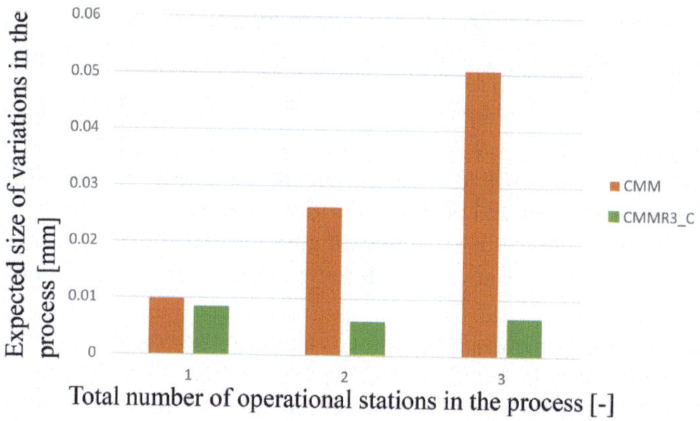

Fig. 3.15 Expected size of variations in the process with three sensing stations (Milisavljevic et al. 2017)

uncertainty is included in DCER3_Q and DCER3_C models, and (5) the dimensional quality is improved with use of two sensing stations in the process regarding DCER3_Q model and the dimensional quality is maintained within desirable boundaries with use of three sensing stations in the process regarding DCER3_C model. It can be concluded that once uncertainty is inserted in DCE the results are improved regarding process cost and quality. Furthermore, the most favorable solutions we achieve with DCER3_Q so that process cost is reduced, quality is improved, and all constraints are satisfied when two sensing stations and five sensors are used in the process. Dimensional quality is improved once uncertainty is inserted in DCE when the simplifications made in DCE are quality orientated.

3.6 Synopsis

In this chapter, we present a method for adaptable and concurrent design of NMS, a critical component of the computational framework, Design for Dynamic Management; see Fig. 3.1. In this chapter Illustrative Example 1 from Chap. 2 is first modeled using the Stream of Variation Approach and its validity for use in the computation framework is illustrated. Using this model, the design problem is then formulated as a compromise Decision Support Problem (cDSP) and solutions that satisfice system requirements such as observability and controllability while meeting overall design goals such as cost are determined. The design strategy is developed to address the following questions:

(1) *If there is flexibility in selecting sensors and tooling at design time, is it possible to select a particular configuration such that all single faults can be diagnosed? Can such faults be accommodated?*

(2) *If there is flexibility in selecting sensors and tooling at design time, are there configurations that allow different system properties and goals to be satisfied?*

Based on the results in this chapter our observation is that the NMS is diagnosable if we modify the configuration of sensors, that is, their number, position, and distribution in the process. The NMS will be controllable if we modify the configuration of tools, that is, their number and position. Further, the cost of the process was shown to be affected by the choice of sensing configuration. The quality of the process is improved, that is, size of variations is reduced with appropriate choice of tooling and sensing configurations.

We illustrate a way in which flexibility can be incorporated in the design process with the approach laid out in this chapter; see Fig. 3.16. By considering the feasible design space and studying the requirements for NMS and the performance resulting from the choice of design parameters, one can perform trade-off analysis and select a design that is cost effective and capable of meeting the needs of flexible manufacturing.

In the next chapter, we present the second component of the DFDM framework, namely, operability analysis wherein we address system functionality with a change in the requirements; see Fig. 3.16. Two operability models for system analysis are

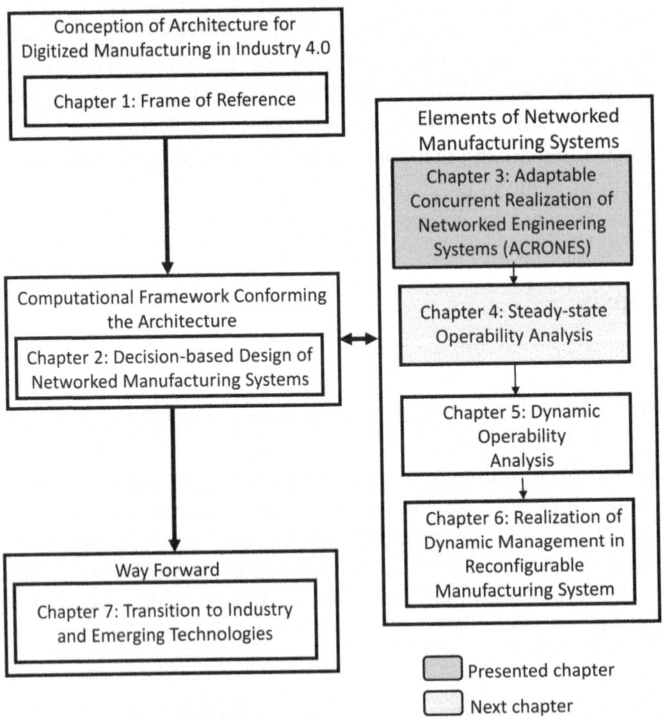

Fig. 3.16 Organization of the monograph—presented and next chapter

used to study the performance of the system prior to the design being finalized; see Operability in Fig. 3.1. In Chap. 4 we present a Steady-State Operability Model (SSOM) for analyzing system functionality when a system is in steady state and in Chap. 5 we cover dynamic operability.

Glossary

Aspiration Space Space framed by designer's wishes

Combined Model Integration of process decision models and performance observation models

Control System Inspection system with sensing stations and sensors that preform measurements

Cost-Quality Trade-off Trade-off between the cost of the process and the quality of the product in order to find satisfying solutions

cDSP Compromise Decision Support Problem

Feasible Design Space Space framed by constraints and bounds

Mechanical System Production line with operational stations and tools that preform machining or assembling operations

NMS Networked Manufacturing Systems

Operational Station Attributes Number of operational stations in the production line

Performance Observation Models Models used measure quality of product

Process Controllability Capability of the process to mitigate the errors and drive the system from an arbitrary state to a desired state along specified state trajectory

Process Decision Models Model used to simulate the process

Process Diagnosability Capability of the process to detect faults and identify their cause

PT Programmable tooling

Sensor Attributes Type, number, position, and distribution of sensors in the inspection system

Sensing Cost Sensing cost relates to the expense of building sensing stations, using Programmable Tooling (PT) control actions, and penalties for reducing the number of sensing stations

Sensors Distribution Scheme Distribution of sensors in the process. There are three types of distribution (1) end of line where sensors are installed in the end station, (2) distributed where sensors are installed on some sensing stations in the process, and (3) saturated distribution where sensors are installed on each sensing station in the process

Sensing Penalties Penalties due to a reduced number of sensing stations in the process.

Sensing Station Attributes Type, and number of sensing station in the inspection system

Sensing Stations Stations with installed sensors

SoV Stream of Variation model
Tool Attributes Type, number, and position of tools in the production line
Uncertainty Inherent randomness or unpredictability of a system, model parameters uncertainty, and model structure uncertainty

References

Altiok, T. (1997). *Performance analysis of manufacturing systems*. Berlin: Springer Verlag. ISBN: 978-1-4612-7341-7 (Print) 978-1-4612-1924-8 (Online).

Apley, D. W., & Shi, J. (1998). Diagnosis of multiple fixture faults in panel assembly. *Journal of Manufacturing Science and Engineering, 120*, 793–801.

Buzacott, J. A., & Shantikumar, J. G. (1993). Stochastic models of manufacturing systems. Englewood Cliffs, New Jersey: Prentice Hall. ISBN 0-13-847567-9

Ding, Y., Kim, P., Ceglarek, D., & Jin, J. (2003). Optimal sensor distribution for variation diagnosis in multistation assembly processes. *IEEE Transactions on Robotics and Automation, 19*(4), 543–556.

Ding, Y., Ceglarek, D., & Shi, J. (2000). Modeling and diagnosis of multistage manufacturing processes: Part I—State space model. In *Proceedings of the 2000 Japan/USA Symposium on Flexible Automation* (pp. 23–26).

Ding, Y., Ceglarek, D., & Shi, J. (2002a). Fault diagnosis of multistage manufacturing processes by using state space approach. *ASME Transactions, Journal of Manufacturing Science and Engineering, 124*, 313–322.

Ding, Y., Ceglarek, D., & Shi, J. (2002b). Design evaluation of multistation manufacturing processes by using state space approach. *ASME Transactions, Journal of Mechanical De-sign, 124*(4), 408–418.

Ding, Y., Ceglarek, D., & Shi, J. (2002c). Diagnosability analysis of multi-station manufacturing processes. *ASME Transactions, Journal of Dynamics Systems, Measurement, and Control, 124*, 1–13.

Gershwin, S. B. (1994). *Manufacturing systems engineering*. Englewood Cliffs, New Jersey: Prentice Hall. ISBN 0-13-560608-X

Hopp, W. J., & Spearman, M. L. (1996). *Factory physics: Foundations of factory management*. Chicago, IL: Irwin/McGraw Hill. ISBN 1577667395.

Izquierdo, L. E., Shi, J., Hu, S. J., & Wampler, C. W. (2007). Feedforward control of multistage assembly processes using programmable tooling. *Transactions of NAMRI/SME, 35*, 295–302.

Jiao, Y. (2012). Compensation-oriented quality control in multistage manufacturing processes (Doctoral dissertation).

Jiao, Y., & Djurdjanovic, D. (2010). Joint allocation of measurement points and controllable tooling machines in multistage manufacturing processes. *IIE Transactions, 42*(10), 703–720.

Jin, J., & Shi, J. (1999). State space modeling of sheet metal assembly for dimensional control. *ASME Transactions Journal of Manufacturing Science and Engineering, 121*(4), 756–762.

Krishnamurthy, A., Suri, R., & Vernon, M. (2004). Re-examining the performance of MRP and Kanban material control systems for multi-product flexible manufacturing systems. *International Journal of Flexible Manufacturing Systems, 16*(2), 123–150.

Liu, H., Chen, W., Kokkolaras, M., Papalambros, Y. P., & Kim, M. H (2006). Probabilistic analytical target cascading: A moment matching formulation for multilevel optimization under uncertainty. *Journal of Mechanical Design, 128*, 991–1000.

Mantripragada, R., & Whitney, D. E. (1999). Modeling and controlling variation propagation in mechanical assemblies using state transition models. *IEEE Transactions on Robotics and Automation, 15*(1), 124–140.

Marston, M., Allen, J. K., & Mistree, F. (2000). The decision support problem technique: Integrating descriptive and normative approaches. *Engineering Valuation & Cost Analysis* (Special Issue on Decision-Based Design: Status and Promise), *3*, 107–129.

Milisavljevic, J. (2015). *Accounting for uncertainty in the realization of multistage manufacturing processes.* Master's Thesis, The School of Aerospace and Mechanical Engineering, University of Oklahoma, Norman, Oklahoma.

Milisavljevic-Syed, J., Allen, J. K., Commuri, S., & Mistree, F. (2018). A method for the concurrent design and analysis of networked manufacturing systems under uncertainty. *Engineering Optimization*, *51*(4), 699–717.

Milisavljevic, J., Commuri, S., Allen, J. K., & Mistree, F. (2017). Concurrent design exploration method (CDEM) of networked manufacturing systems. In *ASME International Design Engineering Technical Conferences*, Cleveland, Ohio. Paper number DETC2017-67557.

Mistree, F., Hughes, O. F., & Bras, B. A. (1993). The compromise decision support problem and the adaptive linear programming algorithm. In M. P. Kamat (Eds.), *Structural optimization: Status and promise* (pp. 247–286). Washington, DC: AIAA.

Mistree, F., Hughes, O. F., & Bras, B. (1992). Compromise decision support problem and the adaptive linear programming algorithm. In M. P. Kamat (Ed.), *Structural optimization: Status and promise* (pp. 251–290). Washington, DC: AIAA.

Shi, J., & Zhou, S. (2009). Quality control and improvement for multistage systems: A survey. *IIE Transactions, 41*(9), 744–753.

Smith, W. F., Milisavljevic, J., Sabeghi, M., Allen, J. K., & Mistree, F. (2014). Accounting for uncertainty and complexity in the realization of engineered systems. In *International Conference on Complex Systems Design & Management*, Paris, France, November 12–14, Paper number 55.

Smith, W. F., Milisavljevic, J., Sabeghi, M., Allen, J. K., & Mistree, F. (2015). The realization of engineered systems with considerations of complexity. In *ASME Design Automation Conference*, Boston, MA, August 2–5. Paper number DETC2015-46211.

Spearman, M. L., & Zazanis, M. A. (1992). Push and pull production systems: Issues and comparisons. *Operations Research, 40*(3), 521–532.

Suri, R. (1998). Quick response manufacturing: A companywide approach to reducing lead times. Portland, OR: Productivity Press. ISBN 1563272016

Womack, J. P., & Jones, D. T. (2003). Lean thinking: Banish waste and create wealth in your corporation. New York, NY: Simon and Schuster. ISBN 0-74-324927-5

Xiao, A. (2003). *Collaborative multidisciplinary decision making in a distributed environment.* Ph.D. Dissertation, The G.W. Woodruff School of Mechanical Engineering, Georgia Institute of Technology, Atlanta, GA.

Chapter 4
Steady-State Operability Analysis

Nomenclature

μ	Measure function that calculates the size of the corresponding space
u_{AOS}	Input points of AOS
u_{AIS}	Input points of AIS
d^N	Disturbances, at their nominal values
y	Output points of DOS
d_{AIS}	Expected disturbances in AIS
d_{DIS}	Expected disturbances in DIS
P_i	Total number of fixture points
M_i	Total number of sensors
M_{pi}	Potential number of sensors
$P_{i_{x,z}}$	Potential position of tools
$M_{i_{x,z}}$	Potential position of sensors
$M_{i,k}$	Potential distribution of sensors per station
$\sum V$	Sensor noise covariance
$\sum W$	Disturbances covariance
C	Overall cost
y_k	Dimensional variations at station
$M_{i,k}$	Distribution of sensors per station
M_S	Total number of sensing stations
X_1	Natural uncertainty (sensor noise)
X_2	Natural uncertainty (process disturbance)
σ	Standard deviation
ΔX_1	Variation in fixture position
ΔX_2	Variation in sensor position

J. Milisavljevic-Syed et al., *Architecting Networked Engineered Systems*,
https://doi.org/10.1007/978-3-030-38610-8_4

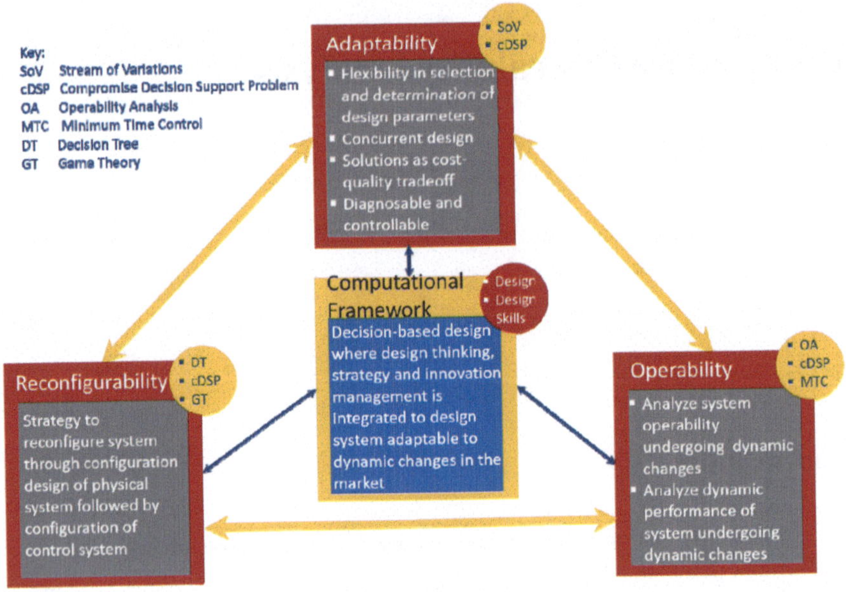

Fig. 4.1 Computational framework, design for dynamic management (Milisavljevic-Syed et al. 2019b)

In Chap. 2, we introduced a computational framework, Design for Dynamic Management (DFDM), for decision-based design of flexible, operable, and rapidly configurable manufacturing processes; see Fig. 4.1. In Chap. 3, we covered the first component of the computational framework, namely, adaptable and concurrent design of Networked Manufacturing Systems (NMS). In Chap. 3 we assumed that the numerical models of the NMS as well as the design parameters are known a priori and do not change after the implementation of the NMS. However, computational models are only abstractions of the physical process and are inherently inaccurate. Similarly, parameters are also subject to errors and variations. Therefore, while steady-state operability analysis, that is, operability with fixed models and parameter values is important, it is equally important to investigate the dynamic operability, that is, the operability of the NMS under variations in assumed models and parameter values. In this chapter, we present the second component of the computational framework, namely, operability analysis of the designed NMS when subjected to dynamic changes and/or disturbances; see Operability in Fig. 4.1. Here, the design is evaluated for changes in inputs or external disturbances that may occur at run-time. The design of NMS that maximizes the 'Operability Index' while minimizing the cost is illustrated through examples.

In this chapter our focus is on analyzing the steady-state operability of NMS. In Sect. 4.1, we present a critical review of the literature. In Sect. 4.2, we present steady-state operability analysis in the design of NMS and our contributions. In Sect. 4.3, we present the Steady-State Operability Model (SSOM) and the implementation of the compromise Decision Support Problem (cDSP) construct and we show how steady-state operability analysis facilitates dynamic management. In Sect. 4.4, we illustrate the efficacy of the method by applying it to a 2-D panel stamping process involving N-stations (Illustrative Example 1 Sect. 2.1.3). In Sect. 4.5, we present the numerical analysis of results and provide some insight into operability indices (servo operability, nominal operability, and steady-state operability) used in the design of NMS.

4.1 Steady-State Operability Analysis: A Critical Review

In control theory, the relationship between a process and its control is well understood. Systems that are linear can be stabilized and controlled over any region of operation. Control of such systems was investigated as early as the 1940's (Ziegler and Nichols 1943). However, control of highly nonlinear systems is not easy even when the region of operation is limited. Georgakis et al. (2003) show that effective control of a process is affected by the design of the process. The quote "A well designed plant is also a well-controlled one" is recognized by many researchers. According to Georgakis and co-authors (Georgakis et al. 2003), considering operability issues early in the design stage is important for chemical manufacturers to minimize variations in their products. Operability Analysis is the method that can be used for systematic exploration of the beneficial as well as a detrimental interactions between process and control design. A precise measure of operability is necessary in order to effectively accomplish this interaction.

> An operability measure should quantify the inherent ability of the process to move from one steady state to another and to reject any of the expected disturbances in a timely fashion with the limited control action available. (Georgakis et al. 2003).

Methods for operability analysis are classified into two categories: linear-based and nonlinear-based.

Linear Methods for Operability Analysis

Linear control methods are commonly used to control multiple-input-multiple-output (MIMO) systems governed by linear dynamics. (Georgakis et al. 2003). Developing a controller as the inverse of the process is useful if the process is required to track a reference signal. The controller and the process in cascade connection would then behave like a system with unit gain and the output of the system would track the input (reference signal). However, for such a technique to be successful, the inverse function of the process must exist. Morari (1983) identified the relationship between the invertibility of the transfer function matrix of a system and its resilience. Thus, the factors that prevent inversion of the process are (1) right-half-plane (RHP) zeros, (2) time delays, (3) constraints on the input variables, and (4) model uncertainty. In addition to Morari's invertibility of the transfer function there are other methods based in linear systems theory that are worth mentioning:

1. Singular Value Decomposition (see for example, Moore 1986; Grosdidier and Morari 1986; Cao et al. 1996) in addressing various aspects of control problems (control variable pairing, optimal sensor location, robust controller design, and resiliency);
2. Relative Gain Array (RGA) gives a quantitative measure of control-loop interaction for multivariable systems (for example, McAvoy 1983; Grosdidier et al. 1985; Zhu et al. 1997);
3. Relative Disturbance Gain (Stanley et al. 1985);
4. Block Relative Gain (Manousiouthakis et al. 1986);
5. Relative Sensitivity (Arkun 1988); and
6. Closed-loop Disturbance Gain (Hovd and Skogestad 1992).

While a well-designed controller is important for ensuring system performance, it is often required to ascertain the best possible performance that can be obtained. Swartz (Swartz 1996) introduced 'plant operability assessment' where solution obtained represents an upper bound on the performance of all linear stabilizing feedback controllers. Chenery (1997) introduced the linear controllability analysis based on optimal LTI control, while Lewin (1996) studied disturbance direction and frequency to study the bounds on performance.

Nonlinear Methods for Operability Analysis

Nonlinear methods are developed to utilize nonlinear models. The methods worth mentioning are:

1. Flexibility Index (FI) for quantifying the steady-state operability of nonlinear processes (Swaney and Grossmann 1985a). While some authors do not make any assumptions (Grossmann and Floudas 1987), others assume that the limiting points lie in the uncertain-parameter vertex directions (Swaney and Grossmann 1985b). Extended version of FI approach was also developed for assessing the feasibility and flexibility of the dynamic systems (Dimitriadis and Pistikopoulos 1995);
2. Maintaining controllability of the plant with a sufficient number of manipulated variables with enough range to keep the plant under nominal operating conditions when the disturbances affect the process (Fisher et al. 1988);
3. Operability of the plant in the presence of disturbances through optimization (Bahri et al. 1996). Further, addressing the dynamic operability using dynamic mixed-integer nonlinear programming (Bahri et al. 1996);
4. An integrated design and control approach under parametric uncertainty and disturbances where flexibility aspects were incorporated in a multiperiod design subproblem coupled with a feasibility analysis of dynamic systems (Mohideen et al. 1996);
5. Operability characteristics through dynamic simulations of the SISO control structures (Lyman et al. 1996);
6. Operability of CSTRs with exothermic reactions (Russo and Bequette 1995, 1998); and

7. Methods for the inherent steady-state operability of linear continuous processes (Vinson and Georgakis 1998, 2000), nonlinear processes (Subramanian and Georgakis 2000, 2001), extended to dynamic operability analysis (Uztiirk and Georgakis 2001), and examination operability characteristics of non-square systems (Subramanian et al. 2001).

To remain competitive manufacturing enterprises design manufacturing processes which are adaptable to dynamic market changes. Associated challenges are unexpected disturbances in the manufacturing process, variation in product design, and changes in the scale of the product. To meet these challenges, we propose to fully examine the dynamic and control characteristics of a system at the design stage. It is evident that the current approaches for the design of the NMS, have certain limitations, and unresolved difficulties that we identify as the gaps; see Table 1.2:

- ***Gap 1***: Expand operability analysis to fit any engineering system, such as manufacturing process, assembling process, steel making process, heat exchanger, etc., rather than plant design (Milisavljevic 2018).
- ***Gap 2***: Frame the operability and disturbance spaces, and analyze system functionality with the change in the requirements in addition to disturbances (Georgakis et al. 2003; Fisher et al. 1988; Bahri et al. 1996; Lyman et al. 1996; Russo et al. 1995, 1998; Subramanian et al. 2000, 2001); and
- ***Gap 3***: Expand steady-state operability analysis to the design of complex systems (multi-variable and different goals problems) and analyze dynamic performance of the system due to change in the requirements in addition to disturbances (Mohideen et al. 1996; Uztürk and Georgakis 1998).

In this chapter, we focus on expanding operability analysis to fit any engineering system, framing operability and disturbance spaces, and analyze system functionality with change in the requirements in addition to disturbances; see the highlighted section in Table 4.1.

In Sect. 4.2, we present contributions: (1) the use of operability analysis in design of any engineered system, and (2) use information from ACRONES to frame operability and disturbance space and preform operability analysis.

4.2 Steady-State Operability Analysis in Design of the NMS

Operability analysis is commonly used in design of chemical plants; see Sect. 4.1. In this monograph, we extend this approach to the design of any engineered system. The approach is then illustrated through its application in the panel stamping process; see Sect. 4.4.

Typically, a design engineer has to have certain domain knowledge or sufficient information in order to define (frame) operability and disturbance spaces (Georgakis et al. 2003). However, by connecting Adaptable Concurrent Realization of Networked Engineering System (ACRONES) with the method for operability analysis in the computational framework we provide support to design engineer to frame

Table 4.1 Overview of foundational papers for identifying key unresolved difficulties in the design of networked manufacturing systems regarding system operability

	Research Gaps	Expand operability analysis to fit any engineering system Frame the operability/disturbance spaces Analyze system functionality with change in the requirements									Expand to design of complex systems and analyze dynamic performance	
Analysis	Feasibility Analysis of Dynamic Systems											*
	Operability Analysis of Dynamic Systems	*	*	*	*	*	*	*	*	*	*	
	Diagnosability and Controllability											
	Analyze of Error Propagation											
	Production Function and Inventory Control											
	Production Volume and Product Variability											
Methods	Operability Characteristics of Non-Square Systems									*		
	Dynamic Operability Analysis										*	
	Operability Characteristics through Dynamic Simulations				*							
	Integrated Design and Control Approach under Parametric Uncertainty											*
	Optimization-based Operability			*								
	Maintaining Controllability of the Process in the Presence of Disturbance		*									
	Steady-State Operability of Nonlinear Processes	*				*	*	*	*			
	Dynamic Modeling of Manufacturing System											
	Stream of Variation Modeling											
	Stochastic Modeling of Manufacturing Systems											
	Push and Pull Strategy											
Aspects	Operability	*	*	*	*	*	*	*	*	*	*	*
	Adaptability											*
Paper		Georgakis, C., et al., 2003	Fisher, W. R., et al., 1988	Bahri, P. A., et al., 1996	Lyman, P. R., et al., 1996	Russo, L. P., et al., 1995	Russo, L. P., et al., 1998	Subramanian, S., et all., 2000	Subramanian, S., et all., 2001	Subramanian, S., et all., 2001	Uztiirk, D., et al., 1998	Mohideen, M. J., et all., 1996

operability space and disturbance space without domain knowledge or sufficient information; see Fig. 4.3. We obtain a feasible solution space, by exercising the ACRONES, which we use to define operability space and disturbance space, and to perform the operability analysis of the system; see Sect. 4.3.

4.3 Steady-State Operability Analysis in Design for Dynamic Management

Operability analysis is a component of the computational framework, Design for Dynamic Management, where we analyze the functionality of a system undergoing dynamic changes. In Chap. 3, we obtain the ranges of solutions by exercising the ACRONES that we use as the input in the method for operability analysis. There are two different models in the method for operability analysis (1) the Steady-state Operability Model (SSOM) that we present in this chapter, and (2) the Dynamic Operability Model (DOM) that we present in Chap. 5.

In the DFDM computational framework, operability analysis is connected with the ACRONES method; see adaptability Fig. 4.1. We obtain the ranges of solutions by exercising ACRONES. The output from ACRONES is the input for the steady-state operability model, where we define operability space and disturbance space; see operability in Fig. 4.1. We determine the functionality of the system through operability analysis by exercising SSOM. The output information from the SSOM goes to the DOM in order to analyze dynamic performance of the system; see operability in Fig. 4.1. However, if we cannot identify functional solutions of a system, then the next step is to reconfigure the system; see the fourth quadrant in Fig. 4.1. We perform operability analysis with the Steady-State Operability Model (SSOM) in order to explore the functional system design; see Sect. 4.3.1. Furthermore, we verify SSOM with 2-D panel stamping process as Illustrative Example 1; see Sect. 4.4. In Sect. 4.5, we present the results and discussion.

4.3.1 Steady-State Operability Model (SSOM)

The use of Steady-State Operability Model (SSOM) for analysis of the functional properties of NMS allows the design engineer to assess if the original models used for the components of NMS are satisfactory or if a variant of the design has to be considered. The aim is to minimize a distance between aspiration space framed by the Desired Input Space (DIS) and the Desired Output Space (DOS), and feasible design space framed by the Achieved Input Space (AIS) and the Achieved Output Space (AOS) and locate solutions of the functional system design; see Fig. 4.2. Furthermore, we connect the SSOM model with the ACRONES method where preliminary design information is generated in order to help a designer frame the operability spaces (AIS, AOS); see Fig. 4.3. The approach is described in detail in the following sections.

With SSOM we analyze (1) the extent to which the design requirements are met, (2) nominal operability of the process, and (3) operability of a system under disturbances/variations in requirements. In Fig. 4.4, we present the solution algorithm of the SSOM model.

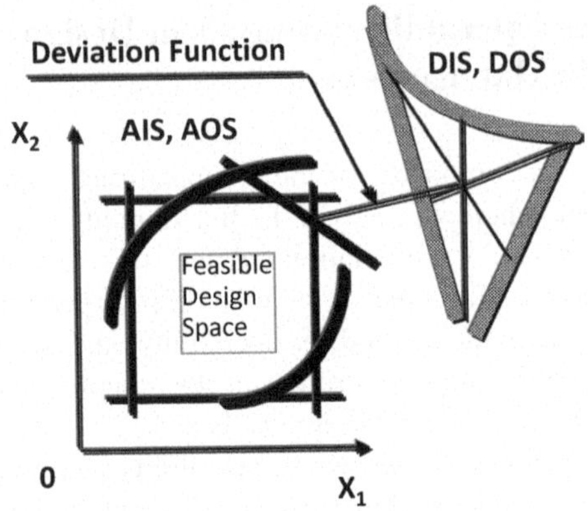

Fig. 4.2 Solution space exploration

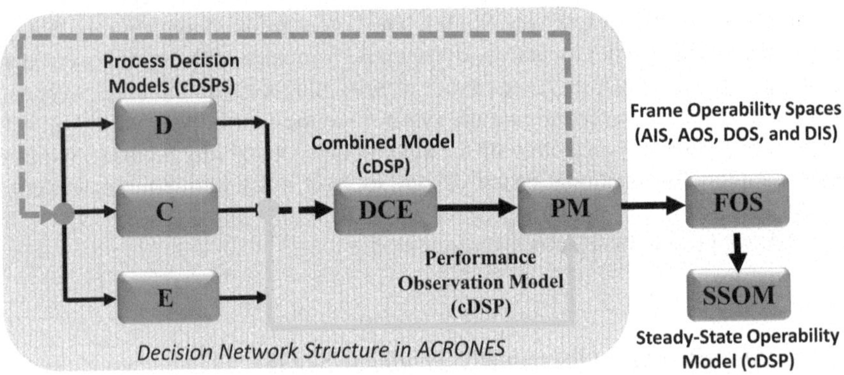

Fig. 4.3 Connecting the ACRONES with the SSOM

The ACRONES + SSOM method affords an elegant and efficient way to explore the solution space and identify possible solutions for a functional system; see Fig. 4.4. The solution scheme includes the following steps:

Step 1: Obtain information from ACRONES and frame the operability spaces; see Sect. 4.4.1.
Step 2: Examine the Servo Operability Index in the Output Space (SOIOS) for the variant system design of a system; see Eq. (4.1), or the Servo Operability Index in the Input Space (SOIIS) for the original design of a system; see Eq. (4.2). The SOIOS and SOIIS values are between 0 and 1 for operable systems. SOIOS less than '1' implies that the expectations of the designer are greater than what the process can deliver and that redesign is required. If the SOIIS less than 1 there is a need to increase the available ranges of some of the inputs and redesign is required.

Fig. 4.4 Solution Scheme (Milisavljevic-Syed et al. 2018)

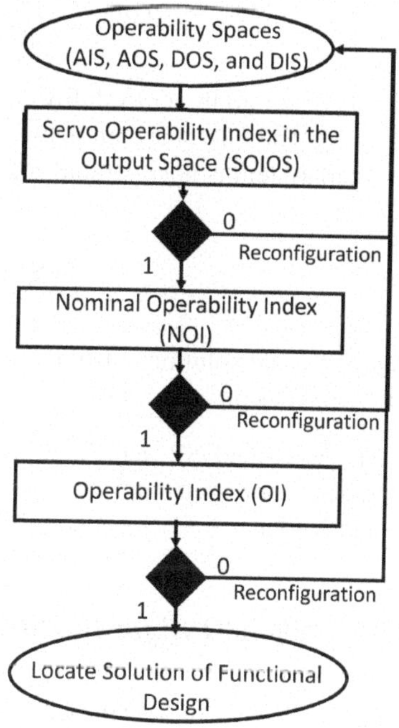

$$\mathrm{SOIOS} = \frac{\mu\left[\mathrm{AOS_u}\left(\mathrm{d^N}\right) \cap \mathrm{DOS}\right]}{\mu[\mathrm{DOS}]};\ 0 \leq \mathrm{SOIOS} \leq 1 \tag{4.1}$$

$$\mathrm{SOIIS} = \frac{\mu\left[\mathrm{AIS} \cap \mathrm{DIS_y}\left(\mathrm{d^N}\right)\right]}{\mu\left[\mathrm{DIS_y}\left(\mathrm{d^N}\right)\right]};\ 0 \leq \mathrm{SOIIS} \leq 1 \tag{4.2}$$

Step 3: Examine the nominal operability of the process in order to determine the anticipated ranges of disturbances. In original system design, the region of disturbances is based on the available input space (AIS) that needs to be higher than the desired input space (DIS); see Eq. (4.3). In variant system design, the region of disturbances is based on the Tolerable Disturbance Space (TDS) that needs to be higher than the Expected Disturbance Space (EDS); see Eq. (4.4). The Nominal Operability Index NOI values are between 0 and 1. If the value is below 1 for the NOI of the original system design then the range of AIS needs to be relaxed and redesign is required. If the value is below 1 for the NOI of the variant system design then the range of TDS needs to be relaxed and redesign is required.

$$\mathrm{NOI} = \frac{\mu\left[\mathrm{AIS} \cap \mathrm{DIS_d}\left(\mathrm{y^N}\right)\right]}{\mu\left[\mathrm{DIS_d}\left(\mathrm{y^N}\right)\right]};\ 0 \leq \mathrm{NOI} \leq 1 \tag{4.3}$$

$$\text{NOI} = \frac{\mu[\text{TDS} \cap \text{EDS}]}{\mu[\text{EDS}]};\ 0 \le \text{NOI} \le 1 \tag{4.4}$$

Step 4: Examine the Operability Index (OI) of the system in the steady-state; see Eq. (4.5). The OI values are between 0 and 1, where a value below 1 implies that designed system is not operable and redesign is required.

$$\text{OI} = \frac{\mu[\text{AIS} \cap \text{DIS}]}{\mu[\text{DIS}]} \tag{4.5}$$

Step 5: Locate solutions of functional design. If OI is equal to 1 than solutions of functional design are located.

The use of cDSP and ACRONES in the steady-state operability is a novel approach. For detailed information about the computational framework, Design for Dynamic Management; see Sect. 2.4.

4.3.2 The CDSP for the Steady-State Operability Model

In this section, we present the word formulation of the cDSP construct for the Steady-State Operability model supported by mathematical expressions. In Table 4.2, we present the cDSP construct and the cDSP construct for the SSOM analysis. The cDSP construct for the SSOM is the following:

Table 4.2 The cDSP construct for the steady-state operability analysis

cDSP		cDSP for the SSOM
Given		
• An alternative that is to be improved through modification • Assumptions used to model the domain of interest • The system parameters • All other relevant information		Information generated with the ACRONES help designer frame the operability spaces and process disturbances: • Desired Input Space (DIS); see Eq. (4.6) • Achieved Input Space (AIS); see Eq. (4.13) • Desired Output Space (DOS); see Eq. (4.10) • Achieved Output Space (AOS); see Eq. (4.6) • Tolerable Disturbance Space (TDS); see Eq. (4.18) • Expected Disturbance Space (EDS); see Eq. (4.21)
n	Number of system variables	
p + q	Number of system constraints	
p	Equality constraints	
q	Inequality constraints	
m	Number of system goals	
$g_i(X)$	System constraint function $g_i(X) = C_i(X) - D_i(X)$	
$f_k(d_i)$	Function of deviation variables to be minimized at priority level k for the preemptive case	

(continued)

Table 4.2 (continued)

cDSP		cDSP for the SSOM
W_i	Weight for the Archimedean case	
Find		
• The values of the independent system variables (they describe the physical attributes of an artifact)		• System variables that describe SOIOS, SOIIS, NOI, OI
X_j	$j = 1, \ldots, n$	
• The values of the deviation variables (they indicate the extent to which the goals are achieved)		• The values of the deviation variables (they indicate the extent to which the goals SOIOS, SOIIS, NOI, OI are achieved)
d_i^-, d_i^+	$i = 1, \ldots, m$	
Satisfy		
• The system constraints that must be satisfied for the solution to be feasible • There is no restriction placed on linearity or convexity	**Existing system design** • Servo operability index in the Output Space; see Eq. (3.1) • Nominal operability Index; see Eq. (3.4)	**New system design** • Servo operability index in the input space; see Eq. (3.2) • Nominal operability Index; see Eq. (3.3)
$g_i(X) = 0$ $g_i(X) \geq 0$	$i = 1, \ldots, p$ $i = p + 1, \ldots, p + q$	
• The system goals that must achieve a specified target value as far as possible • There is no restriction placed on linearity or convexity		
$A_i(X) + d_i^- - d_i^+ = G_i$	$i = 1, \ldots, m$	• Operability Index (OI); see Eq. (3.5)
• The lower and upper bounds on the system		
$X_j^{min} \leq X_j \leq X_j^{max}$; $d_i^-, d_i \geq 0$ and $d_i^- \cdot d_i^+ = 0$	$j = 1, \ldots, n$	
Minimize		
The deviation function that is a measure of the deviation of the system performance from that implied by the set of goals and their associated priority levels or relative weights: • Case A: Preemptive (lexicographic minimum) $Z = [f_1(d_i^-, d_i^+), \ldots, f_k(d_i^-, d_i^+)]$ • Case b: Archimedean $Z = \sum_{i=1}^{m} W_i(d_i^- + d_i^+)$; $\sum W_i = 1;\ W_i \geq 0$		The deviation function that is a measure of the deviation of the system performance from that implied by the set of goals (SOIOS, SOIIS, NOI, OI) and their associated priority levels or relative weights

- **Given** that information about alternatives, assumptions, system parameters, and other relevant information is generated by exercising the ACRONES, the designer has enough information to frame the operability spaces (DIS, AIS, DOS, AOS) and process disturbances (TDS, EDS).
- **Find** independent system variables that describe the physical attributes of operability spaces and process disturbances that describes goals, and the values of the deviation variables that indicate the extent to which the goals (SOIOS, SOIIS, NOI, OI) are achieved.
- **Satisfy** system constraints, system goals (SOIOS, SOIIS, NOI, OI), and system bounds.
- **Minimize** deviation function that is a measure of the deviation of the system performance from the desired one (DOS, DIS) to the actual one (AOS, AIS).

In Sect. 4.4, we present the empirical structural validity of the steady-state operability analysis method by verifying the appropriateness of the chosen example problem intended to test the design method.

4.4 Illustrative Example: Panel Stamping Process

In this section, we analyze the appropriateness of the illustrative example, a 2-D panel stamping process, for the Steady-State Operability method; see the Quadrant 2 in Fig. 1.15. In Sect. 4.5, we analyze the usefulness of the results of the steady-state operability analysis method; see Quadrant 3 in Fig. 1.15.

A panel stamping process represents an integral part of many manufacturing processes, especially in the assembly of automobile bodies. The panel stamping process represents the simplification of the auto body stamping process (Apley and Shi 1998). Further, the assumption is that the panels constitute the auto body parts with 100% dimensional accuracy. The dimensional quality of the completed product is highly dependent on the level of accuracy with which the panels are fixtured during the stamping process. Typically, coordinate sensors that measure the position of a part feature are installed at the end-of-line of the process.

Four different scenarios are considered to demonstrate the design process:

1. Panel assembling process in three stations with an end-of-line sensing configuration; see Fig. 4.5a;
2. Panel assembling process in three stations with a distributed sensing configuration; see Fig. 4.5b;
3. Panel assembling process in four stations with an end-of-line sensing configuration; see Fig. 4.6a; and
4. Panel assembling process in four stations with a distributed sensing configuration; see Fig. 4.6b.

For more information about the panel stamping process see Ding et al. (2002). In Sect. 4.4.1, we present the panel stamping process design specifications and model

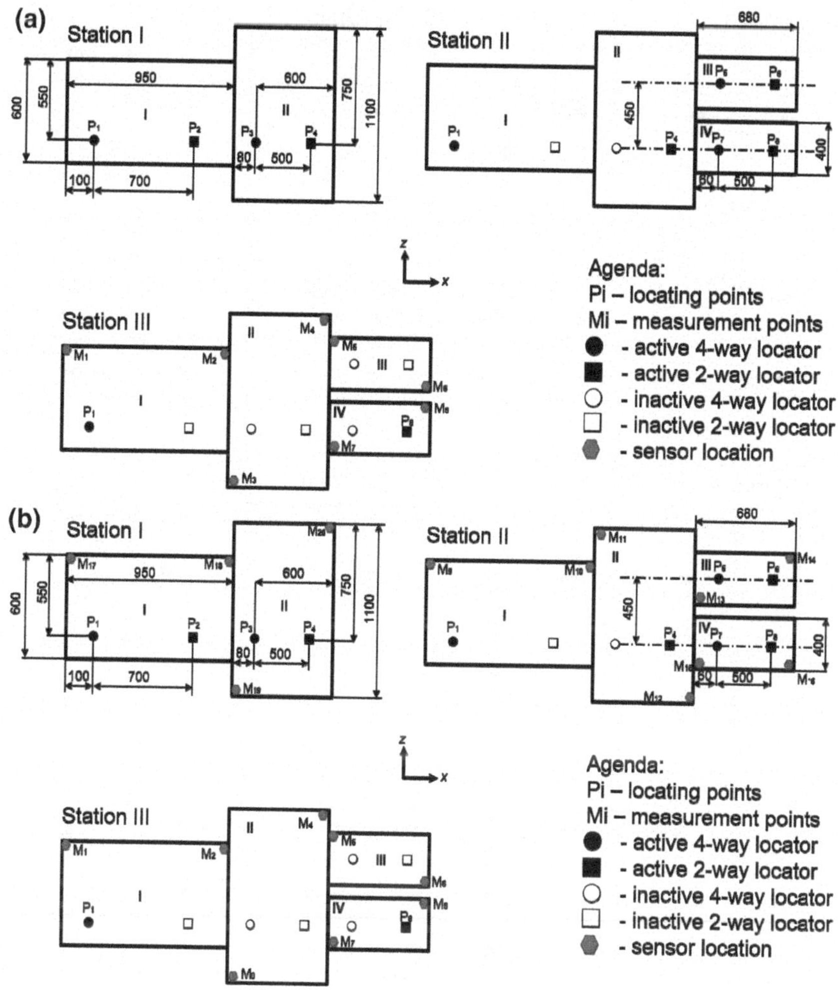

Fig. 4.5 Two-dimensional panel stamping process in three stations (Ding et al. 2002)

equations. In Sect. 4.4.2, we present the cDSP formulation for the Steady-State Operability Analysis of a panel stamping process.

4.4.1 *Model of the Panel Stamping Process and Design Specifications*

In this section, we present design specifications and operability spaces in design of a panel stamping process (Table 4.3).

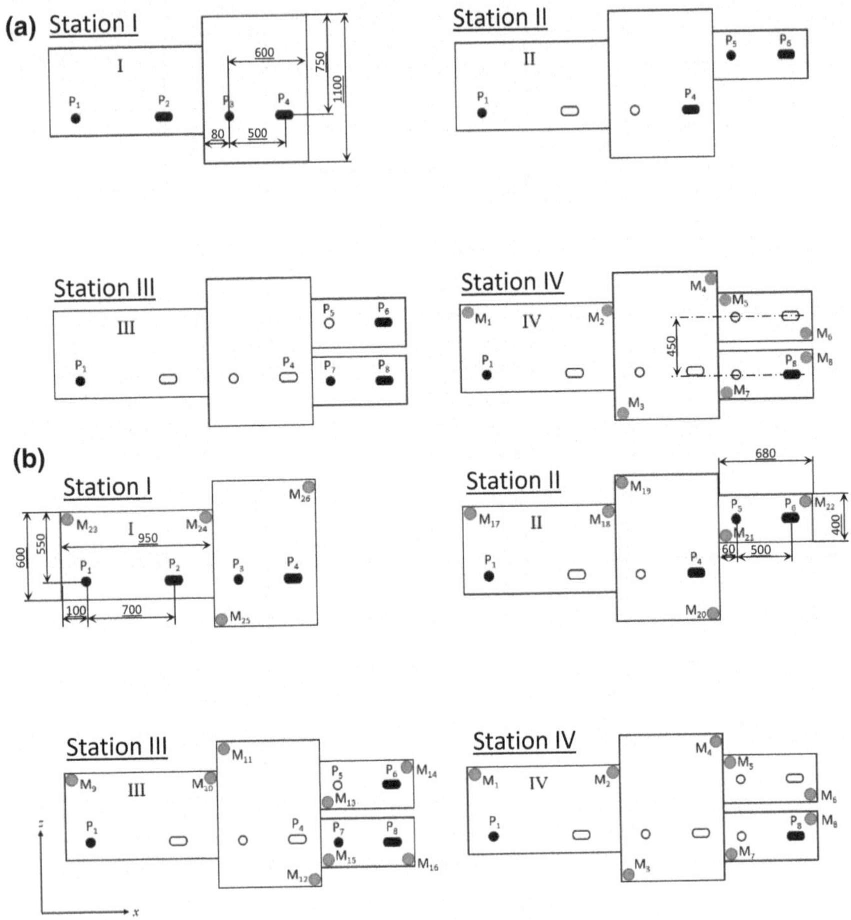

Fig. 4.6 Two-dimensional panel stamping process in four stations

Definition of Operability Spaces in Design of Panel Stamping Process

Achievable Output Space (AOS): Achievable Output Space (AOS) is the collection of output points achieved by solving the model for the entire Available Input Space (AIS). AOS is a function of u and d, $AOS_u(d^N)$ where AOS is calculated by considering all the points inside the AIS, denoted by the subscript u, when the disturbances are at their nominal values, d^N. AOS are the output values achieved by solving comprehensive model (DCE) of panel stamping process.

In Eq. (4.6), we present the mathematical formulation of AOS.

$$AOS = \bigcup_{u \in AIS} AOS_u(d^N) \tag{4.6}$$

Table 4.3 Panel stamping process design parameters

	Coordinate in *x*-direction [mm]	Coordinate in *z*-direction [mm]
Fixture points		
P_1	63.0	50.0
P_2	800.0	50.0
P_3	1019.0	50.0
P_4	1312.0	50.0
P_5	1688.0	470.0
P_6	2005.0	470.0
P_7	1688.0	50.0
P_8	2005.0	50.0
Sensor points		
M1	0.0	600.0
M2	950.0	0.0
M3	950.0	−300.0
M4	1630.0	800.0
M5	0.0	0.0
M6	950.0	600.0
M7	950.0	−300.0
M8	1630.0	800.0
M9	1630.0	300.0
M10	2310.0	700.0
M11	1630.0	−150.0
M12	2310.0	200.0
M13	0.0	0.0
M14	950.0	600.0
M15	950.0	−300.0
M16	1630.0	800.0
M17	1630.0	300.0
M18	2310.0	700.0
M19	1630.0	−150.0
M20	2310.0	200.0

Thus in panel stamping process, input points of AOS we express as a function of the identified variables; see Eq. (4.7).

$$u = f\left(P_i, M_i, P_{i_{x,z}}, M_{i_{x,z}}\right) \tag{4.7}$$

Table 4.4 Available input space (AIS)

Symbol	Description	Lower bound	Upper bound	Units
P_i	Total number of fixture points	0	8	[–]
M_{Pi}	Potential number of sensors	0	20	[–]
$P_{i_{x,z}}$	Potential position of tools	0	2005	[mm]
$M_{i_{x,z}}$	Potential position of sensors	0	2310	[mm]
$M_{i,k}$	Potential distribution of sensors per station	0	8	[–]
$\sum V$	Sensor noise covariance	$0.0009 \cdot I$	$0.009 \cdot I$	[mm^2]
$\sum W$	Disturbances covariance	$0.0001 \cdot I$	$0.001 \cdot I$	[mm^2]

Table 4.5 Desired Output Space (DOS)

Symbol	Description	Lower bound	Upper bound	Units
C	Overall cost	0	n	[$]
y_k	Dimensional variations at station	−0.8	0.8	[mm]
M_i	Total number of sensors	0	n	[–]
$M_{i,k}$	Distribution of sensors per station	0	n	[–]
M_S	Total number of sensing stations	0	n	[–]

Table 4.6 Expected disturbance space (EDS)

Symbol	Description	Lower bound	Upper bound	Units
$X_1 \pm 3\sigma$	Natural uncertainty (sensor noise)	−0.0013	0.0013	[mm]
$X_2 \pm 3\sigma$	Natural uncertainty (process disturbance)	−0.0014	0.0014	[mm]
$P_{i_{x,z}} \pm \Delta X_1$	Model parameter uncertainty (variation in fixture position)	0 ± 15	2005 ± 15	[mm]
$M_{i_{x,z}} \pm \Delta X_2$	Model parameter uncertainty (variation in sensor position)	0 ± 25	2310 ± 25	[mm]

In Eq. (4.8), we present the disturbances, at their nominal values, as a function of sensor noise and process disturbances covariance.

$$d^N = f\left(\sum V, \sum W\right) \tag{4.8}$$

In Eq. (4.9), we present the mathematical formulation of AOS in the panel stamping process.

$$AOS = \bigcup_{u \in AIS} AOS_{(P_i, M_i, P_{i_{x,z}}, M_{i_{x,z}})}\left(\sum V, \sum W\right) \tag{4.9}$$

Desired Output Space (DOS): Desired Output Space (DOS) is desired operating window for the process outputs. DOS is defined by a designer as their wish for achieving certain output space.

In Eq. (4.10), we present the mathematical formulation of DOS.

$$DOS = \bigcup DOS_d(y) \tag{4.10}$$

Thus in panel stamping process, output points of DOS are expressed as a function of the identified variables; see Eq. (4.11).

$$y = f\left(M_i, M_{i,k}, M_{Si}, C, y_k\right) \tag{4.11}$$

In Eq. (4.12), we present the mathematical formulation of DOS in the panel stamping process.

$$DOS = \bigcup DOS_d\left(M_i, M_{i,k}, M_{Si}, C, y_k\right) \tag{4.12}$$

Available Input Space (AIS): Available Input Space (AIS) is represented by the range over which the inputs of the process can change. The AIS is the input points available by exercising ACRONES or through prior designer knowledge, for example, design experience.

In Eq. (4.13), we present the mathematical formulation of AIS.

$$AIS = \bigcup AIS_d(u) \tag{4.13}$$

Thus in a panel stamping process, the input points of AIS are expressed as a function of the variables identified with ACRONES; see Eq. (4.14).

$$u = f\left(P_i, M_i, P_{i_{x,z}}, M_{i_{x,z}}, M_{i,k}, \sum V, \sum W\right) \tag{4.14}$$

In Eq. (4.15), we present the mathematical formulation of AIS in the panel stamping process.

$$AIS = \bigcup AIS_d\left(P_i, M_i, P_{i_{x,z}}, M_{i_{x,z}}, M_{i,k}, \sum V, \sum W\right) \tag{4.15}$$

Desired Input Space (DIS): Desired Input Space (DIS) is the set of input values required to reach the entire DOS. The total Desired Input Space defined as the union of $DIS_d(y)$ for all y in DOS, or the union of $DIS_y(d)$ for all d in EDS. For example, in panel stamping process if we want to reach a certain output (cost) of process we need to adjust points in the DIS accordingly.

In Eq. (4.16), we present the mathematical formulation of DIS.

$$DIS = \bigcup_{y \in DOS} DIS_d(y) = \bigcup_{d \in EDS} DIS_y(d) \tag{4.16}$$

In the panel stamping process, input points that need to be adjusted $u \in AIS$; see Eq. (4.11), in order to reach the output points $y \in DOS$; see Eq.(4.8). In Eq. (4.17), we present the mathematical formulation of DIS in the panel stamping process.

$$\begin{aligned} DIS &= \bigcup_{y \in DOS} DIS_d\left(M_i, M_{i,k}, M_{Si}, C, y_k\right) \\ &= \bigcup_{d \in EDS} DIS_y\left(\mathrm{x}_1 \pm 3\sigma, \mathrm{x}_2 \pm 3\sigma P_{i_{x,z}} \pm \Delta \mathrm{X}_1, M_{i_{x,z}} \pm \Delta \mathrm{X}_2\right) \end{aligned} \tag{4.17}$$

Tolerable Disturbance Space (TDS): Tolerable Disturbance Space (TDS) is the region of disturbances that can be tolerated with the available inputs AIS, keeping the system at the nominal operating point. TDS is determined only for the existing system design. In the panel stamping process, TDS is determined by exercising ACRONES.

In Eq. (4.18), we present the mathematical formulation of TDS.

$$TDS = \bigcup_{d \in TDS} AIS_u(d) \tag{4.18}$$

Thus in panel stamping process, expected disturbances d in AIS are expressed as a function of the variables identified with ACRONES; see Eq. 4.19.

$$d = f\left(\mathrm{x}_1 \pm 3\sigma, \mathrm{x}_2 \pm 3\sigma P_{i_{x,z}} \pm \Delta \mathrm{X}_1, M_{i_{x,z}} \pm \Delta \mathrm{X}_2\right) \tag{4.19}$$

In Eq. (4.20), we present the mathematical formulation of TDS in the panel stamping process.

$$TDS = \bigcup_{d \in TDS} AIS_u\left(\mathrm{X}_1 \pm 3\sigma, \mathrm{X}_2 \pm 3\sigma P_{i_{x,z}} \pm \Delta \mathrm{X}_1, M_{i_{x,z}} \pm \Delta \mathrm{X}_2\right) \tag{4.20}$$

Expected Disturbance Space (EDS): Expected Disturbance Space (EDS) in steady-state operability reflects on uncertainties in model parameters. In panel stamping process, EDS reflects on natural uncertainty due to sensor noise and process disturbances, and model parameter uncertainty such as variations in fixture and sensors position.

In Eq. (4.21), we present the mathematical formulation of EDS.

$$EDS = \bigcup_{d \in EDS} DIS_u(d) \tag{4.21}$$

Thus in panel stamping process, expected disturbances d in DIS we express as a function of the variables identified with ACRONES; see Eq. (4.22).

$$d = f\left(\mathrm{X}_1 \pm 3\sigma, \mathrm{X}_2 \pm 3\sigma P_{i_{x,z}} + \Delta \mathrm{X}_1, M_{i_{x,z}} \pm \Delta \mathrm{X}_2\right) \tag{4.22}$$

In Eq. (4.23), we present the mathematical formulation of EDS in the panel stamping process.

$$EDS = \bigcup_{d \in EDS} DIS_u\left(\mathrm{X}_1 \pm 3\sigma, \mathrm{X}_2 \pm 3\sigma P_{i_{x,z}} \pm \Delta \mathrm{X}_1, M_{i_{x,z}} \pm \Delta \mathrm{X}_2\right) \tag{4.23}$$

In Sect. 4.4.2, we present the cDSP formulation for the steady-state operability analysis example.

4.4.2 The CDSP Formulation for the Steady-State Operability Analysis of a Panel Stamping Process

Given

Design Parameters			
Number of stations	N	3	[–]
Number of parts	n_p	4	[–]
Potential fixture faults	m_r	18	[–]
Degrees of freedom of 2-D rigid part	DOF	3	[–]
Desired process diagnosability index	μ_D	100	[%]
Desired process controllability index	μ_C	100	[%]
Operability Spaces			
Available Input Space	*AIS*	[–]	Table 4.4
Desired Output Space	*DOS*	[–]	Table 4.5
Expected Disturbance Space	*EDA*	[–]	Table 4.6
Design Alternatives			
3-stage process with end-of-line sensing	Fig. 4.5a		
3-stage process with distributed sensing	Fig. 4.5b		
4-stage process with end-of-line sensing	Fig. 4.6a		
4-stage process with distributed sensing	Fig. 4.6b		
Assumptions			
Workpieces are 2-D rigid parts			
There are only potential fixture faults in the process			
Local directions of fixture points in x-direction are adjustable in n-stations			
Sensors in x- and z-direction are positioned at the corner points			
No three sensors are collinear in x- and z- direction			
Fixture points $P_1, P_2, P_3, P_4, P_7, P_8$ are collinear in z- direction			

(continued)

(continued)

Fixture points P_5, P_6 are collinear in z- direction
Process diagnosability is full
Process controllability is full
System goals weight coefficients are based on a designer decision
Operability spaces information are obtained from ACRONES

Find

System Variables		
Achievable Output Space	AOS	$AOS \subset AIS$
Desirable Input Space	$DIS_d(y^N)$	$DIS \subset DOS$
Total Desirable Input Space	DIS	$DIS = \bigcup_{y \in DOS} DIS_d(y) = \bigcup_{d \in EDS} DIS_y(d) = \bigcup_{y \in DOS; d \in EDS} DIS_d(y)$
Tolerable Disturbance Space	TDS	$TDS = \bigcup_{d \in TDS} AIS_u(d)$

Deviation Variables		
Under achievement of the Goal 1	d_1^-	[–]
Under achievement of the Goal 2	d_2^-	[–]
Under achievement of the Goal 3	d_3^-	[–]

Satisfy

System Constraints				
Fixture points P P_i and P_j cannot be the same points (i, j = 1, …, 8)	$P_{i,j}$	$P_{i,x} \neq P_{j,x}$ $P_{i,z} \neq P_{j,z}$	C_1	[mm]
Sensors points M_i and M_j cannot be the same points (i, j = 1, …, 20)	$M_{i,j}$	$M_{i,x} \neq M_{j,x}$ $M_{i,z} \neq M_{j,z}$	C_2	[mm]
Sensors have to take corner positions	M_i	$M_{i,x} \equiv \min \vee \max$ $M_{i,z} \equiv \min \vee \max$	C_3	[mm]
No three sensors can be collinear in x- and z- direction	$M_{i,j,r}$	$M_{i,x} \neq M_{j,x} \neq M_{r,x}$ $M_{i,z} \neq M_{j,z} \neq M_{r,z}$	C_4	[mm]
Fixture points $P_1, P_2, P_3, P_4, P_7, P_8$ must be collinear in z- direction	$P_{i,z}$	$P_{1,z} = P_{2,z} = P_{3,z} = P_{4,z} = P_{7,z} = P_{8,z}$	C_5	[mm]
Fixture points P_5, P_6 must be collinear in z- direction	$P_{i,z}$	$P_{5,z} = P_{6,z}$	C_6	[mm]

(continued)

(continued)

Product of deviation variables has to be equal to zero	d_i^-, d_i^+	$d_i^- \times d_i^+ = 0$	C_7	[–]
Deviation variables have to be positive	d_i^-, d_i^+	$d_i^-, d_i^+ \geq 0$	C_8	[–]
System Goals				
Maximize Servo Operability in the Output Space*		$\frac{\mu[AOS_u(d^N) \cap DOS]}{\mu[DOS]} + d_1^- - d_1^+ = 1$	G_1	[–]
Maximize Servo Operability Index in the Input Space**		$\frac{\mu[AIS \cap DIS_y(d^N)]}{\mu[DIS_y(d^N)]} + d_1^- - d_1^+ = 1$	G_1	[–]
Maximize Nominal operability Index*		$\frac{\mu[TDS \cap EDS]}{\mu[EDS]} + d_2^- - d_2^+ = 1$	G_2	[–]
Maximize Nominal operability**		$\frac{\mu[AIS \cap DIS_d(y^N)]}{\mu[DIS_d(y^N)]} + d_2^- - d_2^+ = 1$	G_2	[–]
Maximize System Operability		$\frac{\mu[AIS \cap DIS]}{\mu[DIS]} + d_3^- - d_3^+ = 1$	G_3	[–]
System Bounds				
Total number of sensors	M_i	0–20	B_1	[–]
Distribution of sensors per station	$M_{i,k}$	0–8	B_2	[–]
Total number of sensing stations	M_{Si}	0–n	B_3	[–]
Dimensional variations at station	y_k	−0.8 to 0.8	B_4	[mm]
Overall cost	Cost	0– n	B_5	[$]
Servo Operability Index in the Output Space or Servo Operability in the Input Space	*SOIOS* or *SOIIS*	0–100	B_6	[%]
Nominal operability Index	*ROI*	0–100	B_7	[%]
Operability Index	*OI*	0–100	B_8	[%]

Minimize

Deviation Functions		
Preemptive formulation where we are minimizing the goal function	$Z = \min(d_i^-)$	[–]

In Sect. 4.5, we present the empirical performance validity where we verify the ability to produce useful results for the chosen example problem.

4.5 Results and Discussion

Steady State Operability Analysis that incorporates ACRONES and DFDM was introduced in Sect. 4.3 and the model used for analysis was given in Fig. 4.3. In this section, we present the results and discuss the steady operability of the system. The SSOM is illustrated using an example of the automotive panel stamping process where three and four stations are used; see Figs. 4.5 and 4.6.

4.5.1 Design of a Three Stage Panel Stamping Process

Servo Operability Index in the Output Space (SOIOS)

In the case of variant system design we want to see what portion of the Desired Output Space (DOS) can be achieved using the Available Input Space (AIS). The goal is to maximize the Servo Operability Index in the Output Space (SOIOS). The SOIOS is full (equal to 1) if the Available Output Space (AOS) and the Desired Output Space (DOS) are identical, to realize the entire DOS in the design. If SOIOS is partial (less than 1) then redesign is required. In Table 4.7, we present the system configuration obtained by exercising the ACRONES that reach full SOIOS for a three stage panel stamping process with different sensing configurations. In Fig. 4.7a–c, we present a distributed sensing configuration. In Fig. 4.7d–f, we present the end-of-line sensing configuration.

In design of the three stage panel stamping process with distributed and end-of-line sensing AOS is framed by exercising ACRONES and DOS is framed by designer wishes. It can be seen that DOS is much smaller than AOS since thew designer's goal

Table 4.7 Full servo operability index in the output space (SOIOS) solution space

Sensing configuration	Available output space			Desirable output space			Full servo operability space		
	Mi [–]	Ms [–]	Cost [$]	Mi [–]	Ms [–]	Cost [$]	Mi [–]	Ms [–]	Cost [$]
Distributed	4 ÷ 20	2 ÷ 3	7 ÷ 23	0 ÷ 7	0 ÷ 3	0 ÷ 10	4 ÷ 7	2 ÷ 3	7 ÷ 10
End-of-line	0 ÷ 8	0 ÷ 1	3 ÷ 11	0 ÷ 2	0 ÷ 1	0 ÷ 5	0 ÷ 2	0 ÷ 1	3 ÷ 5

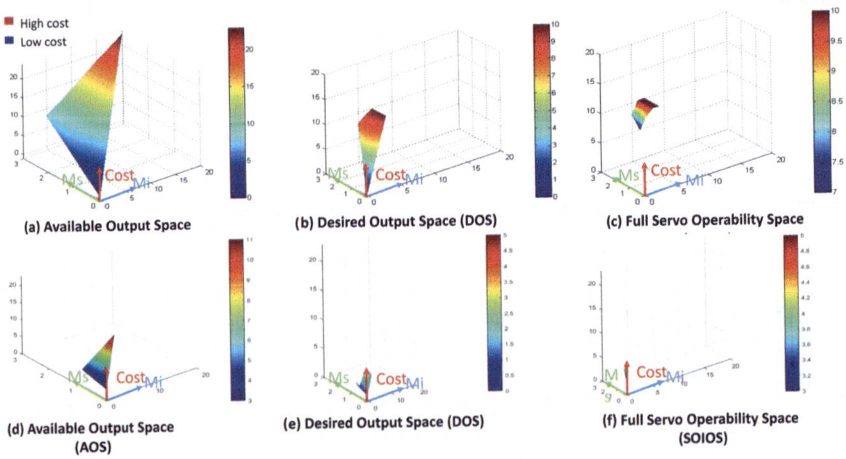

Fig. 4.7 Servo operability index in the output space (SOIOS) solution space for distributed and end-of-line sensing configuration

is to reduce the cost of the system; see Fig. 4.7e and d. Full SOIOS is the intersection between AOS and DOS; see Fig. 4.7f. In Table 4.7, we present the solution space of SOIOS, AOS, and DOS, where full SOIOS is $7 \leq \text{Cost} \leq 10$, number of sensors $4 \leq M_i \leq 7$, and sensing stations $2 \leq M_s \leq 3$.

In Table 4.7, we present the solution space of the SOIOS in the design of the three stage panel stamping process with end-of-line sensing. The full SOIOS solution space is the cost of the process $3 \leq \text{Cost} \leq 5$, number of sensors $0 \leq M_i \leq 2$, and sensing stations $0 \leq M_s \leq 1$.

Nominal Operability Index (NOI)

In the case of variant system design, we determine the anticipated ranges of disturbances with the goal of maximizing maximize the Nominal Operability Index (NOI). The NOI is full (equal to 1) if the Expected Disturbance Space (EDS) (natural uncertainty and model parameter uncertainty) of the system is within the Tolerable Disturbance Space (TDS). If NOI is partial (less than 1) than redesign is required.

TDS is framed by a designer's experience of multistage system design recommendations where EDS is framed by exercising ACRONES and full NOI is framed as intersection of EDS and TDS. In presence of natural uncertainty in the design of the panel stamping process with distributed sensing it can be seen that TDS is larger than EDS; Fig. 4.8a, b. Further, it can be seen that full NOI is at the intersection between EDS and TDS; see Fig. 4.8c.

In Table 4.8, we present the system configuration that reaches full NOI for three stage panel stamping process with distributed and end-of-line sensing configuration in the presence of natural uncertainty.

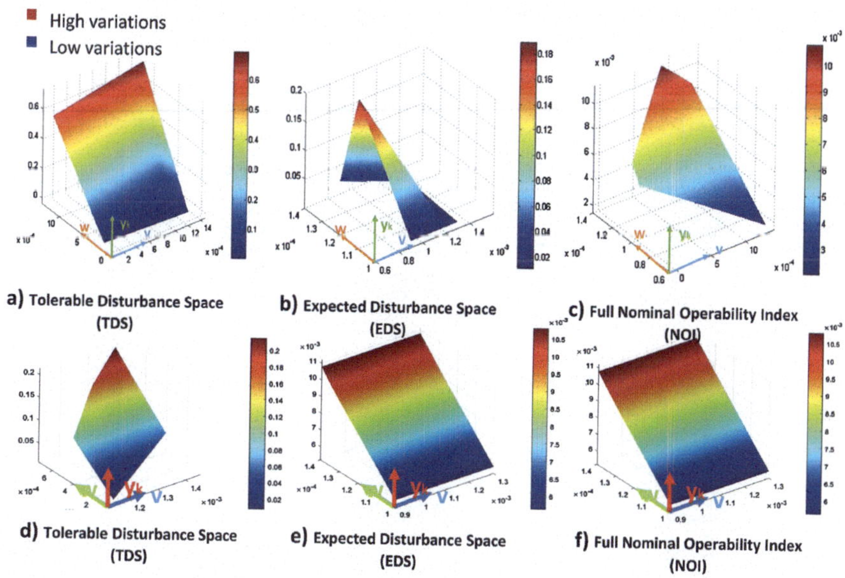

Fig. 4.8 Nominal operability index (NOI) solution space in the presence of natural uncertainty for distributed and end-of-line sensing configuration

Table 4.8 Full nominal operability Index (NOI) solution space for natural uncertainty

Sensing configuration	i	Tolerable disturbance space			Expected disturbance space			Full nominal operability index		
		V [mm]	W [mm]	y_k [mm]	V [mm]	W [mm]	y_k [mm]	V [mm]	W [mm]	y_k [mm]
Distributed	1	0.00124	0.00091	0.38600	0.00090	0.00010	0.01130	0.00090	0.00010	0.00563
	2	0.00000	0.00011	0.00682	0.00090	0.00013	0.18900	0.00000	0.00011	0.00682
	3	0.00147	0.00122	0.69300	0.00090	0.00014	0.02190	0.00090	0.00014	0.01080
	4	0.00103	0.00094	0.41200	0.00120	0.00010	0.01130	0.00103	0.00010	0.00563
	5	0.00003	0.00114	0.60600	0.00130	0.00010	0.01120	0.00003	0.00010	0.00565
	6	0.00126	0.00050	0.11700	0.00120	0.00013	0.01890	0.00120	0.00013	0.00938
	7	0.00131	0.00000	0.00206	0.00130	0.00014	0.02190	0.00130	0.00000	0.00206
End-of-line	1	0.00121	0.00037	0.06705	0.00090	0.00010	0.00563	0.00090	0.00010	0.00563
	2	0.00114	0.00046	0.09705	0.00090	0.00013	0.00939	0.00090	0.00013	0.00939
	3	0.00144	0.00036	0.06178	0.00090	0.00014	0.01080	0.00090	0.00014	0.01080
	4	0.00135	0.00034	0.07603	0.00120	0.00010	0.00563	0.00103	0.00010	0.00563
	5	0.00127	0.00060	0.16670	0.00130	0.00010	0.00563	0.00130	0.00010	0.00563
	6	0.00116	0.00012	0.00821	0.00120	0.00013	0.00938	0.00120	0.00013	0.00938
	7	0.00138	0.00068	0.21492	0.00130	0.00014	0.01079	0.00130	0.00014	0.01079

In the presence of natural uncertainty in the design of the panel stamping process with end-of-line sensing it can be seen that TDS is larger than EDS; see Fig. 4.8d, e. Further, we can see that NOI is equal to EDS; see Fig. 4.8f.

Operability Index

The operability index provides us information about whether system can regulate itself in the presence of disturbance (natural uncertainty) and if the system is functional in the presence of disturbance.

Available Input Space (AIS) is framed by exercising the ACRONES and the Desired Input Space (DIS) is framed by designer wishes, experience or assumptions, where full operability of the system is intersection of AIS and DIS. In Table 4.9, we present solution space of operable system in the presence of disturbances (natural uncertainty).

In presence of natural uncertainty in the design of a panel stamping process with distributed sensing it can be seen that AIS is larger than DIS; see Fig. 4.9a, b. Further, it can be seen that full OI is in intersection between AIS and DIS; see Fig. 4.9c.

In the presence of natural uncertainty in design of the panel stamping process with end-of-line sensing it can be seen that AIS is larger than DIS; see Fig. 4.9d, e. Further, it is seen that full OI is at the intersection between AIS and DIS; see Fig. 4.9f.

4.5.2 Design of a Four Stage Panel Stamping Process

Servo Operability Index in the Output Space (SOIOS)

We observe three attributes in operability spaces for the servo operability index in the output space, namely, the number of sensors, sensing stations, and related cost of the process. In Table 4.10, we present the system configuration that reaches full SOIOS for four stage panel stamping process with different sensing configurations. In Fig. 4.10a–c, we present the distributed sensing configuration. In Fig. 4.10d–f, we present the end-of-line sensing configuration.

In case in design of the four-stage panel stamping process with distributed sensing AOS is larger than the DOS; see Fig. 4.10a, b. Full servo operability of the system is discovered at the intersection between AOS and DOS; see Fig. 4.10c.

In the design of the four stage panel stamping process with end-of-line sensing if we want cost of design to be $7 \leq \text{Cost} \leq 10$ then the number of sensors and the number of sensing stations in the system configuration must satisfy the following constraints: $0 \leq M_i \leq 3$ and $0 \leq M_s \leq 1$; see Table 4.10.

Nominal Operability Index (NOI)

TDS is framed by a designer's experience in multistage system design where EDS is framed by exercising the ACRONES and full NOI is identified at the intersection of these spaces. In the design of a four stage panel stamping process with distributed

Table 4.9 Full operability index (OI) solution space for natural uncertainty

Sensing configuration	i	Available input space			Desired input space			Full operability index		
		V [mm]	W [mm]	y_k [mm]	V [mm]	W [mm]	y_k [mm]	V [mm]	W [mm]	y_k [mm]
Distributed	1	0.00124	0.00091	0.14850	0.00232	0.00085	0.03501	0.00124	0.00085	0.33686
	2	0.00096	0.00041	0.20160	0.00198	0.00189	0.17000	0.00096	0.00041	0.07990
	3	0.00147	0.00122	0.13560	0.00143	0.00111	0.05928	0.00143	0.00111	0.57414
	4	0.00103	0.00094	0.17080	0.00346	0.00140	0.09394	0.00103	0.00094	0.41192
	5	0.00143	0.00114	0.34740	0.00229	0.00055	0.01445	0.00143	0.00054	0.13632
	6	0.00126	0.00050	0.01650	0.00116	0.00075	0.02668	0.00115	0.00050	0.11693
	7	0.00131	0.00062	0.44800	0.00054	0.00330	0.52085	0.00054	0.00062	0.17969
End-of-line	1	0.00090	0.00010	0.00563	0.00232	0.00085	0.33996	0.00124	0.00085	0.33686
	2	0.00090	0.00013	0.00939	0.00198	0.00186	1.65396	0.00096	0.00041	0.07990
	3	0.00090	0.00014	0.01080	0.00143	0.00111	0.57668	0.00143	0.00111	0.57414
	4	0.00120	0.00010	0.00563	0.00346	0.00140	0.91224	0.00103	0.00094	0.41192
	5	0.00130	0.00010	0.00563	0.00229	0.00055	0.14032	0.00143	0.00054	0.13632
	6	0.00120	0.00013	0.00938	0.00116	0.00075	0.25964	0.00115	0.00050	0.11693
	7	0.00130	0.00014	0.01079	0.00054	0.00330	5.07071	0.00054	0.00062	0.17969

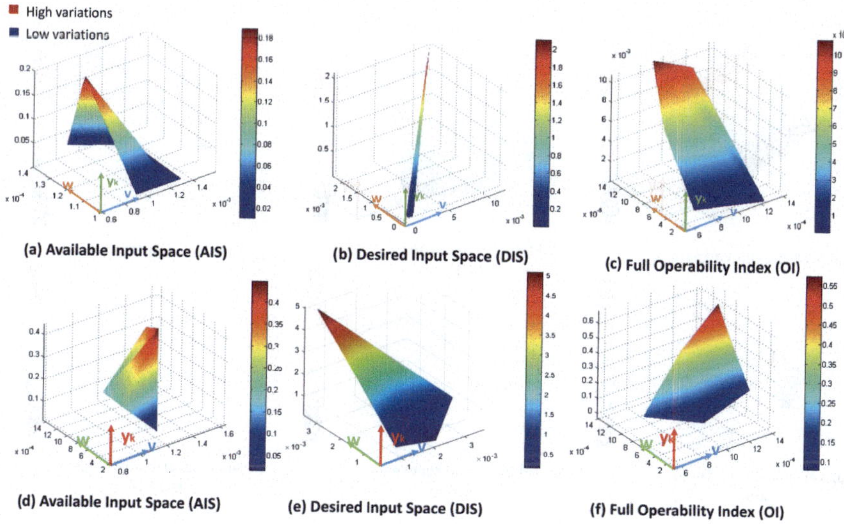

Fig. 4.9 Operability index (OI) solution space in presence of natural uncertainty for distributed and end-of-line sensing configuration

Table 4.10 Full servo operability index in the output space (SOIOS) solution space

Sensing configuration	Available output space			Desirable output space			Full servo operability space		
	Mi [–]	Ms [–]	Cost [$]	Mi [–]	Mi [–]	Ms [–]	Cost [$]	Ms [–]	Mi [–]
Distributed	4 ÷ 20	3 ÷ 4	11 ÷ 27	0 ÷ 15	0 ÷ 4	0 ÷ 15	4 ÷ 15	3 ÷ 4	11 ÷ 15
End-of-line	4 ÷ 6	1	7 ÷ 13	0 ÷ 5	0 ÷ 1	0 ÷ 10	5 ÷ 6	1	7÷ 10

sensing in the presence of natural uncertainty it can be seen that TDS is larger than EDS; Fig. 4.11a, b. Full NOI is identified as intersection of TDS and EDS; Fig. 4.11c.

In Table 4.11, we present the system configuration that reaches full NOI for the four stage panel stamping process with distributed and end-of-line sensing configurations in the presence of natural uncertainty.

The four stage panel stamping process with end-of-line sensing is designed in such way that sensor noise, V, and process disturbances, *W*, for combinations i (i = 1–7) of the EDS are within the range of sensor noise V and process disturbances W of the TDS; Fig. 4.11d, e. In other words, EDS is a subspace of TDS. The NOI is the same as EDS; see Fig. 4.11f.

Operability Index

Operability for variant system design is examined by discovering a region that intersects AIS and DIS. The Desired Input Space (DIS) is necessary to reach all points in desired output space where the value of the output space is not at the nominal value.

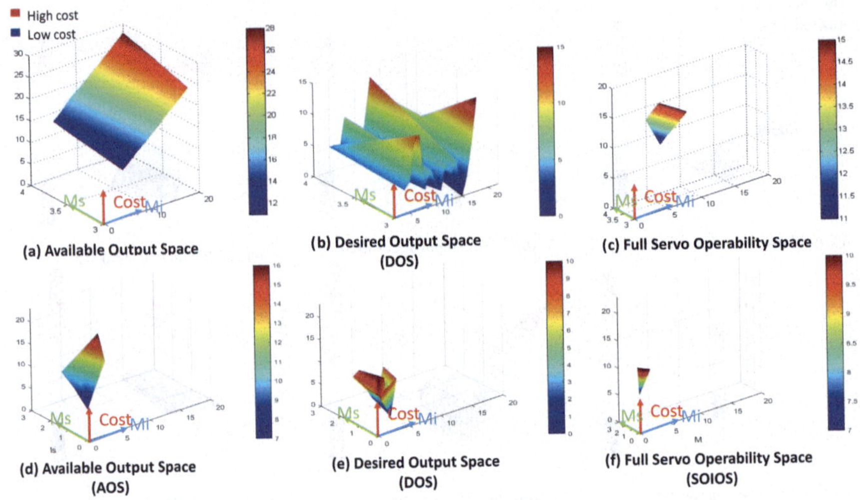

Fig. 4.10 Servo operability index in the output space (SOIOS) solution space for distributed and end-of-line sensing configuration

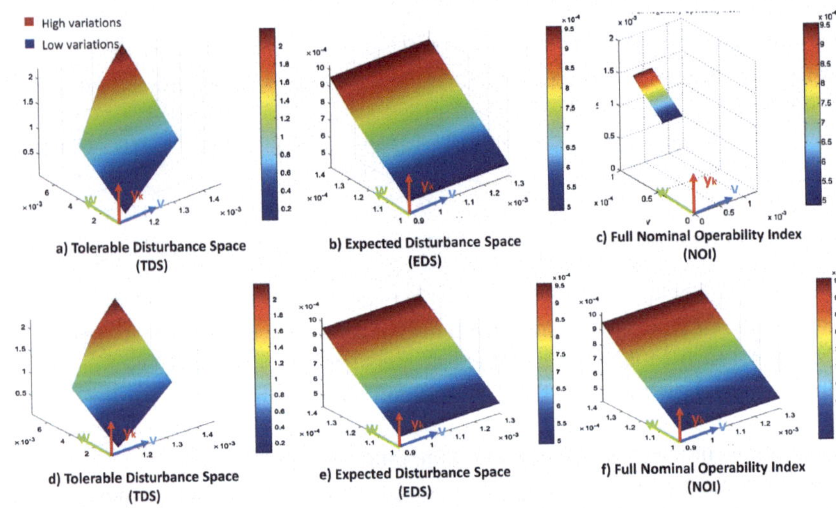

Fig. 4.11 Nominal operability index (NOI) solution space in presence of natural uncertainty for distributed and end-of-line sensing configuration

The available Input Space (AIS) is the input of the process which is able to change over a certain range. AIS are the input points available by exercising ACRONES or through prior designer experience.

AIS is framed by exercising the ACRONES; see Fig. 4.12a. DIS is framed by designer wishes, experience or assumptions, and it is greater than AIS; see Fig. 4.12b.

Table 4.11 Full nominal operability index (NOI) solution space for natural uncertainty

Sensing configuration	i	Tolerable disturbance space			Expected disturbance space			Full nominal operability index		
		V [mm]	W [mm]	y_k [mm]	V [mm]	W [mm]	V [mm]	W [mm]	y_k [mm]	y_k [mm]
Distributed	1	0.00121	0.00372	0.65992	0.00090	0.00010	0.00049	0.00090	0.00010	0.00049
	2	0.00114	0.00455	0.98834	0.00090	0.00013	0.00082	0.00090	0.00013	0.00082
	3	0.00144	0.00355	0.60259	0.00090	0.00014	0.00095	0.00090	0.00014	0.00095
	4	0.00135	0.00399	0.75948	0.00120	0.00010	0.00049	0.00103	0.00010	0.00049
	5	0.00127	0.00597	1.70455	0.00130	0.00010	0.00049	0.00130	0.00010	0.00049
	6	0.00116	0.00121	0.07047	0.00120	0.00013	0.00083	0.00120	0.00013	0.00083
	7	0.00138	0.00678	2.19876	0.00130	0.00014	0.00096	0.00130	0.00014	0.00096
End-of-line	1	0.00121	0.00372	0.65992	0.00090	0.00010	0.00049	0.00090	0.00010	0.00049
	2	0.00114	0.00455	0.98834	0.00090	0.00013	0.000823	0.00090	0.00013	0.00082
	3	0.00144	0.00355	0.60259	0.00090	0.00014	0.000954	0.00090	0.00014	0.00095
	4	0.00135	0.00399	0.75948	0.00120	0.00010	0.000492	0.00103	0.00010	0.00049
	5	0.00127	0.00597	1.70455	0.00130	0.00010	0.000492	0.00130	0.00010	0.00049
	6	0.00116	0.00121	0.07047	0.00120	0.00013	0.000826	0.00120	0.00013	0.00083
	7	0.00138	0.00678	2.19876	0.00130	0.00014	0.000957	0.00130	0.00014	0.00096

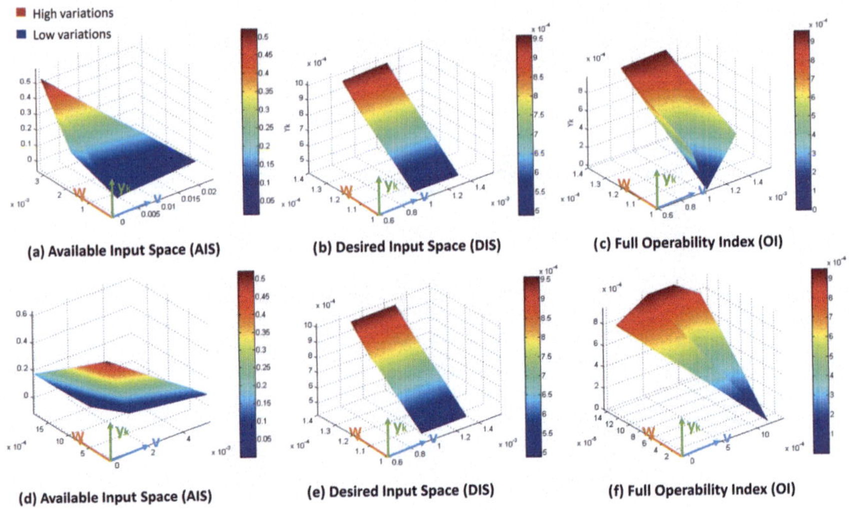

Fig. 4.12 Operability index (OI) solution space in presence of natural uncertainty for distributed and end-of-line sensing configuration

Full operability of the system is identified at the intersection of AIS and DIS. In Fig. 4.12c, we present the solution space in the presence of disturbances (natural uncertainty) under which the system is operable.

AIS is smaller than DIS; see Fig. 4.12d, e. Full operability of the system is identified as intersection of AIS and DIS. In Fig. 4.12f, we present the solution space in presence of disturbances (natural uncertainty) where the system is operable (Table 4.12).

4.6 Synopsis

In this chapter, 'Steady-State Operability Analysis', which is a critical component of the DFDM Framework is presented; see Fig. 4.1. In this analysis, the validity of the empirical models used to represent the components of NMS is first verified. The ACRONES method described in Chap. 3 is then used to determine the desired and actual space of feasible inputs then calculate the operability indices for the design choices. Similarly, the effect of external changes and uncertainties are quantified in terms of the operability indices. The use of these indices in making design choices is illustrated using the example of a panel stamping process (Fig. 4.13).

In the Chap. 5, the second part of the operability analysis, namely, Dynamic Operability Analysis is presented and its use, in analyzing the dynamic performance of NMS, is demonstrated using illustrative examples.

Table 4.12 Full operability index (OI) sclution space for natural uncertainty

Sensing configuration	i	Available input space			Desired input space			Full operability index		
		V [mm]	W [mm]	y_k [mm]	V [mm]	W [mm]	V [mm]	W [mm]	y_k [mm]	y_k [mm]
Distributed	1	9.51E − 03	1.56E − 03	1.17E − 01	2.32E − 03	8.50E − 04	3.50E − 02	2.32E − 03	8.50E − 04	3.47E − 02
	2	8.62E − 03	1.32E − 03	8.39E − 02	1.00E − 05	1.88E − 03	1.70E − 01	1.00E − 05	1.32E − 03	8.33E − 02
	3	6.23E − 03	1.45E − 03	1.01E − 01	1.43E − 03	1.00E − 05	5.93E − 02	1.43E − 03	1.00E − 05	0.00E + 00
	4	1.22E − 02	1.25E − 03	7.55E − 02	3.46E − 03	1.40E − 04	9.39E − 02	3.46E − 03	1.40E − 04	7.50E − 02
	5	1.33E − 02	1.06E − 02	5.35E + 00	2.29E − 02	5.50E − 04	1.45E − 02	1.33E − 02	5.50E − 04	1.49E − 02
	6	1.22E − 02	1.32E − 03	8.42E − 02	1.16E − 03	1.40E − 04	2.67E − 02	1.16E − 03	1.40E − 04	9.50E − 04
	7	1.39E − 02	1.44E − 02	9.85E − 00	5.40E − 04	3.30E − 03	5.21E − 01	5.40E − 04	3.30E − 03	5.21E − 01
End-of-line	1	9.51E − 03	1.56E − 03	1.17E − 01	4.32E − 03	8.50E − 04	3.47E − 02	4.32E − 03	8.50E − 04	3.48E − 02
	2	8.62E − 03	1.32E − 03	8.39E − 02	2.00E − 05	1.88E − 03	1.69E − 01	2.00E − 05	1.32E − 03	8.33E − 02
	3	6.23E − 03	1.45E − 03	1.01E − 01	1.43E − 03	1.11E − 03	0.00E + 00	1.43E − 03	1.11E − 03	2.90E − 02
	4	1.22E − 02	1.25E − 03	7.55E − 02	5.45E − 03	1.00E − 05	9.40E − 02	5.45E − 03	1.00E − 05	0.00E + 00
	5	1.33E − 02	1.06E − 02	5.35E + 00	2.28E − 03	5.40E − 04	1.52E − 02	2.28E − 03	5.40E − 04	1.40E − 02
	6	1.22E − 02	1.32E − 03	8.42E − 02	3.15E − 03	7.40E − 04	9.50E − 04	3.15E − 03	7.40E − 04	2.63E − 02
	7	1.39E − 02	1.44E − 02	9.85E − 00	5.40E − 04	3.00E − 04	5.20E − 01	5.40E − 04	3.00E − 04	4.33E − 03

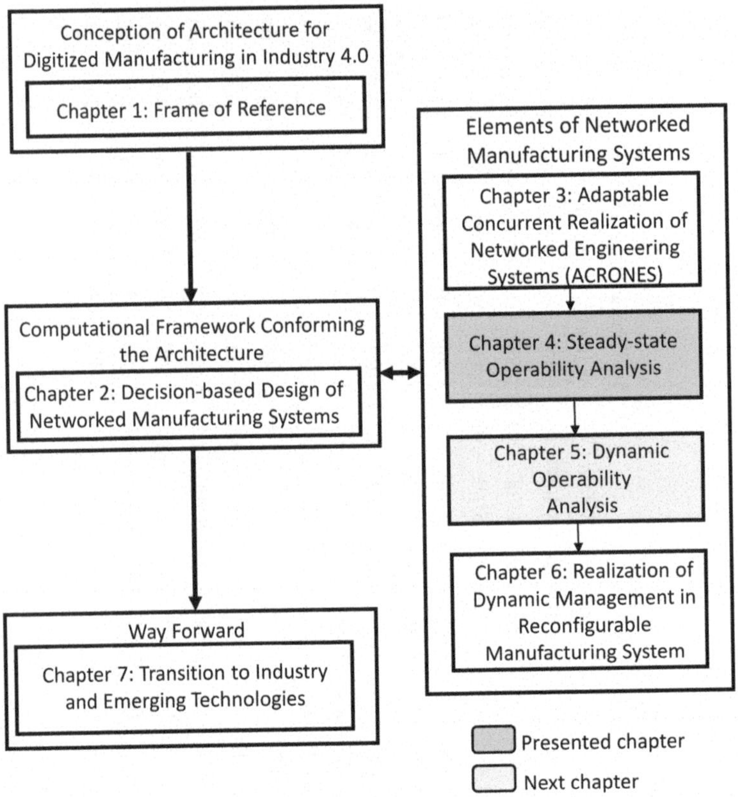

Fig. 4.13 Organization of the monograph—presented and next chapter

Glossary

ACRONES Adaptable Concurrent Realization of Networked Engineering System
AIS Achieved Input Space
AOS Achieved Output Space
cDSP compromise Decision Support Problem
CSTR Continuous single tank reactor
DIS Desired Input Space
Disturbance Space Robust solution space obtained by exercising the ACRONES
DOS Desired Output Space
FI Flexibility index
Functional System Operable system
MIMO Multiple input multiple output
NMS Networked Manufacturing Systems
NOI Nominal Operability Index
OI Operability Index

Operability Space Solution space obtained by exercising the ACRONES
Original System Design New system design without existing specifications
RGA Relative Gain Array
RHP Right-Half-Plane
SISO Single input single output
SOIIS Servo Operability Index in the Input Space
SOIOS Servo Operability Index in the Output Space
SSOM Steady-State Operability Model
Steady-State Operability The functionality of the system after it recovers from disruptions and reaches steady state, in relation to its normal operation
Variant System Design System design with existing specifications

References

Apley, D. W., & Shi, J. (1998). Diagnosis of multiple fixture faults in panel assembly. *Journal of Manufacturing Science and Engineering, 120*, 793–801.

Arkun, Y. (1988). Relative sensitivity: A dynamic closed-loop interaction measure and design tool. *AIChE Journal, 34*(4), 672–675.

Bahri, P. A., Bandoni, J. A., & Romagnoli, J. A. (1996). Effect of disturbances in optimizing control: Steady-state open-loop Backoff problem. *AIChE Journal, 42*(4), 983–994.

Cao, Y., Biss, D., & Perkins, J. (1996). Assessment of input-output controllability in the presence of control constraints. *Computers & Chemical Engineering, 20*(4), 337–346.

Chenery, S. D. (1997). Process controllability analysis using linear and nonlinear optimization, London, UK, Imperial College of Science, Technology, and Medicine, Ph.D. monograph.

Dimitriadis, V. D., & Pistikopoulos, E. N. (1995). Flexibility analysis of dynamic systems. *Industrial and Engineering Chemistry Research, 34*(12), 4451–4462.

Ding, Y., Shi, J., & Ceglarek, D. (2002). Diagnosability analysis of multi-station manufacturing processes. *Journal of Dynamic Systems, Measurement, and Control, 124*, 1–13.

Fisher, W. R., Doherty, M. F., & Douglas, J. M. (1988). The interface between design and control. 1. Process controllability. *Industrial and Engineering Chemistry Research, 27*(4), 597–605.

Georgakis, C., Uztürk, D., Subramanian, S., & Vinson, D. R. (2003). On the operability of continuous processes. *Control Engineering Practice, 11*(8), 859–869.

Grosdidier, P., Morari, M., & Holt, B. R. (1985). Closed-loop properties from steady-state gain information. *Industrial and Engineering Chemistry Fundamentals, 24*(2), 221–235.

Grosdidier, P., & Morari, M. (1986). Analysis of interactions using structured singular values. In *IEEE American Control Conference*.

Grossmann, I. E., & Floudas, C. A. (1987). Active constraint strategy for flexibility analysis in chemical processes. *Computers & Chemical Engineering, 11*, 675–693.

Hovd, M., & Skogestad, S. (1992). Simple frequency-dependent tools for control system analysis, structure selection and design. *Automatica, 28*(5), 989–996.

Jin, J., & Shi, J. (1999). State space modeling of sheet metal assembly for dimensional control. *ASME Transactions Journal of Manufacturing Science and Engineering, 121*(4), 756–762.

Lewin, D. (1996). A simple tool for disturbance resiliency diagnosis and feedforward control design. *Computers & Chemical Engineering, 20*(1), 13–25.

Lyman, P. R., Luyben, W. L., & Tyreus, B. D. (1996). Method for assessing the effect of design parameters on controllability. *Industrial and Engineering Chemistry Research, 35*(10), 3484–3497.

Manousiouthakis, V., Savage, R., & Arkun, Y. (1986). Synthesis of decentralized process control structures using the concept of block relative gain. *AIChE Journal, 32*(6), 991–1003.

McAvoy, T. J. (1983). *Interaction analysis: Principles and applications*. NC: Instrument Society of America Research Triangle Park.

Milisavljevic, J. (2018). Architecting networked engineering systems. Doctoral Dissertation, The School of Aerospace and Mechanical Engineering, University of Oklahoma, Norman, Oklahoma.

Milisavljevic-Syed, J., Commuri, S., Allen, J. K., & Mistree, F. (2018). A method for the concurrent design and analysis of networked manufacturing systems. *Engineering Optimization*, pp. 1–19.

Milisavljevic-Syed, J., Commuri, S., Allen, J. K., & Mistree, F. (2019b). Concurrent design exploration method for realizing networked manufacturing systems for industry 4.0. In *52nd CIRP Conference on Manufacturing Systems*, PROC-D-18-00262.

Mohideen, M. J., Perkins, J. D., & Pistikopoulos, E. N. (1996). Optimal design of dynamic systems under uncertainty. *AIChE Journal, 42*(8), 2251–2272.

Morari, M. (1983). Design of resilient processing plants—III: A general framework for the assessment of dynamic resilience. *Chemical Engineering Science, 38*(11), 1881–1891.

Moore, C. (1986). Application of singular value decomposition to the design, analysis, and control of industrial processes. In *American Control Conference*, IEEE.

Russo, L. P., & Bequette, B. W. (1995). Impact of process design on the multiplicity behavior of a jacketed exothermic CSTR. *AIChE Journal, 41*(1), 135–147.

Russo, L. P., & Bequette, B. W. (1998). Operability of chemical reactors: Multiplicity behavior of a jacketed styrene polymerization reactor. *Chemical Engineering Science, 53*(1), 27–45.

Stanley, G., Marino-Galarraga, M., & McAvoy, T. J. (1985). Shortcut operability analysis 1: The relative disturbance gain. *Industrial & Engineering Chemistry Process Design and Development, 24*, 1181–1188.

Subramanian, S., & Georgakis, C. (2000). Steady-state operability characteristics of reactors. *Computers & Chemical Engineering, 24*(2–7), 1563–1568.

Subramanian, S., Uztürk, D., & Georgakis, C. (2001). An optimization-based approach for the operability analysis of continuously stirred tank reactors. *Industrial and Engineering Chemistry Research, 40*(20), 4238–4252.

Swaney, R. E., & Grossmann, I. E. (1985a). An index for operational flexibility in chemical process design. Part I: Formulation and theory. *AIChE Journal, 31*(4), 621–630.

Swaney, R. E., & Grossmann, I. E. (1985b). An index for operational flexibility in chemical process design. Part II: Computational Algorithms. *AIChE Journal, 31*(4), 631–641.

Swartz, C. (1996). A computational framework for dynamic operability assessment. *Computers & Chemical Engineering, 20*(4), 365–371.

Uztiirk, D., & Georgakis, C. (1998). An optimal control perspective on the inherent dynamic operability of processes. In: *Annual AIChE Meeting*, Miami, FL. Paper 217a.

Uztürk, D., & Georgakis, C. (2001). Inherent dynamic operability of processes I: Definitions and analysis in the SISO case. *Industrial and Engineering Chemistry Research, 41*(3), 421–432.

Vinson, D. R., & Georgakis, C. (1998). A new measure of process output controllability. *IFAC Proceedings Volumes, 31*(11), 663–672.

Vinson, D. R., & Georgakis, C. (2000). A new measure of process output controllability. *Journal of Process Control, 10*(2–3), 185–194.

Zhu, Z. X., Lee, J., & Edgar, T. F. (1997). Steady state structural analysis and interaction characterization for multivariable control systems. *Industrial and Engineering Chemistry Research, 36*(9), 3718–3726.

Ziegler, J., & Nichols, N. (1943). Process lags in automatic control circuits. *Trans. ASME, 65*(5), 433–443.

Chapter 5
Dynamic Operability Analysis

Nomenclature

t_f^d	Desired response time
t_f	Response time
t_f^*	Shortest desired response time
y_{sp}	Change in the requirements
d	Change in disturbance
u	Input points of AOS
y	Output points of DOS
F_{i-1}	Feed flow rate
F_i	Flow rate
V_i	Volume
t	Time
C_{Ai}	Concertation of A
C_{Ai-1}	Feed concertation of A
k_i	Reaction rate constant
ρ	Density of A
c_p	Heat capacity of A
T_i	Reactor temperature
T_{i-1}	Feed temperature
ΔH	Heat of reaction
U	Overall heat-transfer coefficient
T_{Ci}	Jacket temperature
T_{C0}	Coolant feed temperature
A_i	Heat-transfer area
ρ_C	Density of coolant
c_{pC}	Heat capacity of coolant
V_{Ci}	Volume of the jacket
F_{Ci}	Coolant flow rate
E	Activation energy

J. Milisavljevic-Syed et al., *Architecting Networked Engineered Systems*,
https://doi.org/10.1007/978-3-030-38610-8_5

R	Reference or nominal value
A_{si}	Side heat-transfer area
A_{bi}	Bottom heat-transfer area
A_{si}^R	Reference side heat-transfer area
V_i^R	Reference volume
x_{1i}	Normalized reactor holdup
x_{2i}	Concentration of reactor A
x_{3i}	Reactor temperature
x_{4i}	Coolant temperature
x_{4i}	Coolant temperature
q_i	Normalized flow rate
q_{Ci}	Normalized coolant flow rate
α_i	Ratio of coolant flow rate and flow rate and
μ_i	Ratio of flow rate and coolant flow rate
δ_{si}	Thickness of side heat-transfer area
δ_{bi}	Thickness of bottom heat-transfer area
k_0	Arrhenius constant
ΔH	Heat of reaction
$ODF = 1$	Overdesign factor in the reactor volume obtain DIS
T_i	Strong constraint
F_{Ci}	Assumption
k	Reactor rate constant at reactor temperature
$Q_{\max}/Q$	Controllability index of Luyben
ΔT	Temperature difference between jacket and reactor
λ_{ij}	The jth eigenvalue of ith reactor (i is omitted for single reactors)
i	Reactor number

In Chap. 2, we presented the DFDM framework for the design of Networked Manufacturing Systems (NMS). In Chap. 3, ACRONES, a method to concurrently design mechanical and control components of NMS is presented. The performance of the NMS is addressed in Chap. 4 where operability analysis is used to determine the steady state performance of the NMS. Once the components of the NMS are designed, any deviations in inputs or the performance of component modules will affect the performance of the NMS. Chapter 4 addressed this issue by determining the operability index of the NMS when the steady state performance is considered. In this chapter, we address the second issue in the performance of NMS, that is "how does the dynamic operability of the NMS alter when either inputs or process parameters change?" In order to answer this question, we first propose a metric called "Dynamic Operability Index" which we then use to study the dynamic performance of the NMS.

In Sect. 5.2, we discuss how dynamic operability analysis fits in the computational framework, Design for Dynamic Management (DFDM). In Sects. 5.2.1 and 5.2.2, we present the Dynamic Operability Model (DOM) and introduce the compromise Decision Support Problem (cDSP) construct for DOM. In Sect. 5.3, we consider a Continuous Stirred Tank Reactor (CSTR) as an Illustrative Example 2; see Sect. 2.1.3. Further, we verify the appropriateness of this problem to test the design method. In

Sect. 5.4, we present empirical performance validity and verify the ability to produce useful results for the chosen example problem and provide insight into the use of dynamic operability in the design of NMS.

5.1 Dynamic Operability Analysis: A Critical Review

The connection between the design of the process and control of the process is investigated as early as the early 40's (Ziegler and Nichols 1943). A poorly designed process can impose serious limitations on the ability of the controls to ensure desired performance (Georgakis et al. 2003). The quote, "Well designed plant is also a well-controlled one" is recognized by many authors. In fact, considering operability issues early in the design stage is an additional motivation for chemical suppliers to minimize variations of their products according to Georgakis and co-authors (Georgakis et al. 2003). The bridge between design and control is introduced by operability analysis where systematic exploration of the beneficial as well as a detrimental interaction between process and control designs is taking place. A precise measure of operability is necessary in order to effectively accomplish this interaction.

> An operability measure should quantify the inherent ability of the process to move from one steady state to another and to reject any of the expected disturbances in a timely fashion with the limited control action available. (Georgakis et al. 2003)

Operability analysis methods are classified into two categories: those based on linear systems and those based on nonlinear systems.

Linear-Based Operability Analysis Methods

Linear methods are developed to address the problems of multiple-input-multiple-output (MIMO) systems, without any explicit account of the limited range available to the input variables (Georgakis et al. 2003). Morari (1983) identified the relationship between the invertibility of the transfer function matrix of a system and its resilience, where factors that prevent inversion of the process are (1) right-half-plane (RHP) zeros, (2) time delays, (3) constraints on the input variables, and (4) model uncertainty. In addition to Morari's invertibility of the transfer function matrix there are other linear-based methods that are worthy of mentioning:

1. Singular Value Decomposition (see for example, Moore 1986; Grosdidier and Morari 1986; Cao et al. 1996) in addressing various aspects of control problems (control variable pairing, optimal sensor location, robust controller design, and resilience);
2. Relative Gain Array (RGA) gives a quantitative measure of control-loop interaction for multivariable systems (for example, McAvoy 1983; Grosdidier et al. 1985; Zhu et al. 1997);
3. Relative Disturbance Gain (Stanley et al. 1985);
4. Block Relative Gain (Manousiouthakis et al. 1986);

5. Relative Sensitivity (Arkun 1988); and
6. Closed-loop Disturbance Gain (Hovd and Skogestad 1992).

Swartz (1996) introduces an operability assessment where solution obtained represents an upper bound on the performance of all linear stabilizing feedback controllers. Chenery (1997) introduce linear controllability analysis, based on optimal LTI control. Lewin (1996) proposes a function of disturbance direction and frequency.

Nonlinear-Based Operability Analysis Methods

Nonlinear methods are developed to utilize nonlinear models. Some commonly used methods are:

1. Flexibility Index (FI) for quantifying the steady-state operability of nonlinear processes (Swaney and Grossmann 1985a). Further, the FI approaches are based on one of the following: assumption that the limiting points lie in the uncertain-parameter vertex directions (Swaney and Grossmann 1985b), any assumptions (Grossmann and Floudas 1987), and an extended version of the FI approach for assessing the feasibility and flexibility of dynamic systems (Dimitriadis and Pistikopoulos 1995);
2. Maintaining controllability of the plant with sufficient numbers of manipulated variables with enough range to keep the plant under nominal operating conditions when the disturbances affect the process (Fisher et al. 1988);
3. Operability of a plant in the presence of disturbances through optimization (Bahri et al. 1996). Further, dynamic operability using dynamic mixed-integer nonlinear programming has been developed (Bahri et al. 1996);
4. A concurrent design and control approach under parametric uncertainty and disturbances where flexibility aspects are incorporated in a multiperiod design subproblem coupled with a feasibility analysis of dynamic systems (Mohideen et al. 1996);
5. Operability characteristics through dynamic simulations of the SISO control structures (Lyman et al. 1996);
6. Operability of C8TRs with exothermic reactions (Russo and Bequette 1995, 1998); and
7. Methods for the inherent steady-state operability of linear continuous processes (Vinson and Georgakis 1998, 2000), nonlinear processes (Subramanian and Georgakis 2000, Subramanian et al. 2001), extended to dynamic operability analysis (Uztiirk and Georgakis 2001), and examination operability characteristics of non-square systems (Subramanian et al. 2001).

A manufacturing system must be adaptable to dynamic market changes for it to remain competitive. The design of these systems must take into account the unexpected disturbances in the manufacturing process, variation in product design, and product scale change. In order to address these challenges, there is a need to fully examine the dynamic and control characteristics of a system at the design stage. From the current approaches for the design of NMS (see Table 5.1), it is evident that there are limitations and unresolved difficulties that have to be addressed during the design

Table 5.1 Overview of foundational papers for identifying key unresolved difficulties in the design of NMS system operability

	Gaps	Expand operability analysis to fit any engineering system, frame the operability/disturbance spaces analyze system functionality with change in the requirements									Expand to design of complex systems and analyze dynamic performance	
	Paper	Georgakis et al. (2003)	Fisher et al. (1988)	Bahri et al. (1996)	Lyman et al. (1996)	Russo and Bequette (1995)	Russo and Bequette (1998)	Subramanian and Georgakis (2000)	Subramanian et al. (2001)	Subramanian al. (2001)	Uztürk and Georgakis (1998)	Mohideen et al. (1996)
Analysis	Feasibility analysis of dynamic Systems											*
	Operability analysis of dynamic Systems	*	*	*	*	*	*	*	*	*	*	
	Diagnosability and controllability											
	Analyze of error propagation											
	Production function and inventory control											
	Production volume and product variability											
Methods	Operability characteristics of n on-square systems									*		
	Dynamic operability analysis										*	
	Operability characteristics through dynamic simulations				*							
	Integrated design and control approach under parametric uncertainty											*
	Optimization-based operability			*								
	Maintaining controllability of the process in the presence of disturbance		*									
	Steady-state operability of nonlinear Processes	*				*	*	*	*			
	Dynamic modeling of manufacturing system											
	Stream of variation modeling											
	Stochastic modeling of manufacturing systems											
	Push and pull satrategy											
Aspects	Operability	*	*	*	*	*	*	*	*	*	*	*
	Adaptability											*

process. These are identified in terms of the following gaps; see Tertiary Question 2 (TQ2_2) in Table 1.2:

- ***Gap 1***: Expand operability analysis to address the design and analysis of any engineering system, such as manufacturing process, assembling process, steel making process, heat exchanger, etc., rather than plant design (Milisavljevic 2018).
- ***Gap 2***: Frame the operability and disturbance spaces, and analyze system functionality with change in the requirements in addition to disturbances (Georgakis et al. 2003; Fisher et al. 1988; Bahri et al. 1996; Lyman et al. 1996; Russo and Bequette 1995, 1998; Subramanian and Georgakis 2000; Subramanian et al. 2001); and

- ***Gap 3***: Expand dynamic operability analysis to the design of complex systems (multi-variable and different goals problems) and analyze dynamic performance of the system due to change in the requirements in addition to disturbances (Mohideen et al. 1996; Uztürk and Georgakis 1998).

In this chapter, we focus on expanding dynamic operability analysis to the design of complex systems and analyzing dynamic performance of the system; see highlighted section in Table 5.1. Furthermore, in order to analyze the conditions under which the system transitions from one state to another and maintains stability even in the presence of changes, we propose dynamic operability analysis as part of the design of the NMS. With dynamic operability analysis we obtain information about the desired dynamic performance and stability of the system, and operating ranges of achievable performance of process.

In Sect. 5.2, we present how dynamic operability analysis fits in the DFDM. Further, we present Dynamic Operability Model (DOM) and cDSP for the DOM; see Sects. 5.2.1 and 5.2.2.

5.2 Dynamic Operability Analysis in Design for Dynamic Management

Dynamic operability analysis is a component of the computational framework, DFDM, where dynamic performance of the system undergoing dynamic change is analyzed. In Sect. 4.5, we obtained the ranges of solutions of the functional design by exercising the Steady-State Operability Model (SSOM). This functional design is the starting point for analyzing the dynamic performance and is the input in the Dynamic Operability Model (DOM).

Operability analysis in the DFDM framework is connected with the Adaptable Concurrent Realization of Networked Engineering Systems (ACRONES) method. With ACRONES, we obtain the ranges of solutions; see Adaptability in Fig. 5.1. The output from ACRONES is the input in the Steady-State Operability Model (SSOM), where we define the operability space and disturbance space; see Operability in Fig. 5.1. SSOM will help determine if the system meets functional requirements when there are no disturbances or variations in inputs (steady state). If the design determined by ACRONES meets the functional requirements in steady state, then we analyze the dynamic performance of the system; see operability in Fig. 5.1. However, if we cannot locate solutions of functional design (in steady or dynamic state) then the next step is to identify an appropriate reconfiguration strategy; see Reconfigurability in Fig. 5.1.

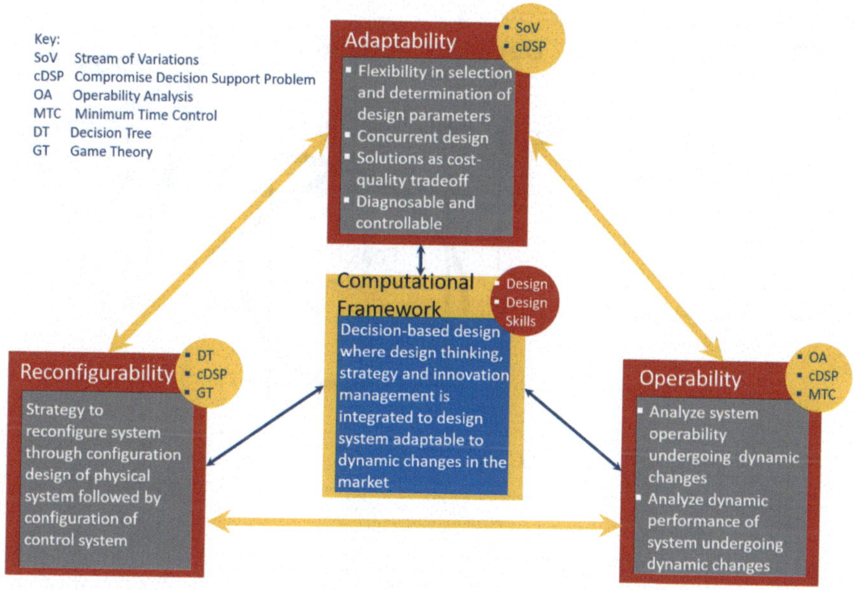

Fig. 5.1 Computational framework, design for dynamic management (Milsavljevic-Syed et al. 2019b)

5.2.1 Dynamic Operability Model

The Dynamic Operability Model (DOM) presented in this chapter is used to obtain insight into the effect of disturbances such as changing requirements or changing functionality of NMS components on the overall performance. The aim is to minimize a distance between the aspiration space framed by the Dynamic Desired Output Space (DDOS) and the desired response time t_f^d and the feasible design space framed the Dynamic Achieved Output Space (DAOS) and response time, t_f; see Fig. 5.2. Furthermore, we connect SSOM with DOM in order to determine system functionality in steady-state and obtain Dynamic Available Input Space (DAIS) that we use to frame DAOS; see Sect. 4.3.

Dynamic operability of the system is defined as the shortest time it takes for a system to adjust to the desired set point after a change in requirements and/or after the occurrence of a system disturbance. The measure of operability is based on the idea that time spent away from the desired set point is due to off-specification products (Milisavljevic-Syed et al. 2018). Different types of controllers, feedback, run-to-run, or minimum-time optimal controllers can be utilized to evaluate measure of operability. In this monograph, we use the minimum-time optimal controller approach proposed by Subramanian and co-authors (Subramanian et al. 2001). Minimum time is defined as the smallest duration that is required for the process to overcome the worst disturbance and/or change in the requirements.

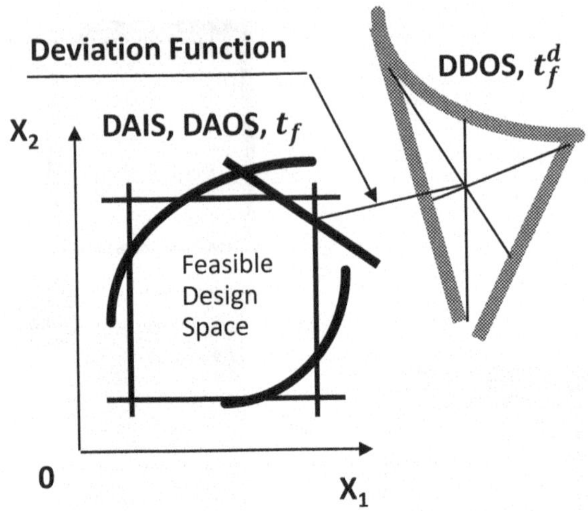

Fig. 5.2 Solution space exploration on over time

The DOM is examined in the shortest time, $t_f^*(y_{sp}, d)$, for system to adjust to change in the requirements, $y_{sp} \in DOS$, and/or disturbances, $d \in EDS$, that can occur over time.

The solution scheme provides an efficient way to explore the solution space and identify possible solutions of dynamic performance of a system due to changes in requirements and/or disturbances; see Fig. 5.3. The solution scheme includes the following steps:

Step 1 Obtain information from the ACRONES and the SSOM to frame Dynamic Available Input Space (DAIS).

Step 2 Examine the Dynamic Desired Operability Spaces (DDOS) that gives the desired dynamic performance of a system; see Eq. 5.1. If the response time of the system, t_f, is less than or equal to desired response time, t_f^d, that is set by a designer, then the system achieves the desired dynamic performance. If the response time of the system, t_f, is greater than the desired response time, t_f^d, that is set by a designer, then system does not have desired dynamic performance and reconfiguration is required.

$$\text{DDOS} = \left\{ \left(t_f, y_{sp}, d\right) \middle| t_f \le t_f^d\left(y_{sp}, d\right), \forall y_{sp} \in DOS, \forall d \in EDS \right\} \tag{5.1}$$

Step 3 Examine the Dynamic Available Operating Space (DAOS) that gives us insights in the dynamic performance achievable by a system for a given choice of DAIS, DOS, EDS; see Eq. 5.2. For detailed information about the DOS and EDS see Sect. 4.4. If the response time of the system, t_f, is less than or equal to shortest desired time for a system to reach and remain within desired performance bounds, t_f^*, set by a designer, then the system has desired dynamic performance. If the response time of the system, t_f, is greater than shortest desired time, t_f^*, set by a designer, then the system does

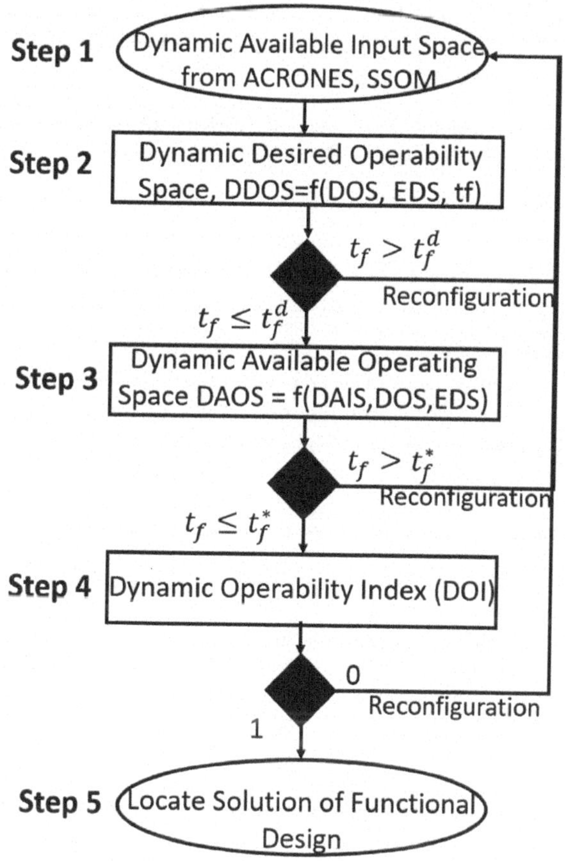

Fig. 5.3 Solution scheme of dynamic operability model

not have the desired dynamic performance and reconfiguration is required.

$$\begin{aligned} \text{DAOS} = d\Big\{\big(t_f, y_{sp}, d\big)\big|t_f \leq t_f^*\big(y_{sp}, d\big), \forall y_{sp} \in DOS, \\ \forall d \in EDS, u \in dAIS\} \end{aligned} \tag{5.2}$$

Step 4 Examine the Dynamic Operability Index (DOI) of the continuous system; see Eq. 5.3. It gives us a portion of the operating ranges achievable within the response time, $t_f^d(y, d)$, for a given DAIS. The DOI takes values between 0 and 1, where a DOI value of 1 implies that the system is fully operable. A DOI value less than 1 implies that designed system is not operable when changes occur in requirements and/or disturbances over time. In such cases, a redesign of the system is required to meet performance requirements.

$$DOI\frac{\mu[DDOS \cap DAOS]}{\mu[DOS \cup EDS]} \tag{5.3}$$

Step 5 If the DOI is equal to 1 then the best solutions are located, that is, the upper bound of the performance is achieved by the selected control (Milisavljevic-Syed et al. 2018).

The use of cDSP and ACRONES in dynamic operability analysis is a novel approach. In Sect. 5.2.2, we present the cDSP for the dynamic operability model.

5.2.2 *The cDSP for the Dynamic Operability Model*

In this section, we present the word formulation of the cDSP construct for the dynamic operability model (DOM) supported by mathematical formulation of the problem. In Table 5.2, the cDSP construct augmented to accommodate dynamic operability analysis is presented. The cDSP construct is as follows:

Table 5.2 The cDSP construct for the dynamic operability analysis

cDSP		cDSP for the DOM
Given		
• An alternative that is to be improved through modification • Assumptions used to model the domain of interest • The system parameters • All other relevant information		Information generated with the ACRONES and examined with SSOM help designer frame the operability spaces (DAIS) and desired response time in case of change and disturbances • Dynamic available input space (DAIS)
n	number of system variables	• Desired response time, t_f^d
p + q	number of system constraints	• Shortest desired time for a system to reach and maintain desired performance, t_f^*
p	equality constraints	
q	inequality constraints	
m	number of system goals	
$g_i(X)$	system constraint function $g_i(X) = C_i(X) - D_i(X)$	
$f_k(d_i)$	function of deviation variables to be minimized at priority level k for the preemptive case	
W_i	weight for the archimedean case	
Find		
• The values of the independent *system variables* (they describe the physical attributes of an artifact). X_j $j = 1, \ldots, n$ • The values of the *deviation variables* (they indicate the extent to which the goals are achieved). D_i^-, d_i^+ $I = 1, \ldots, m$		• System variables that describe DDOS, DAOS, DDAOS, DOEDS, DOI; see Eqs. 5.1–5.3 • The values of the *deviation variables* (they indicate the extent to which the goals DDOS, DAOS, DOI are achieved)

(continued)

Table 5.2 (continued)

cDSP	cDSP for the DOM
Satisfy	
• The *system constraints* that must be satisfied for the solution to be feasible • There is no restriction placed on linearity or convexity $g_i(X) = 0; i = 1,\ldots, p$ $g_i(X) \geq 0; i = p + 1, \ldots, p + q$ • The *system goals* that must achieve a specified target value as far as possible • There is no restriction placed on linearity or convexity $A_i(X) + d_i^- - d_i^+ = G_i; i = 1, \ldots, m$ • The lower and upper *bounds* on the system $X_j^{min} \leq X_j \leq X_j^{max}; \; j = 1, \ldots, n$ $d_i^-, d_i \geq 0$ and $d_i^- \cdot d_i^+ = 0$	• Dynamic desired operability space; see Eq. 5.1 • Dynamic available operating spaces; see Eq. 5.2 • Dynamic operability index; see Eq. 5.3
Minimize	
The *deviation function* that is a measure of the deviation of the system performance from that implied by the set of goals and their associated priority levels or relative weights • Case a: preemptive (lexicographic minimum) $Z = \left[f_1(d_i^-, d_i^+), \ldots, f_k(d_i^-, d_i^+)\right]$ • Case b: archimedean $Z = \sum_{i=1}^{m} W_i(d_i^- + d_i^+);$ $\sum W_i = 1; \; W_i \geq 0$	The *deviation function* that is a measure of the deviation of the system performance from that implied by the set of goals (DDOS, DAOS, DOI) and their associated priority levels or relative weights

- ***Given*** that information about alternatives, assumptions, system parameters, and other relevant information are generated with ACRONES and SSOM, designer have enough information to frame the operability spaces (DAIS).
- ***Find*** independent *system variables* that describe the physical attributes of dynamic operability spaces that describes goals, and the values of the *deviation variables* that indicate the extent to which the goals (DDOS, DAOS, DOI) are achieved.
- ***Satisfy*** *system constraints* responses time, *system goals* (DDOS, DAOS, DOI), and *system bounds.*
- ***Minimize*** *deviation function* that is a measure of the deviation of the system performance from desired one (DDOS) to actual one (DAOS, DAIS).

In Sect. 5.3, we present the empirical structural validity and verify the appropriateness of the chosen example problem, design of CSTR as Illustrative Example 2, intended to demonstrate the design method.

5.3 Illustrative Example: Continuous Stirred Tank Reactors

A Continuously Stirred Tank Reactor (CSTR) is used to exemplify the analysis method presented in this chapter. We examine the interaction between design and control of CSTR with a first-order reaction. The problem we address is the resizing of vessel $\overline{V}$ and achieving system operability for different design choices, D_i, available input space (product flow rate, F, coolant flow rate, F_C, and reactor volume V) and the expected disturbance space (feed flow rate F_0 and feed temperature T_0) given that desired output space (exit concentration C_A) is set. In this monograph, we considered design of single-CSTR (Scenario 1) and two-CSTRs-in-series systems (Scenario 2). The two configurations have the same feed flow rates and conversion specifications. In Fig. 5.4, we present the schematics of these systems. For a liquid-phase exothermic reaction taking place in a jacketed CSTR, the mass and energy balances can be written, with the assumptions of constant physical properties and complete mixing. In Sect. 5.3.1, we present the model equations that are dimensionless with appropriate transformations.

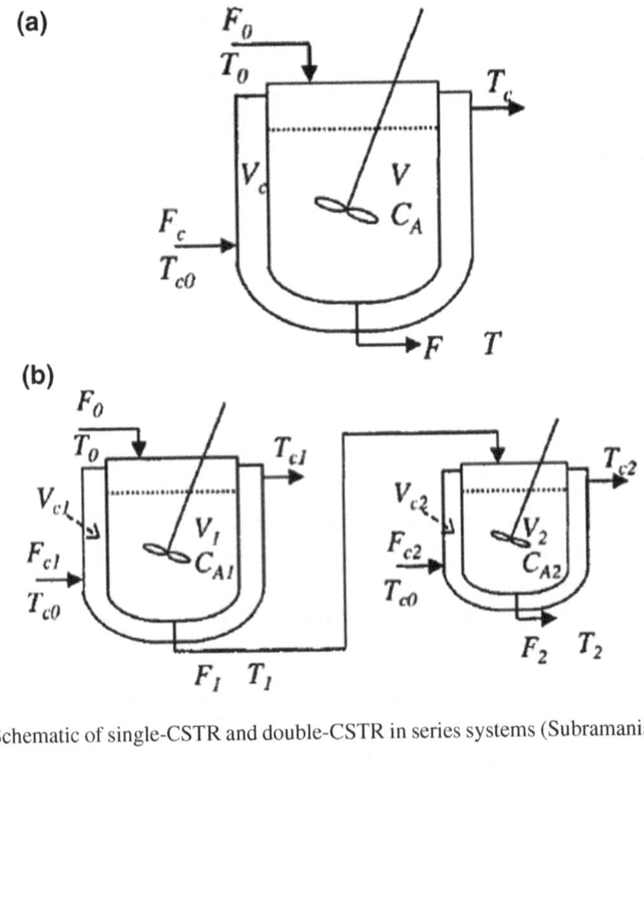

Fig. 5.4 Schematic of single-CSTR and double-CSTR in series systems (Subramanian et al. 2001)

5.3.1 *The CSTR Model Equations and Design Specifications*

In this section, we present the main equations describing the CSTR process, and parameters of the operability and disturbance spaces; see Tables 5.3, 5.4, 5.5 and 5.6. Furthermore, we present the Minimum-Time Optimal Control Problem. In Tables 5.7, 5.8, 5.9, 5.10, 5.11 and 5.12, we present the design procedure and alternatives for a single- and double-CSTR. Lastly, we present the cDSP construct for the operability analysis in design of CSTR.

Table 5.3 CSTR design parameters

Parameter	Value	Units
E	30,000	$\left[\text{Btu lbmol}^{-1}\right]$
$k_{140°\text{F}}$	0.5	$\left[\text{h}^{-1}\right]$
C_{A0}	1.0	$\left[\text{lbmol ft}^{-3}\right]$
U	300	$\left[\text{Btu h}^{-1}{}°\text{F}^{-1}\text{ft}^{-2}\right]$
T_0	70	[°F]
T_{C0}	70	[°F]
ΔH	−30,000	$\left[\text{Btu lbmol}^{-1}\right]$
c_p	0.75	$\left[\text{Btu lb}^{-1}{}°\text{F}^{-1}\right]$
c_{pc}	1.00	$\left[\text{Btu lb}^{-1}{}°\text{F}^{-1}\right]$
M	50.0	$\left[\text{lb lb mol}^{-1}\right]$
ρ	50.0	$\left[\text{lb ft}^{-3}\right]$
ρ_c	62.3	$\left[\text{lb ft}^{-3}\right]$

Table 5.4 Available input space (AIS)

Symbol	Description	Lower bound	Upper bound	Units
F_{Ci}	Coolant flow rate	0	$4F_{Ci}^{R}$	$[\text{ft}^3\text{h}^{-1}]$
F_i	Product flow rate	50	150	$[\text{ft}^3\text{h}^{-1}]$
V_i	Volume of the reaction mixture	$0.3V^{R}$ODF	V^{R}ODF	$[\text{ft}^3]$

Table 5.5 Desired output space (DOS)

Symbol	Description	Lower bound	Upper bound	Units
C_{Ai}	Exit of concentration of reactor A	0.02	0.2	$[\text{lbmol ft}^3]$
T_i	Reactor temperature	100	200	[°F]

Table 5.6 Expected disturbance space (EDS)

Symbol	Description	Lower bound	Upper bound	Units
T_{i-1}	Feed temperature	50	90	[°F]
F_{i-1}	Feed flow rate	50	150	$[\text{ft}^3\text{h}^{-1}]$

Table 5.7 Design specifications for a single-CSTR

Symbol	Description	Equation	Assumptions
F_{i-1}	Feed flow rate		
C_{Ai-1}	Feed concentration of A		
x	Conversion		
C_{Ai}	Reactor exit concentration	$C_{Ai} = C_{Ai-1}(1-x)$	
V_i	Reactor volume	$V_i = \frac{F_{i-1}(C_{Ai-1}-C_{Ai})}{kC_{Ai}}$	Height to diameter ratio of 2
V_{Ci}	Volume of the jacket		Thickness of 4 in.
F_{Ci}	Coolant flow rate	See Eq. 5.16	
T_{Ci}	Jacket temperature	See Eq. 5.14	

Table 5.8 Design alternatives for a single-CSTR

	D1	D2	D3
T	140.00	160.00	180.00
V^R	3800.00	1689.61	790.29
Q^R	2587.50	2512.50	2437.50
F_c^R	81.90	63.88	53.12
ΔT	6.77	11.29	18.17

Table 5.9 Design specifications for a double-CSTR

Symbol	Description	Equation	Assumptions
F_0	Feed flow rate		
C_{A0}	Concentration of component A in the 2nd reactor		
x	Exit concentration		
C_{A2}	Reactor exit concentration in the 2nd reactor	$C_{A2} = C_{A0}(1-x)$	
C_{A1}	Steady-state material balance of the 1st reactor	$C_{A1} = \frac{F_0 C_{A0}}{F_0 + F_1 k_1}$	
V_2	Steady-state material balance of the 2nd reactor	$V_2 = \frac{F_0(C_{A1}-C_{A2})}{k_2 C_{A2}}$	$V_2 = R_V V_1$
V_1	Volume of the 1st reactor	$R_V k_1 k_2 (1-x) V_1^2 + F_0(1-x)(k_1 + R_V k_2) V_1 - F_0^2 x = 0$	
C_{A1}	Reactor exit concentration in the 1st reactor	$C_{A1} = \frac{(V_2 k_2 + F_0)}{F_0} C_{A2}$	
V_{Ci}	Volume of the jacket		
F_{Ci}	Coolant flow rate	From Eq. 5.7	
T_{Ci}	Jacket temperature	From Eq. 5.6	

Table 5.10 Design alternatives for double-CSTR for Rv = 1.0

Variables	D1	D2	D3	D4	D5	D6	D7	D8	D9
T_1	140.00	140.00	140.00	160.0	160.0	160.0	180.0	180.0	180.0
T_1	140.00	160.00	180.00	140.00	160.00	180.00	140.00	160.00	180.00
V_1^R	694.43	454.53	294.72	454.53	308.77	207.75	294.72	207.75	144.42
V_2^R	694.43	454.53	294.72	454.53	308.77	207.75	294.72	207.75	144.42
C_{A1}^R	0.22	0.31	0.40	0.16	0.22	0.30	0.12	0.17	0.22
Q_1^R	2066.68	1820.81	1524.68	2171.60	1991.69	1763.28	2216.46	2087.06	1916.68
Q_2^R	520.82	691.69	912.82	415.90	520.82	674.22	371.04	425.44	520.82
F_{c1}^R	77.73	72.34	63.49	65.26	64.06	60.87	56.79	57.94	58.44
F_{c2}^R	15.85	16.77	18.86	12.70	12.60	13.80	11.48	10.35	10.64
ΔT_1	16.79	19.63	21.94	23.41	27.78	32.03	31.89	37.91	44.37
ΔT_2	4.23	7.46	13.13	4.48	7.26	12.25	5.34	7.73	12.06

Table 5.11 Design alternatives for double-CSTR for Rv = 2.0

Variables	D1	D2	D3	D4	D5	D6	D7	D8	D9
T_1	140.00	140.00	140.00	160.0	160.0	160.0	180.0	180.0	180.0
T_1	140.00	160.00	180.00	140.00	160.00	180.00	140.00	160.00	180.00
V_1^R	484.43	306.61	191.62	327.30	215.39	140.46	219.10	149.30	100.75
V_2^R	968.86	613.21	383.25	654.60	430.79	280.91	438.20	298.60	201.50
C_{A1}^R	0.29	0.39	0.51	0.21	0.29	0.39	0.16	0.22	0.29
Q_1^R	1860.86	1553.15	1205.42	2021.55	1785.86	1499.46	2108.85	1933.83	1710.86
Q_2^R	726.64	959.35	1232.08	565.95	726.64	938.04	478.65	578.67	726.64
F_{c1}^R	73.34	64.44	51.45	64.34	61.27	54.92	57.79	58.45	57.40
F_{c2}^R	22.29	23.55	26.92	17.37	17.76	19.54	14.80	14.17	15.06
ΔT_1	19.22	21.77	23.11	27.12	31.67	35.36	36.97	43.79	50.35
ΔT_2	4.73	8.47	14.88	4.78	8.12	13.94	5.29	8.25	13.47

Table 5.12 Design alternatives for double-CSTR for Rv = 0.5

Variables	D1	D2	D3	D4	D5	D6	D7	D8	D9
T_1	140.00	140.00	140.00	160.0	160.0	160.0	180.0	180.0	180.0
T_1	140.00	160.00	180.00	140.00	160.00	180.00	140.00	160.00	180.00
V_1^R	968.86	654.60	438.20	613.21	430.79	298.00	383.25	280.91	201.50
V_2^R	484.43	327.30	219.10	306.61	215.39	149.30	191.62	140.46	100.75
C_{A1}^R	0.17	0.23	0.31	0.13	0.17	0.23	0.10	0.13	0.17
Q_1^R	2224.18	2035.42	1797.36	2282.55	2149.18	1974.08	2293.78	2200.58	2074.18
Q_2^R	363.32	477.08	640.14	304.95	363.32	463.42	293.72	311.92	363.32
F_{c1}^R	80.16	77.15	71.73	65.40	65.18	63.88	55.78	56.97	58.02
F_{c2}^R	10.98	11.42	12.97	9.29	8.70	9.32	9.13	7.55	7.32
ΔT_1	14.48	17.20	19.85	20.15	24.01	28.16	27.70	32.69	38.46
ΔT_2	3.75	6.40	11.22	4.27	6.44	10.49	5.63	7.36	10.69

Main CSTR Process Equations

We present the main equations that represent the CSTR process; see Eqs. 5.4–5.7. After state variable transformation we present the dimensionless transformed state equations that are used in MTCP calculations; see Eqs. 5.33–5.36.

In Eqs. 5.4–5.7, we present the mass balances and energy balances for both single and double reactors-in-series systems.

$$\frac{dV_i}{dt} = F_{i-1} - F_i \tag{5.4}$$

$$\frac{d(V_i C_{Ai})}{dt} = F_{i-1} C_{Ai-1} - F_i C_{Ai} - V_i k_i C_{Ai} \tag{5.5}$$

$$\rho c_p \frac{d(V_i T_i)}{dt} = \rho c_p F_{i-1} T_{i-1} - \rho c_p F_i T_i + (-\Delta H) V_i k_i C_{Ai} - UA_i (T_i - T_{Ci}) \tag{5.6}$$

$$\rho_C c_{pC} V_{Ci} \frac{dT_{Ci}}{dt} = \rho_C c_{pC} F_{Ci} (T_{C0} - T_{Ci}) + UA_i (T_i - T_{Ci}) \tag{5.7}$$

In Eqs. 5.8–5.10, we present the procedure to calculate these parameters.

$$k_i = k_0 e^{-E/RT_i} \tag{5.8}$$

$$A_i = A_{si} + A_{bi} \tag{5.9}$$

$$A_{si} = A_{si}^R \frac{V_i}{V_i^R} \tag{5.10}$$

In Eqs. 5.11–5.27, we present the state variable transformation in order to make them dimensionless.

$$x_{1i} = \frac{V_i}{V_i^R} \tag{5.11}$$

$$x_{2i} = \frac{C_{Ai}}{C_A^R} \tag{5.12}$$

$$x_{3i} = \frac{T_i - T^R}{T^R} \tag{5.13}$$

$$x_{4i} = \frac{T_{Ci} - T^R}{T^R} \tag{5.14}$$

$$q_i = \frac{F_i}{F^R} \tag{5.15}$$

$$q_{Ci} = \frac{F_{Ci}}{F^R_{Ci}} \tag{5.16}$$

$$\alpha_i = \frac{F^R_{Ci}}{F^R} \tag{5.17}$$

$$\tau = \frac{tF^R}{V^R} \tag{5.18}$$

$$\phi_i = \frac{k_0 e^{-\Upsilon} V^R_i}{F^R} \tag{5.19}$$

$$\gamma = \frac{E}{RT^R} \tag{5.20}$$

$$\mu_i = \frac{V_i}{V_{Ci}} \tag{5.21}$$

$$\upsilon_i = \frac{V^R_1}{V^R_i} \tag{5.22}$$

$$\beta = \frac{(-\Delta H)C^R_A}{\rho c_p T^R} \tag{5.23}$$

$$\xi = \frac{\rho c_p}{\rho c_{pc}} \tag{5.24}$$

$$\delta_{si} = \frac{UA^R_{si}}{\rho c_p F^R} \tag{5.25}$$

$$\delta_{bi} = \frac{UA_{bi}}{\rho c_p F^R} \tag{5.26}$$

$$f(x_{3i}) = exp\left(\frac{\Upsilon x_{3i}}{1 + x_{3i}}\right) \tag{5.27}$$

In Eqs. 5.28–5.31, we present the transformed state equations based on transformation in Eqs. 5.11–5.27.

$$\frac{dx_{1i}}{d\tau} = \upsilon_i(q_{i-1} - q_i) \tag{5.28}$$

$$\frac{dx_{2i}}{d\tau} = \frac{\upsilon_i}{x_{1i}}\left[q_{i-1}(x_{2i-1} - x_{2i}) - \phi_i f(x_{3i}) x_{1i} x_{2i}\right] \tag{5.29}$$

$$\frac{dx_{3i}}{d\tau} = \frac{\upsilon_i}{x_{1i}}\left[q_{i-1}(x_{3i-1} - x_{3i}) + \beta\phi_i f(x_{3i}) x_{1i} x_{2i} - (\delta_{si} x_{1i} + \delta_{bi})(x_{3i} - x_{4i})\right] \tag{5.30}$$

$$\frac{dx_{4i}}{d\tau} = \upsilon_i\left[\alpha_i\mu_i q_{Ci}(x_{4i-1} - x_{4i}) + \xi\mu_i f(x_{3i})x_{1i}x_{2i} - (\delta_{si}x_{1i} + \delta_{bi})(x_{3i} - x_{4i})\right] \tag{5.31}$$

Minimum-Time Optimal Control Problem

In Eq. 5.32, we present the minimum-time optimal control problem that we use to determine system responses over time.

$$t_f^*(y_{sp}, d) = \min_u \int_0^{t_f} dt \tag{5.32}$$

Subject to:

$$\begin{aligned}
&\dot{x} = f(x, u, d);\ x(0) = x_0 \\
&y = g(x, u, d) \\
&h_1(y, \dot{x}, x, \dot{u}, u, d) = 0 \\
&h_2(y, \dot{x}, x, \dot{u}, u, d) \leq 0 \\
&Given\, x_0, y_{sp}, d.
\end{aligned}$$

We undertake a series of steps in order to transform minimum-time problem to a nonlinear programming (NLP) problem. In Eqs. 5.33–5.36, we present first derivatives of the state equations based on the NLP problem.

$$\dot{x}_1 = \frac{0.3}{F^R}(x_1 - x_2) \tag{5.33}$$

$$\dot{x}_2 = \frac{0.3V^R}{x_3}\left[\frac{x_1}{F^R C_A^R}(C_{A0} - x_4) - \frac{0.5x_3x_4}{F^R C_A^R} e^{\left[\frac{E}{x_5 R T^R}(x_5 - T^R)\right]}\right] \tag{5.34}$$

$$\begin{aligned}
\dot{x}_3 = \frac{0.3V^R}{x_3}&\left[\frac{x_1}{T^R F^R}(T_0 - x_5) + \frac{(-\Delta H)k_0 x_3 x_4}{\rho c_p T^R F^R} e^{\left(\frac{x_5 - T^R}{x_5}\right)}\right. \\
&\left. - \frac{U}{\rho c_p T^R F^R}\left(A_s^R \frac{x_3}{V^R} + A_b\right)(x_5 - x_7)\right]
\end{aligned} \tag{5.35}$$

$$\dot{x}_4 = \frac{0.3V^R}{x_3}\left[\frac{x_3x_7}{T^R F^R V_C}(T_{C0} - x_7) + \frac{Ux_3}{\rho c_{pc} T^R F^R V_C}\left(A_s^R \frac{x_3}{V^R} + A_b\right)(x_5 - x_7)\right] \tag{5.36}$$

We used the DYNOPT optimization package to solve the NLP problems resulting from dynamic operability analyses. Results are presented in Sect. 5.4.

System Variable Selection and Design Specifications

In Table 5.3, we present the CSTR design parameters. In Tables 5.4, 5.5 and 5.6, we present the operability space and disturbance space in design of CSTR.

In the design of single or double CSTR we make the following assumptions:

- Rate of close and opening of valves of A: $\left[\frac{dF_i}{dt}\right] \leq 19250\left[\mathrm{ft^3 h^{-2}}\right]$,
- Rate of close and opening of valves at coolant: $\left[\frac{dF_{Ci}}{dt}\right] \leq 19250\left[\mathrm{ft^3 h^{-2}}\right]$,
- Constant physical properties, and
- Complete mixing.

5.3.2 The CSTR Design Scenarios

We consider two design scenarios for CSTR with the same flow rates and conversion specifications:

1. Scenario 1: Single Continuously Stirred Tank Reactor; see Fig. 5.4a,
2. Scenario 2: Double Continuously Stirred Tank Reactors; see Fig. 5.4b.

Design Procedure and Alternatives for a Single-CSTR

In Table 5.7, we present design specifications for a single-CSTR. There are three design choices for Scenario 1. Design choice 1 (D1) where reactor temperature T = 140 [°F], design choice 2 (D2) where T = 160 [°F], and design choice 3 (D3) where T = 180 [°F]; see Table 5.8.

Design Procedure and Alternatives for a Double-CSTR

In Table 5.9, we present design specifications for a double-CSTR. In Scenario 2, there are 27 design choices with 3 different reactor volumes and 3 different reactor temperatures; see Tables 5.10, 5.11 and 5.12. In Table 5.10, we present nine different design choices (D1–D9) where reactors volumes are the same and reactor temperatures are 140 [°F], 160 [°F], and 180 [°F]. In Table 5.11, we present nine other design choices (D1–D9) where the ratio of the reactor volumes is 2 and reactor temperatures are as previously specified. In Table 5.12, we present the last nine design choices (D1–D9) where the ratio of the reactor volumes ratio is 0.5.

For more information about the CSTR see Subramanian and co-authors (Subramanian et al. 2001). In Sect. 5.3.3, we present the cDSP formulation of the Dynamic Operability Analysis of single-CSTR.

5.3.3 *The cDSP Construct for the Operability Analysis of the Single-CSTR*

Given

Design parameters				
Activation energy	E	30,000	$\left[\text{Btu lbmol}^{-1}\right]$	Table 5.3
Reaction rate constant	$k_{140°\text{F}}$	0.5	$\left[\text{h}^{-1}\right]$	Table 5.3
Feed concentration of A	C_{A0}	1.0	$\left[\text{lbmol ft}^{-3}\right]$	Table 5.3
Overall heat-transfer coefficient	U	300	$\left[\text{Btu h}^{-1}{}^{\circ}\text{F}^{-1}\text{ft}^{-2}\right]$	Table 5.3
Feed temperature	T_0	70	[°F]	Table 5.3
Coolant feed temperature	T_{C0}	70	[°F]	Table 5.3
Heat of reaction	ΔH	−30,000	$\left[\text{Btu lbmol}^{-1}\right]$	Table 5.3
Heat capacity of A	c_p	0.75	$\left[\text{Btu lb}^{-1}{}^{\circ}\text{F}^{-1}\right]$	Table 5.3
Heat capacity of coolant	c_{pc}	1.00	$\left[\text{Btu lb}^{-1}{}^{\circ}\text{F}^{-1}\right]$	Table 5.3
Molecular weight of A	M	50.0	$\left[\text{lb lbmol}^{-1}\right]$	Table 5.3
Density of A	ρ	50.0	$\left[\text{lb ft}^{-3}\right]$	Table 5.3
Density of coolant	ρ_c	62.3	$\left[\text{lb ft}^{-3}\right]$	Table 5.3
Operability spaces				
Desired output space	*DOS*		[–]	Table 5.5
Expected disturbance space	*EDA*		[–]	Table 5.6
Available input space	*AIS*		[–]	Table 5.4
Design alternatives				
Single CSTR				Figure 5.4a
Double CSTR				Figure 5.4b
Reactor temperature	T_i	140	[°F]	Table 5.7
Reactor temperature	T_i	160	[°F]	Table 5.7
Reactor temperature	T_i	180	[°F]	Table 5.7
Volume ratio	R_V	0.5–2.0	[–]	
Assumptions				
Rate of close and opening of valves at A	$\frac{dF_i}{dt}$	19,250	$\left[\text{ft}^3\text{h}^{-2}\right]$	
Rate of close and opening of valves at coolant	$\frac{dF_{Ci}}{dt}$	19,250	$\left[\text{ft}^3\text{h}^{-2}\right]$	
Height to diameter ratio of reactor volume	$\frac{h_i}{d_i}$	2	[–]	Table 5.6
Volume of the jacket thickness	w_{Ci}	4	[in]	Table 5.6
Volume of the 2nd reactor	V_2			Table 5.9

Find

System variables		
Achievable output space	AOS	$AOS \subset AIS$
Available input space	AIS	Table 5.4
Desirable input space	$DIS_d(y^N)$	$DIS \subset DOS$
Total desirable input space	DIS	$DIS = \bigcup_{y \in DOS} DIS_d(y) = \bigcup_{d \in EDS} DIS_y(d)$
Tolerable disturbance space	TDS	
Max allowable performance time	t_f^d	$t_f^d(y_{sp}, d)$
Max time for system to achieve stability	t_f^*	$t_f^*(y_{sp}, d)$
Dynamic desired available operability spaces	$DDAOS$	$DDAOS = DDOS \cap DAOS$
Dynamic desired expected deviation space	$DDEDS$	$DDEDS = DOS \cup EDS$
Deviation variables		
Under achievement of the goal 1	d_1^-	[–]
Under achievement of the goal 2	d_2^-	[–]
Under achievement of the goal 3	d_3^-	[–]
Under achievement of the goal 4	d_4^-	[–]

Satisfy

System constraints				
Response time has to be less than or equal to the desired response time		$t_f \leq t_f^d$	C1	[s]
Response time has to be greater than or equal to time for system to achieve stability		$t_f \geq t_f^*$	C2	[s]
Product of deviation variables has to be equal to zero	d_i^-, d_i^+	$d_i^- \times d_i^+ = 0$	C1	[–]
Deviation variables have to be positive	d_i^-, d_i^+	$d_i^-, d_i^+ \geq 0$	C2	[–]

System goals			
Maximize servo operability in the output space[a]	$\frac{\mu[AOS_u(d^N) \cap DOS]}{\mu[DOS]} + d_1^- - d_1^+ = 1$	G1	[–]
Maximize servo operability in the input space[b]	$\frac{\mu[AIS \cap DIS_y(d^N)]}{\mu[DIS_y(d^N)]} + d_1^- - d_1^+ = 1$	G1	[–]
Maximize regulatory operability[a]	$\frac{\mu[TDS \cap EDS]}{\mu[EDS]} + d_2^- - d_2^+ = 1$	G2	[–]
Maximize regulatory operability[b]	$\frac{\mu[AIS \cap DIS_d(y^N)]}{\mu[DIS_d(y^N)]} + d_2^- - d_2^+ = 1$	G2	[–]
Maximize system operability	$\frac{\mu[AIS \cap DIS]}{\mu[DIS]} + d_3^- - d_3^+ = 1$	G3	[–]

(continued)

(continued)

Maximize Dynamic Operability Index		$\frac{\mu[DOS \cup EDS]}{\mu[DDOS \cap DAOS]} + d_4^- - d_4^+ = 1$	G4	[-]
System bounds				
Coolant flow rate	F_{Ci}	$0 - 4\,F_{Ci}^R$	B1	[ft^3h^{-1}]
Product flow rate	F_i	50–150	B2	[ft^3h^{-1}]
Volume of the reaction mixture	V_i	$0.3\,V^R - V^R$	B3	[ft^3]
Exit of concentration of reactor A	C_{Ai}	0.02–0.2		[lbmol ft^3]
Reactor temperature	T_i	100–200	B4	[°F]
Feed temperature	T_{i-1}	50–90	B5	[°F]
Feed flow rate	F_{i-1}	50–150	B6	[ft^3h^{-1}]
Servo operability index in the output space	*SOIOS*	0–100	B7	[%]
Regulatory operability index	*ROI*	0–100	B8	[%]
Operability Index	*OI*	0–100	B9	[%]
Systems response time	t_f	0 – n	B10	[s]
Time for system to achieve stability	t_f^*	min-time optimal control—∞	B11	[s]

Minimize

Deviation functions		
Preemptive formulation where we are minimizing the goal function	$Z = \min\left(d_i^-\right)$	[-]

[a]Variant system design
[b]Original system design

In Sect. 5.4, we present the empirical performance validity and verify the ability to produce useful results for the chosen example problem in terms of operability analysis.

5.4 Results and Discussion

Dynamic Operability Analysis incorporated in DFDM is introduced in Sect. 5.2 and the model we use for dynamic operability analysis is presented in Fig. 5.3. The DOM is illustrated using an example of a continuously stirred tank reactors; see Fig. 5.4. In this section, we present the results in design of single and double CSTR and discuss these in the dynamic operability of these systems.

5.4.1 Design of a Single CSTR

In this section, we analyze the dynamic performance of a CSTR system in order determine feasible solutions for the functional design. To accomplish this, we first compute

Dynamic Desired Operating Space (DDOS), Dynamic Achievable Operating Space (DAOS), and Dynamic Operability Index (DOI).

Dynamic Desired Operating Space (DDOS)

For desired a Dynamic Available Input Space (DAIS) and an Expected Disturbance Space (EDS), the Desired Output Space (DOS) is achieved within desired response time, t_f^d. DAIS is related to normalized volume of the reactor, V, flow rate, F, and coolant flow rate F_c; see Table 5.2. EDS is related to feed temperature, T_0, and feed flow rate, F_0; see Table 5.2. DOS is related to exit concentration C_A in single reactor; see Table 5.13. DAIS and EDS information (range of inputs and rate of inputs change) are obtained by exercising the ACRONES and SSOM; see Sect. 5.2. Desired response time to achieve DOS is obtained through min-time control calculus, see Sect. 5.3.1. The goal is to achieve the desired dynamic performance of a system within maximum allowable response time for three different designs Di (i = 1, 2, 3) enumerated in the Sect. 5.3. For more information about design choices in the design of single-CSTR; see Table 5.3. In Table 5.13, we present the dynamic desired operable space response within maximum allowable response time of 2 h.

Dynamic Achievable Operating Space (DAOS)

For a given choice of DAIS and EDS, DOS is achieved in respect to minimum response time, t_f^*. The goal is to achieve dynamic performance of a system within minimum allowable performance time for all three designs Di (i = 1, 2, 3). In Tables 5.16, 5.17 and 5.18, we present the dynamic achievable operable space response within minimum allowable response time of 1 h.

Table 5.13 Dynamic desired output space (DDOS) time output response of a single-CSTR design

Design	Dynamic available input space			Desired output space	Expected disturbance space		t_f^d [h]
	$\frac{V}{V^R}$ [–]	$\frac{F}{F^R}$ [–]	$\frac{F_C}{F_C^R}$ [–]	C_A [lbmol ft^{-3}]	T_0 [°F]	F_0 [ft^{3}h^{-1}]	
Di	0.3–1.0	0.5–1.5	0–4	0.02–0.2	50–90	50–150	0–2

Table 5.14 Dynamic available output space (DAOS) time output response of a single-CSTR design

Design	Dynamic available input space			Desired output space	Expected disturbance space		t_f^* [h]
	$\frac{V}{V^R}$ [–]	$\frac{F}{F^R}$ [–]	$\frac{F_C}{F_C^R}$ [–]	C_A [lbmol ft^{-3}]	T_0 [°F]	F_0 [ft^{3}h^{-1}]	
D1	0.3	0.976–1.246	1–1.510	0.02–0.039	49.5–70	99.75–100	0–1
D2	0.2	0.7–1.3	0.5–1.5	~0.02	~70	~100	0–0.004
D3	0.0	0.8–1.3	0.0–0.42	0.02–0.0204	62.91–70	~100	0–0.014

Table 5.15 Dynamic desired output space (DDOS) time output response of double-CSTR design with volume ration of 1.0, 2.0, and 0.5

Design	Dynamic available input space						Desired output space		Expected disturbance space		t_f^* [h]
	$\frac{V_1}{V_1^R}$ [ft^3]	$\frac{V_2}{V_2^R}$ [ft^3]	$\frac{F_1}{F_1^R}$ [ft^{3}h^{-1}]	$\frac{F_2}{F_2^R}$ [ft^{3}h^{-1}]	$\frac{F_{c1}}{F_{c1}^R}$ [ft^{3}h^{-1}]	$\frac{F_{c2}}{F_{c2}^R}$ [ft^{3}h^{-1}]	C_{A1} [lbmol ft^{-3}]	C_{A2} [lbmol ft^{-3}]	T_0 [°F]	F_0[ft^{3}h^{-1}]	
$\boldsymbol{D_i}$	0.3–1.0	0.3–1.0	0.5–1.5	0.5–1.5	0.0–4.0	0.0–4.0	0.02–0.2	0.02–0.2	50–90	50–150	0.0–2.0

Table 5.16 Dynamic available output space (DAOS) time output response of double-CSTR design with volume ration of 1.0

Design	Dynamic available input space						Desired output space		Expected disturbance space		t_f^*[h]
	V_1[ft^3]	V_2[ft^3]	F_1[ft^{3}h^{-1}]	F_2[ft^{3}h^{-1}]	F_{c1}[ft^{3}h^{-1}]	F_{c2}[ft^{3}h^{-1}]	C_{A1}[lbmol ft^{-3}]	C_{A2}[lbmol ft^{-3}]	T_0[°F]	F_0[ft^{3}h^{-1}]	
D1	400–651	208–400	~100	~30	70–99	61–63	0.02–0.2	0.02–0.19	~70	~100	0–0.92
D2	200–352	136–200	~100	~80	70–108	70–108	~0.02	~0.02	~70	~100	0–1
D3	200–295	88–200	~100	~80	70–108	18–78	~0.02	0.02–0.2	~70	~100	0–1
D4	200–352	148–200	~100	~80	70–112	0–20	~0.02	0.02–0.2	~70	~100	0–1
D5	200–304	93–200	~100	~80	70–112	0–20	~0.02	0.02–0.2	~70	~100	0–1
D6	200–208	62–200	~100	~80	70–87	6–30	~0.02	0.02–0.2	~70	~100	0–1
D7	200–295	88–200	~100	~80	70–87	6–30	~0.02	0.02–0.2	~70	~100	0–1
D8	200–208	62–200	~100	~80	70–87	6–30	~0.02	0.02–0.2	~70	~100	0–1
D9	144	43–145	~100	~80	70–87	0–20	~0.02	0.02–0.2	~70	~100	0–1

Table 5.17 Dynamic available output space (DAOS) time output response of double-CSTR design with volume ration of 2.0

Design	Dynamic available input space						Desired output space		Expected disturbance space		t_f^* [h]
	V_1[ft^3]	V_2[ft^3]	F_1[ft^{3}h^{-1}]	F_2[ft^{3}h^{-1}]	F_{c1}[ft^{3}h^{-1}]	F_{c2}[ft^{3}h^{-1}]	C_{A1}[lbmol ft^{-3}]	C_{A2}[lbmol ft^{-3}]	T_0[°F]	F_0[ft^{3}h^{-1}]	
D1	400–485	291–700	~50	~100	70–86	68–70	0.02–0.07	0.05–0.2	69–70	~100	0.0–0.5
D2	300–307	184–600	50–51	~80	70–85	68–70	0.02–0.1	0.05–0.2	~70	~100	0.0–0.7
D3	140–192	114–280	50–51	~100	160–170	76–80	~0.02	0.05–0.2	~90	~100	0.0–1.0
D4	200–328	196–400	50–51	~100	140–164	65–69	~0.02	0.05–0.2	~70	~100	0.0–1.0
D5	203–215	151–405	58–59	~108	191–204	59–62	0.02–0.15	0.05–0.2	~73	~108	0.0–0.5
D6	120–141	84–240	~50	~100	160–176	31–40	0.05–0.15	0.05–0.2	~70	~100	0.0–0.5
D7	203–215	151–405	58–59	~108	191–204	59–62	0.02–0.15	0.05–0.2	~73	~108	0.0–0.5
D8	141–149	97–282	54–55	~104	157–171	45–51	0.02–0.2	0.05–0.2	~72	~104	0.0–0.5
D9	101	62–202	51–52	~101	123–133	56–60	~0.02	0.05–0.2	~71	~101	0.0–1.0

Table 5.18 Dynamic available output space (DAOS) time output response of double-CSTR design with volume ration of 0.5

Design	Dynamic available input space						Desired output space		Expected disturbance space		t_f^* [h]
	V_1[ft^3]	V_2[ft^3]	F_1[ft^{3}h^{-1}]	F_2[ft^{3}h^{-1}]	F_{c1}[ft^{3}h^{-1}]	F_{c2}[ft^{3}h^{-1}]	C_{A1}[lbmol ft^{-3}]	C_{A2}[lbmol ft^{-3}]	T_0[°F]	F_0[ft^{3}h^{-1}]	
D1	289–400	231—485	~50	~100	70–188	44	0.19–0.2	0.05–0.2	~70	~100	0.0–0.65
D2	289–400	231–485	~50	~100	70–188	44	0.19–0.2	0.05–0.2	~70	~100	0.0–0.65
D3	135–206	106–220	~51	~101	202–287	35–40	0.02–0.1	0.05–0.2	~71	~101	0.0–1.0
D4	184–320	187–307	~50	~100	200–262	27–30	0.02–0.1	0.05–0.2	~70	~100	0.0–1.0
D5	129–200	103–215	~50	~100	240–261	25–30	0.02–0.1	0.05–0.2	~70	~100	0.0–1.0
D6	97–152	58–150	~54	103–104	130–256	24–31	0.02–0.1	0.05–0.2	~72	~104	0.0–1.0
D7	115–150	62–192	~50	99–100	160–223	12–20	0.02–0.1	0.05–0.2	~72	~104	0.0–1.0
D8	91–152	58–141	~54	103–104	129–229	12–21	0.02–0.1	0.05–0.2	~72	~104	0.0–1.0
D9	60–100	30–101	~50	99–100	120–232	9–20	0.02–0.1	0.05–0.2	~70	~100	0.0–1.0

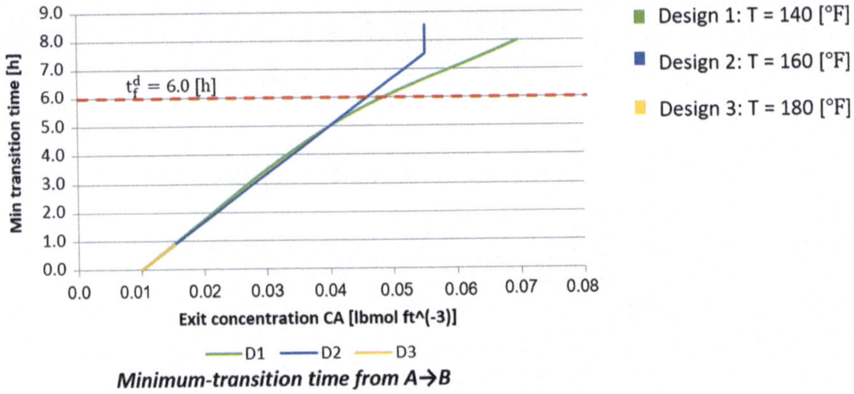

Fig. 5.5 Minimum-transition time for single CSTR designs

Dynamic Operability Index (DOI)

The DOI is calculated from min-time plots. If we set the desired response time to be $t_f^d = 0.01\ [h]$ then we can see that D1 reaches outputs y_{sp} in DOS for 30%, D2 for 60%, and D3 for ~100%; see horizontal dashed line in Fig. 5.5.

The minimum transition times that are needed for moving the reactor from concentration of 0.02 [lbmol ft^{-3}] to nominal concentration of 0.05 [lbmol ft^{-3}] are calculated for D1–D3; see in Fig. 5.5. In Fig. 5.5, we compare the results of three different design of single CSTR. It can be seen that reactor D3 transits faster than D2 and D1. However, reactor D1 transits slower and steadily toward nominal concentration, C_A, than reactors D2 and D3.

We can select different designs of single-CSTR based on designer or customer requirements. For example, if we need the reactor to transit quickly from State A to State B then the best design choice is D3; see Fig. 5.5. On the other hand, if we need the reactor to transit slower and steady from State A to State B than the best design choice is D1.

In Fig. 5.6, we present the minimum-disturbance rejection time calculations in the feed flow rate for D_i. It can be seen that reactor D_3 rejects disturbances faster than D_1 and D_2 because the trajectory of D3 is a straight trajectory that is the shortest distance and shortest time for the system to stabilize. On the other side, design D1 has a trajectory, hyperbola that is not the shortest distance and not the shortest time for a system to stabilize. Based on minimum transition and rejection time it can be concluded that design D_3 gives us better results than D_2 and D_1.

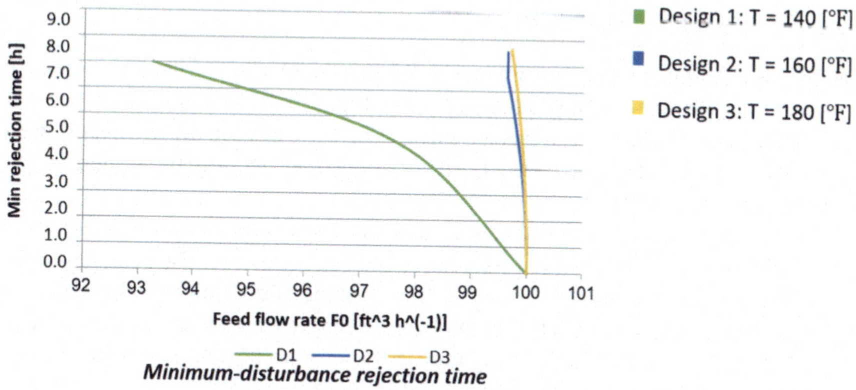

Fig. 5.6 Minimum-disturbance rejection time for single CSTR designs

5.4.2 *Design of a Double-CSTR*

In this section, we analyze dynamic performance of the system in order to locate solutions of functional design of a double-CSTR. We analyze Dynamic Desired Operating Space (DDOS), Dynamic Achievable Operating Space (DAOS), and Dynamic Operability Index (DOI).

Dynamic Desired Operating Space (DDOS)

For desired Dynamic Available Input Space (DAIS) and Expected Disturbance Space (EDS), the Desired Output Space (DOS) is achieved within desired response time, t_f^d. DAIS is design of CSTR with the normalized volumes of Reactors 1 and 2, V_i, flow rate, F_i, and coolant flow rate F_{ci}; see Tables 5.10, 5.11 and 5.12. EDS is related to feed temperature, T_0, and feed flow rate, F_0; see Tables 5.10, 5.11 and 5.12. DOS is related to exit concentration C_A of Reactors 1 and 2; see Tables 5.10, 5.11 and 5.12. The goal is to achieve desired dynamic performance of a system within maximum allowable response time for different designs D_i (i = 1, …, 9). In Table 5.10, we present the design alternatives in design of double-CSTR. In Tables 5.14 and 5.15 we present the dynamic desired operable space responses within maximum allowable response time of 2 h.

Dynamic Achievable Operating Space (DAOS)

For a given choice of DAIS and EDS, DOS is achieved in respect to minimum response time, t_f^*. The goal is to achieve dynamic performance of a system within minimum allowable performance time. In Tables 5.16, 5.17 and 5.18, we present the dynamic achievable operable space response within minimum allowable response time of 1 h.

Dynamic Operability Index (DOI)

The DOI is calculated from min-time plots. We presented results from representative designs D_i for different volumes ratios, R_v; see Fig. 5.7. If desired response time is $t_f^d = 0.6\,[h]$; see the horizontal dashed line in Fig. 5.7; to reach all y_{sp} in DOS then D_8 for Rv = 1.0 has DOI of 100%, D_3 for Rv = 2.0 has DOI of 100%, and D_9 for Rv = 0.5 has DOI of 100%.

The minimum transition time that is needed for moving reactor from concentration of 0.02 [lbmol ft^{-3}] to the nominal concentration of 0.2 [lbmol ft^{-3}] are calculated for D_i for different R_v; see Fig. 5.7. In Fig. 5.7, we compare the results of representative different design of double-CSTR. It can be seen that reactor D_8 transits faster than D_1 and D_6 for Rv = 1.0. Reactor D_3 transits faster than D_6 and D_8 for Rv = 2.0. Lastly, reactor D_9 transits faster than D_1 and D_8 for Rv = 0.5.

We can select different designs of double-CSTR based on designer or customer requirements. For example, if we need the reactor to transit quickly from State A to State B then the best design choice is D8; see Fig. 5.7.

In Fig. 5.8, we present the minimum-disturbance rejection time calculations in the feed flow rate for D_i for different Rv. It can be seen that Reactor D_8 rejects disturbances faster than D_1 and D_6 for Rv = 1.0. Further, Reactors D_3, D_6, D_8 reject disturbances with the same pace for Rv = 2.0. Lastly, Reactor D_9 rejects disturbances faster than D_1 and D_8 for Rv = 0.5.

5.5 Synopsis

In this chapter, we present a method, Dynamic Operability Analysis, that is a critical component of the DFDM framework; see Operability in Fig. 5.1. In Sect. 5.2, we present how the dynamic operability analysis method fits in the DFDM framework. In Sect. 5.3, we investigate structural validity using empirical formulations. A Continuous Stirred Tank Reactor (CSTR) is used to illustrate the design method; see validation of the method Fig. 1.15. In Sect. 5.4, we describe the empirical validation of the design process and compare the performance of the design solutions for the CSTR problem to develop insight into tradeoffs in the design.; see validation of the hypotheses Fig. 1.15.

The approach presented in Chaps. 2–5 will help identify those designs that meet both the steady state as well as dynamic operability requirements for the NMS; see Fig. 5.9. However, it is possible that no design solution exists that would simultaneously satisfy both these requirements. In such a case, it may become necessary to reconfigure the NMS in order to meet these requirements. In Chap. 6, we propose Reconfiguration of Machine Tools (RMT) and Reconfiguration of Inspection Systems (RIS) as one way of reconfiguring the NMS in order to meet the steady state and dynamic operability requirements. The approach is then demonstrated through an illustrative design problem.

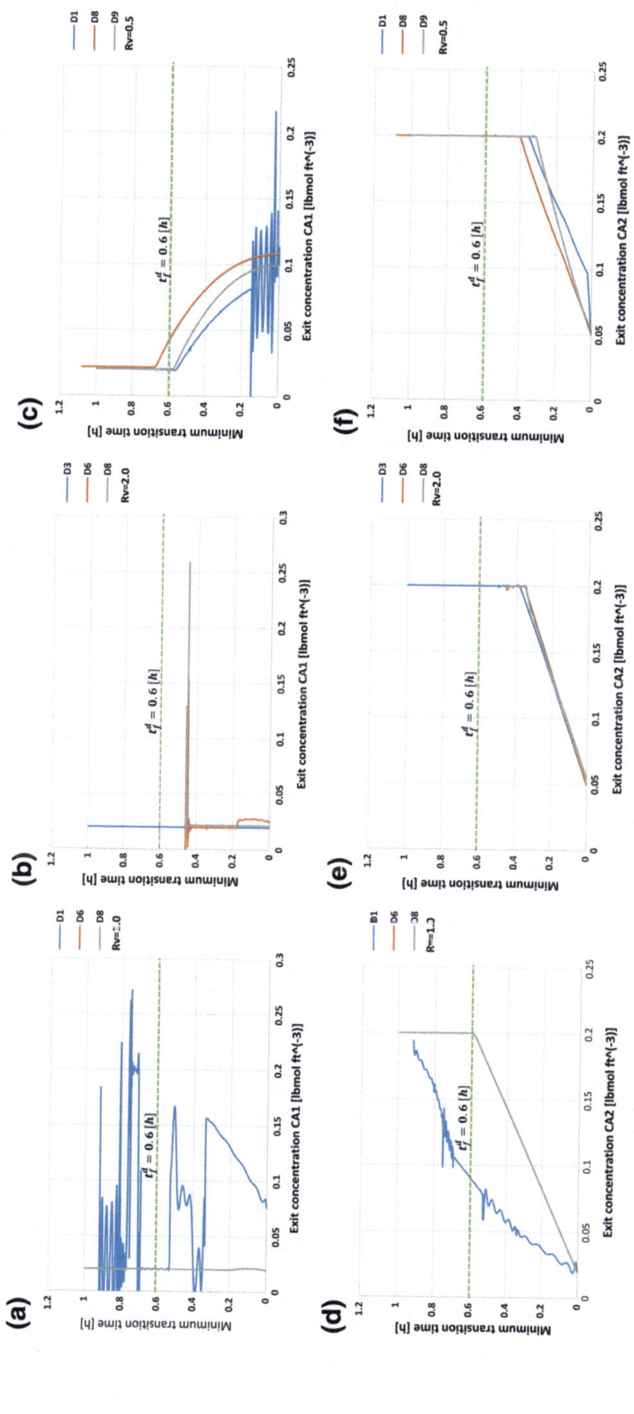

Fig. 5.7 Minimum-transition time for double-CSTR selected designs

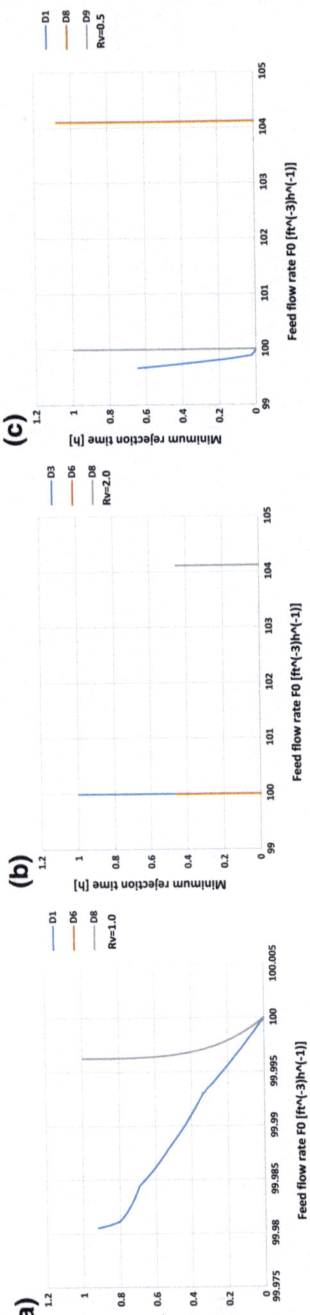

Fig. 5.8 Minimum-disturbance rejection time for double-CSTR selected designs

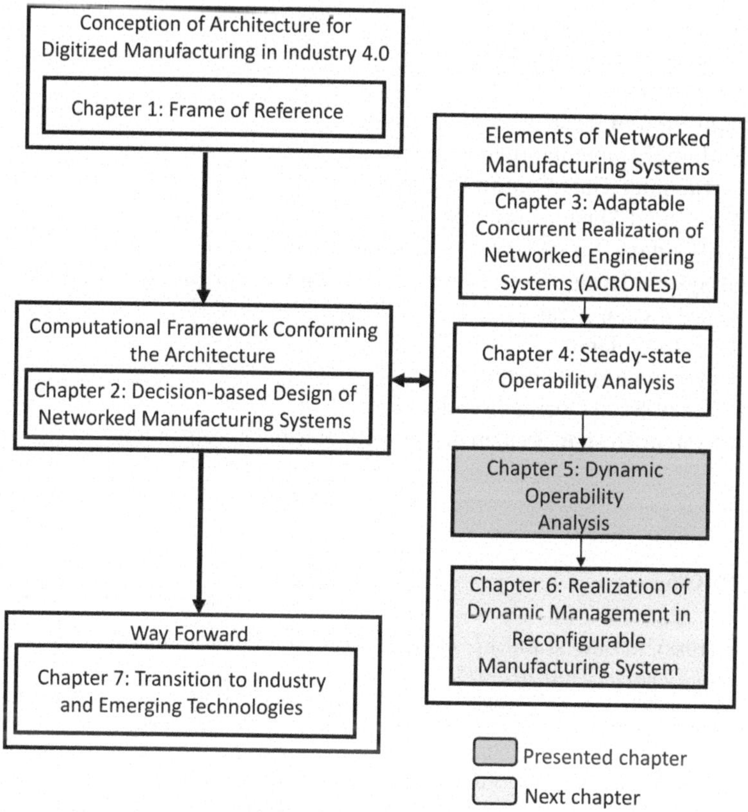

Fig. 5.9 Organization of the monograph—presented and next chapter

Glossary

ACRONES Adaptable Concurrent Realization of Networked Engineering System
AIS Achieved Input Space
AOS Achieved Output Space
cDSP Compromise Decision Support Problem
CSTR Continuous single tank reactor
DAIS Dynamic Available Input Space
DAOS Dynamic Achieved Output Space
DDOS Dynamic Desired Output Space
DFDM Design for Dynamic Management
DIS Desired Input Space
Disturbance Space Space where system can be functional in the presence of disturbance
DOI Dynamic Operability Index
DOM Dynamic Operability Model

DOS Desired Output Space
EDS Excepted Disturbance Space
FI Flexibility index
Functional System System that is operable
MIMO Multiple input multiple output
MTCP Minimum Time Control Problem
NMS Networked Manufacturing Systems
Operability Space Space where system can be functional
Original System Design New system design without existing specifications
RGA Relative Gain Array
RHP Right-half-plane
SISO Single input single output
SSOM Steady-State Operability Model
Variant System Design System design with existing specifications

References

Arkun, Y. (1988). Relative Sensitivity: A Dynamic Closed-Loop Interaction Measure and Design Tool. *AIChE Journal, 34*(4), 672–675.

Bahri, P. A., Bandoni, J. A., & Romagnoli, J. A. (1996). Effect of disturbances in optimizing control: Steady-state open-loop backoff problem. *AIChE Journal, 42*(4), 983–994.

Cao, Y., Biss, D., & Perkins, J. (1996). Assessment of input-output controllability in the presence of control constraints. *Computers & Chemical Engineering, 20*(4), 337–346.

Chenery, S. D. (1997). *Process controllability analysis using linear and nonlinear optimization.* In Ph.D. Monograph. London, UK: Imperial College of Science, Technology, and Medicine.

Dimitriadis, V. D., & Pistikopoulos, E. N. (1995). Flexibility analysis of dynamic systems. *Industrial and Engineering Chemistry Research, 34*(12), 4451–4462.

Fisher, W. R., Doherty, M. F., & Douglas, J. M. (1988). The interface between design and control. 1. Process controllability. *Industrial and Engineering Chemistry Research, 27*(4), 597–605.

Georgakis, C., Uztürk, D., Subramanian, S., & Vinson, D. R. (2003). On the operability of continuous processes. *Control Engineering Practice, 11*(8), 859–869.

Grosdidier, P., Morari, M., & Holt, B. R. (1985). Closed-loop properties from steady-state gain information. *Industrial and Engineering Chemistry Fundamentals, 24*(2), 221–235.

Grosdidier, P., & Morari, M. (1986). Analysis of interactions using structured singular values. In *IEEE American Control Conference.*

Grossmann, I. E., & Floudas, C. A. (1987). Active constraint strategy for flexibility analysis in chemical processes. *Computers & Chemical Engineering, 11,* 675–693.

Hovd, M., & Skogestad, S. (1992). Simple frequency-dependent tools for control system analysis, structure selection and design. *Automatica, 28*(5), 989–996.

Lewin, D. (1996). A simple tool for disturbance resiliency diagnosis and feedforward control design. *Computers & Chemical Engineering, 20*(1), 13–25.

Lyman, P. R., Luyben, W. L., & Tyreus, B. D. (1996). Method for assessing the effect of design parameters on controllability. *Industrial and Engineering Chemistry Research, 35*(10), 3484–3497.

Manousiouthakis, V., Savage, R., & Arkun, Y. (1986). Synthesis of decentralized process control structures using the concept of block relative gain. *AIChE Journal, 32*(6), 991–1003.

McAvoy, T. J. (1983). *Interaction analysis: Principles and applications*. NC: Instrument Society of America Research Triangle Park.
Milisavljevic, J. (2018). Architecting networked engineering systems. Doctoral Dissertation, The School of Aerospace and Mechanical Engineering. University of Oklahoma, Norman, Oklahoma.
Milisavljevic, J., Commuri, S., Allen, J. K., & Mistree, F. (2018). Steady-state operability in design for dynamic management in realization of networked engineering systems. In *ASME Design for Manufacturing and Assembly Conference*. Quebec City, Quebec, Canada. Paper Number DETC2018–85864.
Milisavljevic-Syed, J., Commuri, S., Allen, J. K., & Mistree, F. (2019b). Concurrent design exploration method for realizing networked manufacturing systems for industry 4.0. In *52nd CIRP Annual Conference*, PROC-D-18-00262.
Mohideen, M. J., Perkins, J. D., & Pistikopoulos, E. N. (1996). Optimal design of dynamic systems under uncertainty. *AIChE Journal, 42*(8), 2251–2272.
Morari, M. (1983). Design of resilient processing plants—III: A general framework for the assessment of dynamic resilience. *Chemical Engineering Science, 38*(11), 1881–1891.
Moore, C. (1986). Application of singular value decomposition to the design, analysis, and control of industrial processes. In *American Control Conference, IEEE.*
Russo, L. P., & Bequette, B. W. (1995). Impact of process design on the multiplicity behavior of a jacketed exothermic CSTR. *AIChE Journal, 41*(1), 135–147.
Russo, L. P., & Bequette, B. W. (1998). Operability of chemical reactors: Multiplicity behavior of a jacketed styrene polymerization reactor. *Chemical Engineering Science, 53*(1), 27–45.
Stanley, G., Marino-Galarraga, M., & McAvoy, T. J. (1985). Shortcut operability analysis 1: The relative disturbance gain. *Industrial Engineering Chemistry Process Design and Development, 24,* 1181–1188.
Subramanian, S., & Georgakis, C. (2000). Steady-state operability characteristics of reactors. *Computers & Chemical Engineering, 24*(2–7), 1563–1568.
Subramanian, S., Uztürk, D., & Georgakis, C. (2001). An optimization-based approach for the operability analysis of continuously stirred tank reactors. *Industrial and Engineering Chemistry Research, 40*(20), 4238–4252.
Swaney, R. E., & Grossmann, I. E. (1985a). An Index for operational flexibility in chemical process design. Part I: Formulation and theory. *AIChE Journal, 31*(4), 621–630.
Swaney, R. E., & Grossmann, I. E. (1985b). An index for operational flexibility in chemical process design. Part II: Computational algorithms. *AIChE Journal, 31*(4), 631–641.
Swartz, C. (1996). A computational framework for dynamic operability assessment. *Computers & Chemical Engineering, 20*(4), 365–371.
Uztiirk, D., & Georgakis, C. (1998). An optimal control perspective on the inherent dynamic operability of processes. In *Annual AIChE Meeting*, Miami, FL. Paper 217a.
Uztürk, D., & Georgakis, C. (2001). Inherent dynamic operability of processes I: Definitions and analysis in the SISO case. *Industrial and Engineering Chemistry Research, 41*(3), 421–432.
Vinson, D. R., & Georgakis, C. (1998). A new measure of process output controllability. *IFAC Proceedings Volumes, 31*(11), 663–672.
Vinson, D. R., & Georgakis, C. (2000). A new measure of process output controllability. *Journal of Process Control, 10*(2–3), 185–194.
Zhu, Z. X., Lee, J., & Edgar, T. F. (1997). Steady state structural analysis and interaction characterization for multivariable control systems. *Industrial and Engineering Chemistry Research, 36*(9), 3718–3726.
Ziegler, J., & Nichols, N. (1943). Process lags in automatic control circuits. *Trans. ASME, 65*(5), 433–443.

Chapter 6
Realization of Reconfigurable Manufacturing Systems

Nomenclature

CT	Configuration tree
u	Nodes
v	Edges
U	Node set of the RPL configuration tree
V	Edge set of the RPL configuration tree
u_{ijk}	Element in the node set U
i	Type of node
j	Identifier of the node of the same type
k	Section of the configuration tree to which the node belongs
v	Ordered pair which specifies one edge in the configuration
$u_{i'j'k'}$	Child node of node u_{ijk}
CCT	Capability of the configuration tree CT
u_{1j0}	j-th spindle head in the tool-side branch in the configuration tree
$f_2(u)$	Function which determines whether u_{1j0} belongs to configura tion tree CT
$\sum f(u)$	Sum of the spindle heads in the configuration tree CT
T	Motion of the tool-side branch in the configuration tree
t_{jr}	X-, Y-, Z-, A-, B- and C- axes (from left to right)
H	Motion of the workpiece-side branch in the configuration tree
h_{1r}	X-, Y-, Z-, A-, B- and C- axes (from left to right)
O_j	j-th spindle head-related operation
o_{jr}	Degrees of freedom of motion on the X-, Y-, Z-, A-, B- and C- axes
CCT	Capability of the configuration tree
n	Number of operations
FCT	Functionality of the configuration tree
t_{lr}	Motion in the rth direction of the lth tool-related configuration

J. Milisavljevic-Syed et al., *Architecting Networked Engineered Systems*,
https://doi.org/10.1007/978-3-030-38610-8_6

h_{1r}	Dictates the motion in the rth direction of the workpiece-related configuration
o_{lr}	Motion requirement for the rth direction of the lth operation
$\oplus$	Binary addition; that is, if $t_{lr} = 0$ and $h_{1r} = 0$, $t_{lr} \oplus h_{1r} = 0$, else $t_{lr} \oplus h_{1r} = 1$
FCT'	Flexibility of functionality in the RMT configuration
$n_{\max}$	Maximum number of tools (spindle heads) in the RMT configuration
ECT	Cost of a configuration tree
$\sum_{u \in U_i'} g(u)$	Fixed cost
$\sum_{u \in U_i'} q(u) \times s(u)$	Assembly cost
RCT	Reconfigurability of the RMT configuration tree
UCT	Edge utilization rate of an RMT configuration tree
V	Set of edges in an RMT configuration tree
V'	Set of edges that are the same in the current and previous configuration trees
$card$	Number of elements in the set
$card(V)$	Number of edges in an RMT configuration tree
PCT	Probability of the configuration tree
$p(v)$	Probability that edge v of the current configuration tree exists in the edge set of the next configuration tree
m	Number of feasible configuration trees
d_{m1}	Deviation of the cost of the mth feasible configuration tree
d_{m2}	Deviation of the reconfigurability from the most satisfactory reconfigurability
I	Decision-making function
w_1	Weights of cost
w_2	Weights of reconfigurability
$CRIS$	RIS configuration
cr_i	Ith RIM in the RMS
M	Number of RIM
y_i	Position of the ith RIM, where RIM follows the y_ith RIT
x_i	Number of sensors in the ith RMT, where RIM can detect x_i features
C_p	Processing route
cp_i	Ith operation
N	Number of operations in processing route C_p
$FCRIS$	Detection functionality of the RIS configuration
x_i	Number of sensors in the ith RIM
M	Number of RIS
$CCRIS$	Detection capability of the RIS configuration
y_i	Position of the ith RIM in the RIS
y_0	Distance from the jth operation to the ith RIM

$\sum_{j=y_{i-1}+1}^{j=y_i} (y_i - j)$	Sensitivity to error of the ith RIM
N	Number of RMT in the RMS
$FCRIS$	Constraints on the detection functionality of the RIS configuration design
$H(*)$	Judgement function of the processing surface
$G(*)$	Judgement function of the datum surface
$CCRIS$	Constraint on the detection capability
$D(*)$	Operation
$INT(*)$	Function used to retain the whole number
$ECRIS$	Cost of RIS configuration
c_1	Fixed average cost of installing a RIM
c_2	Fixed average cost of installing a sensor
x_i	Number of sensors in ith RIM
M	Number of RIM in the RIS
$RCRIS$	Diagnosability of the RIS configuration
$XCRIS$	Validity of the RIS configuration design
p_k	Validity of the kth sensor
y_i	Position of the ith RIM in the RIS
$y_i - y_{i-1}$	Expresses the maximum amount of damage done to the product in the process
d_{g1}	Deviation of the cost of the gth feasible RIS configuration
d_{g2}	Represents the diagnosability deviation
I	Decision-making function, which is the sum of d_{g1} and d_{g2}
a_i	Weights of the cost and the diagnosability

The design of Networked Manufacturing Systems (NMS) can be looked at as a two-stage operation where a computational model and a decision support framework are first used to identify the space of feasible solutions. Then, a performance index is selected that allows for dynamic management of NMS to account for uncertainties arising from variations in inputs to the sub-components of the NMS or in the parameters of the models used to represent the NMS. These concepts, introduced as the second component of the computational framework, Design for Dynamic Management (DFDM), are described using illustrative examples. Chaps. 2–6. Selection of robust design solutions enable the NMS to meet design and performance objectives in the presence of modeling errors and variations arising from changing market demands (variations in input space). However, failures in system components or large-scale variations in market demand make it necessary to reconfigure the NMS in order to adapt to the changes and meet production objectives. In this chapter, we present a DFDM-based exploration method for reconfiguration of a Reconfigurable Manufacturing System (RMS). The aim is to identify those configurations that meet design/performance objectives and a strategy to select appropriate solutions when the original design is unable to meet the requirements. The method addresses reconfiguration performed at different levels of granularity, namely module-level, device-level, and system-level reconfiguration. The proposed method is a third component of the DFDM framework; see Reconfigurability in Fig. 6.1.

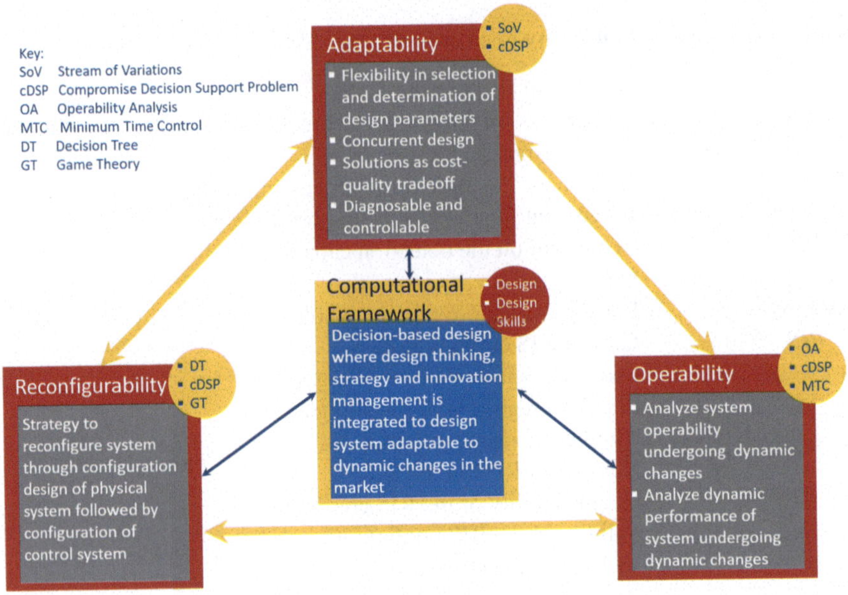

Fig. 6.1 Computational framework, design for dynamic management (Milisavljevic-Syed et al. 2019b)

The DFDM-based reconfiguration method is illustrated through three different examples. In the first example, we consider changes that can be addressed by retooling of the manufacturing system (Reconfigurable Machine Tool (RMT); see Sect. 6.3). In the second example, we consider changes that require the use of different set of sensors (Reconfigurable Inspection Systems (RIS); see Sect. 6.4). In the third example, we consider reconfiguration of system functionality (Reconfigurable Production Subsystem (RPSU); see Sect. 6.5) where individual subsystems are reconfigured. In Sect. 6.6, we present the results and discuss the usefulness of the method.

In this chapter our focus is on the reconfiguration methods that helps meet the design and performance objectives of the NMS. The method involves: (a) selection of a solution that is robust and meets design and performance objectives; (b) identification of associated failure criteria for the NMS and alternate configurations that meet the design/performance goals; (c) using the DFDM framework to reconfigure the NMS and verify that the design objectives are satisfied. In Sect. 6.1, we present current state of the art. In Sect. 6.2, we present the application of reconfiguration method in design of the NMS. In Sect. 6.3, we discuss the application of the DFDM-based reconfiguration method to the modification of a Reconfigurable Machine Tool (RMT). In Sect. 6.4, we discuss the application of the DFDM-based reconfiguration method to the modification of a Reconfigurable Inspection System (RIS). In Sect. 6.5, we present a computational model that is used to validate the performance of a Reconfigurable Manufacturing System (RMS). In Sect. 6.6, we present the results and discuss the usefulness of the method.

6.1 Reconfigurability of the Networked Manufacturing Systems: A Critical Review

Networked Manufacturing Systems (NMS) are complex systems consisting of a series of connected workstations. Conventional manufacturing involve the use of fixed manufacturing lines that are highly reliable and capable of mass-producing products with high quality and low cost. However, such fixed manufacturing lines cannot handle the needs arising from changing market demands in a global economy. Modern manufacturing requires NMS to be capable of learning optimal manufacturing strategies and dynamic reconfiguration of the system to adapt to changing needs. A Reconfigurable Manufacturing System (RMS) allows the NMS to reap the economic benefits of scale production and to quickly adapt to market demands.

In this chapter, reconfiguration of NMS is illustrated through three different examples. In the first example, we consider changes that can be addressed by retooling the manufacturing system (Reconfigurable Machine Tool (RMT); see Sect. 6.3). We define Controllability Indices that are then used to select an optimum tool configuration from a set of possible configurations. In the second example, we consider changes that require the use of a different set of sensors (Reconfigurable Inspection Systems (RIS); see Sect. 6.4). In this case, we define Observability Indices that are then used to select an optimum sensor deployment. In the third example, we consider reconfiguration of system functionality (Reconfigurable Production Subsystem (RPSU)) where individual subsystems are reconfigured. Depending on the complexity of reconfiguration being performed, RMS configuration design can be divided into module-level granularity, equipment-level granularity, and system-level granularity; see Sect. 6.5.

6.1.1 Reconfigurable Machine Tool (RMT)—An Overview

Design of interchangeable configurations is associated with two fields, namely, robotics and manufacturing. In manufacturing, there are three reasons for forcing the reconfiguration: (1) random failure, (2) product variance, and (3) demand fluctuation. Existing approaches for reconfiguration can help RMT adapt to faults or changes in demand that result in only a small deviation from the desired performance (Heisel and Meitzner 2006). The capability of an RMT to adapted to events that cause significant deviation in the performance will make manufacturing enterprises more profitable and improve customer satisfaction (Youssef and El Maraghy 2006; Abdi 2009; Goyal et al. 2012; Wang and Koren 2012). In addition to maintaining desired performance, reconfiguration must satisfy functional constraints such as production rate and delivery time (Dou et al. 2009; Al-Zaher et al. 2013; Puik et al. 2016; Xia et al. 2016).

After critical review of literature, the principal gaps in methods for RMT are as follows:

- Scaling the capability of the system and/or changing its functionality (Andersen et al. 2017);
- Effectively identifying the tasks from the part family (Koren and Kota 1999);
- Clearly modeling the descriptions of different configurations (Mpofu et al. 2008);
- Systematically evaluating the configuration using multiple performance indices (Andersen et al. 2017); and
- Adjusting the configuration to the diverse production scenarios (Gadalla and Xue 2017).
- The details of these gaps are provided in Table 6.1.

Gap 1: *Scaling Capability*

The issue of scaling capability is usually addressed by increasing or reducing the quantity of RMT in the manufacturing system where production capacity is augmented by the addition of parallel production lines (Benderbal et al. 2017; Youssef et al. 2006; Wang et al. 2018; Dou et al. 2009; Goyal et al. 2013; Son et al. 2001; Spicer et al. 2002a, b; Koren and Ulsoy 2002; Attanaik et al. 2010; Ai-Ping et al. 2011). Further, Youssef and co-authors (Youssef et al. 2007), recommend the utilization of multi-machine tools with higher system availability and less cost instead of increasing the number of machine tools. Further, increasing the number of spindle heads increases the capability of the RMT. Deif and co-authors (Deif et al. 2006) and Spicer and co-authors (Spicer et al. 2002a, b) propose a model where the capability is modified using multi-spindle modules.

Gap 2: *Functionality*

Modification of the functionality is achieved by reconfiguring the modules in a NMS. Shabaka and co-authors (Shabaka and ElMaraghy 2007) generate the feasible configuration matched with the required functions according to the axes of motion. Mpofu and co-authors (Mpofu et al. 2008), use the Degree of Freedom (DOF) to determine the required axes of motion for the manufacturing process.

Gap 3: *Selection of Adequate Configuration*

Design requirements given in terms of cost, availability, quality, and utilization can be used as metrics to evaluate different configurations (Mittal and Jain 2014; Kuzgunkaya and ElMaraghy 2007). Heisel and Meitzner (2006) use reconfiguration cost of the RMT as a criterion for evaluating different reconfiguration strategies. Kuzgunkaya and ElMaraghy (2007) and Benderbal and co-authors (Benderbal et al. 2017) take the cost and reconfigurability as indices to manage the RMT location in the system, while Goyal and co-authors (Goyal et al. 2013), take the axes' motion to determine the module assembly in the machine, and Gadalla and co-authors (Gadalla and Xue 2017), take consideration of configuration parameters and reconfiguration process. Ahuett and Molina (Ahuett and Molina 2005) also identified upgradeability and adaptability as main indices for the modular configuration with the consideration of the RMT intended evolution process.

Table 6.1 Overview of important papers for identifying the research motivation in design of RMT configuration (Wang et al. 2018)

Key approaches for reconfigurable configuration design

Paper	Configuration type				Design scenario			Design decision (output)					
	Machine-level		System-level		Environmental foundation	Variance of product design	Fluctuation of product demands	Machine-configuration			System-configuration		
	Single-spindle head	Multi-spindle Head	Series-line	Parallel-line				Module type	Module quantity	Assembly relationship	Machine type	Machine quantity	Machine position
Gadalla et al. (2016)	✓	–	–	–	1	✓	–	–	✓	–	✓	–	–
Benderbal et al. (2017)	✓	–	✓	–	0	✓	–	–	–	–	–	✓	✓
Xu et al. (2017)	✓	–	–	–	1	✓	–	✓	✓	✓	–	–	–
Goyal et al. (2012, 2013)	✓	–	✓	–	0	✓	✓	✓	✓	–	–	✓	–
Mpofu et al. (2008)	✓	–	–	–	1	✓	–	✓	✓	✓	–	–	–
Wang, et al. (2012)	–	–	–	✓	1	–	✓	–	–	–	✓	✓	✓
Son et al. (2010)	–	–	–	✓	0	–	✓	–	–	–	✓	✓	✓
Pattanaik et al. (2007)	–	✓	–	–	1	✓	✓	✓	✓	–	–	–	–
Dou et al. (2009)	–	–	–	✓	1	–	✓	–	–	–	✓	✓	✓
Abdi et al. (2009)	✓	–	–	–	0	✓	✓	–	–	–	✓	–	–
Youssef et al. (2007),	–	✓	–	✓	0	✓	–	✓	✓	✓	✓	✓	✓
Liu et al. (2008)	✓	–	–	–	0	–	✓	✓	✓	–	–	–	–

(continued)

Table 6.1 (continued)

Key approaches for reconfigurable configuration design													
Paper	Configuration type				Design scenario			Design decision (output)					
	Machine-level		System-level					Machine-configuration			System-configuration		
	Single-spindle head	Multi-spindle Head	Series-line	Parallel-line	Environmental foundation	Variance of product design	Fluctuation of product demands	Module type	Module quantity	Assembly relationship	Machine type	Machine quantity	Machine position
Nan et al. (2012)	✓	–	–	–	0	–	✓	–	–	–	✓	–	–
Ai-Ping et al. (2011)	✓	–	–	–	1	–	✓	–	–	–	✓	–	–
Chen et al. (2005)	✓	–	–	–	0	–	✓	✓	✓	✓	–	–	–
Wang et al. (2018)	✓	✓	–	–	**1**	✓	✓	✓	✓	✓	–	–	–

Key approaches for reconfigurable configuration design						
Constraints		Goals				Research gap
Scaling the capability	Converting the function	Min cost		Max reconfigurability		
		Direct cost	Extra cost	Utilization of the existing resource	Availability of the future reconfiguration	
–	✓	✓	✓	–	–	Limited to the several manufacturing function. Lack of models on the reconfigurability. Ignoring the design for changing demands

(continued)

Table 6.1 (continued)

Key approaches for reconfigurable configuration design						
Constraints		Goals				Research gap
Scaling the capability	Converting the function	Min cost		Max reconfigurability		
		Direct cost	Extra cost	Utilization of the existing resource	Availability of the future reconfiguration	
–	✓	✓	–	✓	✓	Ignoring the implementation effort. Concentrating on the system layout. Ignoring the reusability of the existing resource
–	✓	–	–	✓	–	Lack of the cost analysis. Unable to satisfy the changing demands
✓	✓	✓	–	–	✓	Given configuration candidates. Lack of considering the extra cost.. Increasing the complexity with the large number of machines
–	✓	✓	–	✓-	–	Unavailable for the large candidate library without the quantitative model
✓	✓	✓	–	✓	–	Lack of the research on the flexible reconfiguration for dynamic market. Limited to the functional conversion. Ignoring the operational cost
✓	–	✓	✓	–	–	Lack of the reconfiguration perspective.
–	✓	✓	–	–	–	Lack of the specific analysis of capability and functionality requirement

(continued)

Table 6.1 (continued)

Key approaches for reconfigurable configuration design						Research gap
Constraints		Goals				
Scaling the capability	Converting the function	Min cost		Max reconfigurability		
		Direct cost	Extra cost	Utilization of the existing resource	Availability of the future reconfiguration	
–	✓	✓	–	✓	✓	Lack of details in RMT configuration
✓	✓	✓	✓	✓	–	Lack of details in RMT configuration
✓	✓	✓			✓	Fixed configuration candidate. Lack of details in RMT configuration
–	✓	✓	✓	✓	–	Unable to satisfy the changing demands
–	✓	✓	✓	–	–	Lack of the reconfigurability in design; Lack of considering the capability changes
–	✓	–	–	–	–	Unable to deal with a large number of candidates. Lack of considering the capability changes
–	✓	–	–	✓	✓	Unable to satisfying the changing demands Similarity Analysis is Fuzzy
✓	✓	✓	✓	✓	✓	

0 stands for the creating entirely new systems satisfying the requirements
1 stands for the modifying the existing systems to fit the new requirements

Gap 4: Modeling the Design Process Aimed at Reconfiguration and Improving the Effectiveness of the Design Process

Gadalla and Xue (2017) studied the effectiveness of the decision-making process during design of system reconfiguration and proposed a model-based approach for reconfiguration that takes into account the configuration parameters and the reconfiguration process. Wang and Koren (2012), focus on scalability planning for reconfiguration. Benderbal and co-authors (Benderbal et al. 2017) develop a multi-objective method to manage the machine location while minimizing the transition effort and maximize the responsiveness.

From the literature review, we glean shortcomings of current RMT configuration methods such as the lack of the multi-spindle configuration and periodic reconfiguration, and the descriptive models; see Table 6.1. Further, we identify the following gaps:

- How to describe the RMT configuration in a clear, dynamic and computational way?
- How to generate feasible configurations while simultaneously satisfying capability and functionality requirements?
- How to address the trade-off between cost and reconfigurability and verify that the selected configuration meets as many of the design goals as possible?
- How to explore the design space to solutions for different scenarios?

6.1.2 Reconfigurable Inspection System (RIS)—An Overview

Cyber-Physical System (CPS) are comprised of virtual and physical processes, where the available data-accessing and data-processing services via the Internet of Things (IoT) result in an interconnected system (Lee et al. 2008). Increasing use of sensors and communication networks also results in continuous data flow across these systems. The digital Twin and Digital Thread, a virtual representation the processes in the system and the data flows, are necessary to assess the performance of these systems (Gabor et al. 2016). CPS is expected to facilitate data-oriented dynamic management of multistage manufacturing processes to enhance the efficiency of the manufacturing process (Gabor et al. 2016; Lee et al. 2015).

In pace with the rapid development of information technology, digitized manufacturing needs to be equipped with the capability to automatically detect the current status of the manufacturing process. Such detection can help in the identification of process errors in a timely manner and accommodate these errors through reconfiguration of appropriate entities in the manufacturing line, and thereby maintain product quality (Kore 2013). In RMS, RIS consisting of multiple Reconfigurable Inspection Machines (RIM), is used to check the quality of product (Koren and Shpitalni 2010). RIS can be modeled using on Stream of Variation (SoV) method and used to diagnose the root cause of errors (Barhak et al. 2005). As long as the errors are detectable through the output of the RIS, this method can be used to quickly detect errors and their sources. Configuration design of RIS provides customized detection ability sensitive to process errors and enables the collection of enough data to support the identification of root causes of errors.

6.1.3 Reconfigurable Manufacturing System (RMS)

Manufacturing processes are affected by both external factors (product variances or demand fluctuations) and internal factors (random failures of equipment), that lead to deterioration in product quality (Bruccoleri et al. 2006). Digitization of the process is necessary to implement reconfiguration strategies that will improve the manufacturing efficiency, enhance production quality, and increase the responsiveness to changes in demand (Mehrabi et al. 2000). Wang and co-authors (Wang et al. 2018) propose reconfiguration of modules in a manufacturing process in order to adapt to changing requirements.

A manufacturing system consists of multiple modules with different functions and configurations. Reconfiguration of these systems must consider the factors driving the need for reconfiguration, type of reconfiguration that is required (module-level, device-level, or system-level) and the effect of such reconfiguration on the cost and performance of the overall system. The different types of reconfiguration are addressed below.

A. *Module Granularity Design (MGD)*

Modular design of a manufacturing system reduces the overall cost and time-to-market, and improves the reliability and availability of the manufacturing system. Reconfiguration of modular systems is straightforward and involves changing out the module and thereby changing the data and material flows between modules. However, the design engineer has to anticipate the types of reconfigurations needed ahead of time and specify these modules at design time.

B. *Device Granularity Design (DGD)*

The design unit of the DGD is the equipment, and its space-time scale is larger than the MGD. The input to the DGD is the equipment library and process requirements, and the output is the subsystem configuration design. The essence of the DGD mainly revolves around the process requirements, adding, deleting, or moving the equipment, resulting in changes in the configuration of the subsystem.

C. *System Granularity Design (SGD)*

SGD is the concurrent design of multiple subsystems, and it needs to be represented on a larger spatial scale. The SGD input is the subsystem configuration design, such as the production system, the control system, etc., and the market demands. The output is the overall system configuration design. The SGD is based on the characteristics of the overall system, and the design ensures that the reconfigured system utilizes its resources more efficiently.

6.2 Reconfiguration Strategies in Design of the Networked Manufacturing Systems

In Sects. 6.2.1–6.2.3, we present three different types of reconfiguration that can be used in the design of NMS to achieve (1) customized production, (2) comprehensive detection, and (3) dynamic reconfiguration.

6.2.1 *Module-Level Configuration for Customized Production*

The question we address is "*How to effectively reorganize the Reconfigurable Machine Tool (RMT) modular configuration to fit customized production by identifying the number, type and assembly relationship of modules?*" The solution to the question posed can be arrived at using the following elements of the design method: (1) RMT configuration tree, (2) production capability, (3) production functionality, (4) cost, and (5) reconfigurability in the RMT configuration design in order to facilitate customized productions. The following assumptions are made in order to answer the question posed: (1) a module library is given, (2) there are limited tools in one RMT, (3) cost is associated with fixed and assembly cost, and (4) effort of reconfiguration is related to the distance of moving modules.

- ***Gap* 1: *RMT Configuration Tree*.** A tree structure diagram that is used to describe the steps in the reconfiguration of the RMT system.
- ***Gap* 2: *Production Capability*.** The capability to perform multiple parallel operations at the same time. Production capability we measure as the number of tools in the RMT configuration.
- ***Gap* 3: *Production Functionality*.** The ability to process the machining features. Production functionality we measure as the relative movement between the tool and the workpiece in the RMT configuration.
- ***Gap* 4: *Cost*.** Reducing economic expenditure as much as possible during the reconfiguration.
- ***Gap* 5: *Reconfigurability*.** Reuse the module as much as possible during the reconfiguration, mainly include utilization and similarity.

The RMT configuration design method is comprised of two phases. In the first phase, we explore the feasible design space of RMT configuration by using production capacity and production functions as indicators. In the second phase, we explore the satisfied RMT configuration tree by taking the cost and reconfigurability as indicators. The design procedure is illustrated through an example in Sect. 6.3.

6.2.2 Device-Level Configuration for Comprehensive Detection

The question we address is "*How to effectively reorganize the RIS modular layout for comprehensive detection by identifying the required type, position and number of sensors in the RIM?*" The solution to this question can be arrived at by evaluating the following properties of the overall system being designed: (1) detection capability, (2) detection functionality, (3) cost, and (4) diagnosability of the RIS configurations. The assumptions made in the design of RIS configurations are (1) there is a detailed sensor library, (2) one sensor can only detect one kind of error, and (3) detection time is related to the distance between the RMT and corresponding the RIM. The gaps in the design of these reconfigurable systems are identified below:

- ***Gap* 1**: ***Detection Capability***. The ability to detect operational features of all operations in the process route, measured by the Reconfigurable Inspection Machines (RIM) and sensors.
- ***Gap* 2**: ***Detection Functionality***. The ability to timely detect the speed of the processing error, measured by the RIM position.
- ***Gap* 3**: ***Cost***. Reducing economic expenditure as much as possible during the reconfiguration, mainly include the RIM costs and sensor cost.
- ***Gap* 4**: ***Diagnosability***. Data collection performance is improved as much as possible in reconfiguration, mainly includes timeliness and comprehensiveness.

The RMT configuration design method is comprised of two phases. In the first phase, we use detection capacity and detection functions as indicators to explore the feasible design space of the RIS layout. In the second phase, we consider the cost and diagnosability as indicators to explore the satisfied RIS layout. The design procedure is illustrated through an example in Sect. 6.4.

6.2.3 System-Level Configuration for Dynamic Reconfiguration of the System

The question we address is "*What is the interaction between the RMT and the RIS in order to achieve dynamic reconfiguration of the RMS*?" In order to determine this interaction, it is necessary to integrate multiple decision-making strategies into a management model to support dynamic reconfiguration of RMS. Hence, we identify the need for (1) information collection, (2) activity interaction, and (3) exploration of the overall design plan in the RMS configuration design in order to achieve dynamic reconfigurability.

- ***Gap* 1**: ***Information Collection***. A system design needs a standardized model to promote parallel design among multiple activities.
- ***Gap* 2**: ***Activity Interaction***. A system design requires simulation of interactions to reduce unnecessary optimization iterations.

- ***Gap* 3: *Explore the Overall Design Network*.** A system design needs to construct a decision network describing the RMS configuration design and realize the integration of the activity model.

The application of this method to the design of a Reconfigurable Manufacturing System (RMS) is illustrated through an example in Sect. 6.5.

6.3 Designing Reconfigurable Machine Tools (RMT)—An Example

In Sect. 6.3.1, we present a method for Reconfigurable Machine Tool (RMT) configuration design. In Sect. 6.3.2, we present the first step, construction of the configuration tree. In Sect. 6.3.3, we present the second step, construction of the constraint model. In Sect. 6.3.4, we present the last step, the construction of the objective model.

6.3.1 Development of RMT Configuration Design

Current design methods for RMT configuration design are based on modularization theory. Padayachee and other authors, (Koren and Kota 1999; Son et al. 2010; Padayachee et al. 2009; Padayachee and Bright 2012) identify a need for RMT design development regarding geometry, size, accuracy among other parameters. Hence, diverse modules for RMT configuration design are developed. As a result, manufacturing enterprises have individual module libraries to support the design, which are not necessarily available to other manufacturing enterprises. The RMT configurations can be generated by different manufacturing enterprises from different module libraries; see Fig. 6.2. Consequently, modules from different libraries have different parameters (for example, geometry), and different languages to describe two different configurations concurrently in the design process is quite complex.

The complexity in describing two different configurations concurrently in the design process can be overcome by explaining how to carry out the process of design using critical indices and developing a common method for description of the configuration (Roozenburg and Eekels 1995). The first step is to develop a common model and make the design method more comprehensive and practical for use with different module libraries. The most important part of the common design method is the process of mapping; see Fig. 6.3.

In Fig. 6.3, we present the common model that has five blocks (1) design requirement, (2) module libraries, (3) common model, (4) design process, and (5) RMT configuration. The first block is the specified module library mapped into the common model. The second block is based on the design requirement and the common model is used to design a RMT. The third block is the solution of the design process obtained and expressed with the common model. The fourth block is the RMT design

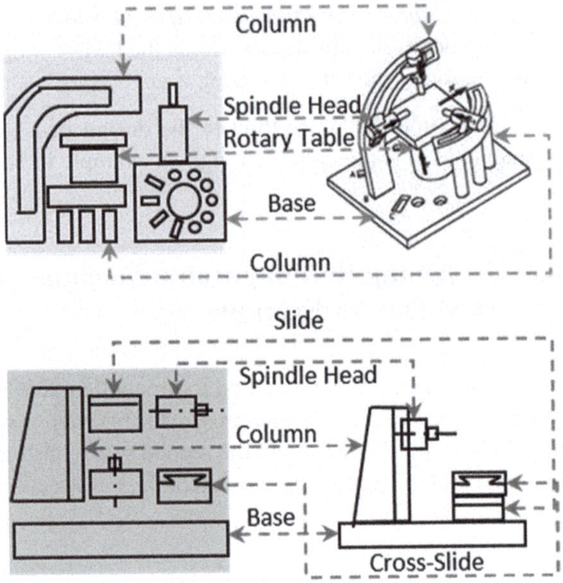

Fig. 6.2 The RMT configuration design from different module libraries (Wang et al. 2018)

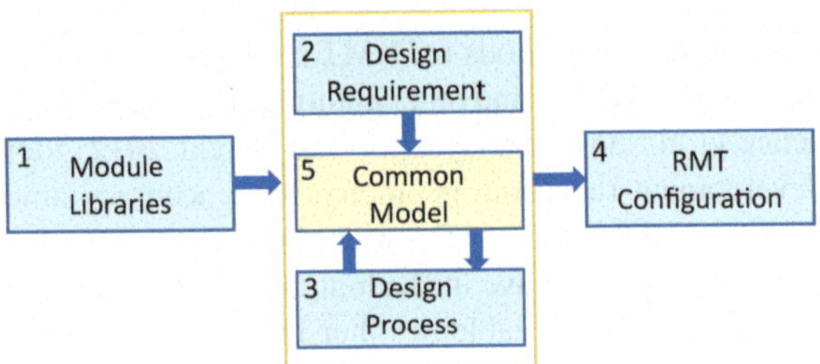

Fig. 6.3 Schematic of the design with the common model (Wang et al. 2018)

in the common model mapped into an RMT configuration with the required module library.

Wang and co-authors (Wang et al. 2018) introduce the concept of a configuration tree as a common model in order to ensure that the method can work for any module library. A configuration tree is a functional model we use to describe RMT configuration based on a tree structure diagram. The tree structure diagram is a data structure made up of finite nodes and edges, as in graph theory. In a tree, the nodes are hierarchically arranged by edges. The RMT design configuration consists of modules, that are integrated into an assembly relationship. Due to the similar structures of the tree and RMT configuration, RMT configurations are developed in the form of a tree. When decomposing the RMT structural configuration, the hierarchy of the design process goes from the most general to the most specific processes, and from configuration to tool-related configuration combined with workpiece-related configuration and modules. Tool-related configuration is used to achieve the motion of the tool, while the workpiece-related configuration is the motion of workpiece which must be configure as well. Hence, each node in the configuration tree represents a module in the RMT configuration. The edges represent the relationship between nodes, that is, the assembly relationship among the modules.

The branches are the subsets of nodes and edges in the configuration tree, and they represent sub-configurations. The branch with the tool-related configurations we call the tool-side branch, and the branch with the workpiece-related configurations we call the workpiece-side branch. Thus, the configuration tree diagram is used to represent the entire RMT configuration. In Fig. 6.4, we present the mapping relationship between the RMT configuration and the configuration tree.

Given a function, the mapping from RMT configuration to configuration tree is one-to-one. Therefore, the configuration tree can be used to describe the RMT configuration design with various module libraries. Furthermore, by using a configuration tree it is possible to partially automate the RMT design process based on its data structure. In Fig. 6.5, we present the proposed RMT configuration design method that includes three stages (1) configuration description, (2) configuration formation, and (3) configuration evaluation.

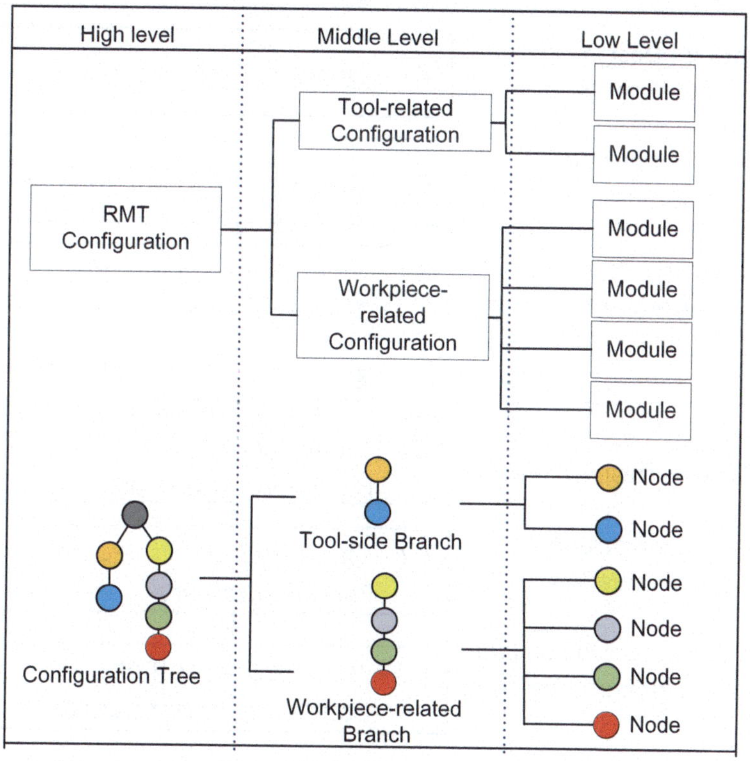

Fig. 6.4 Mapping relationship RMT configuration and configuration tree (Wang et al. 2018)

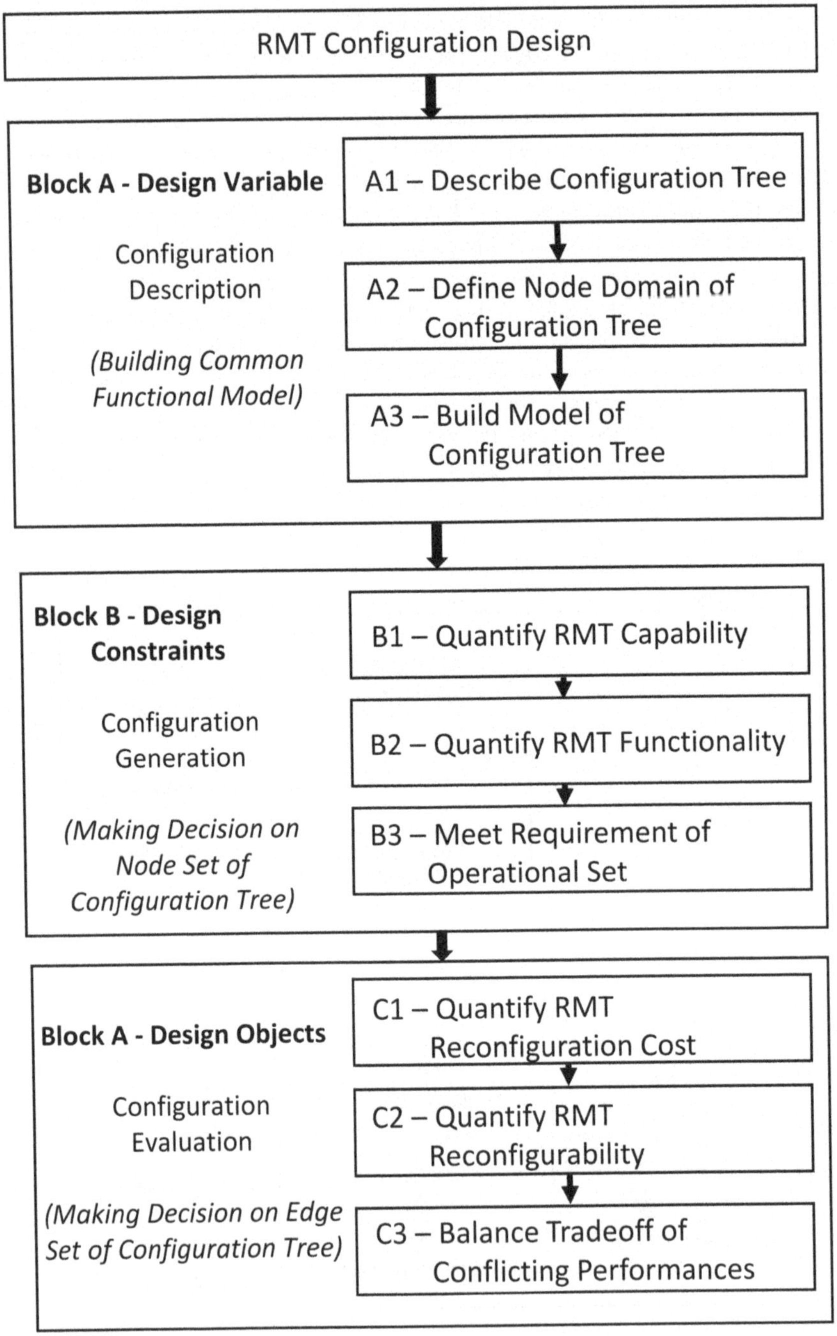

Fig. 6.5 Design method for the RMT configuration (Wang et al. 2018)

- ***Configuration Description, Block A,*** Fig. 6.5: This step is the foundation of the proposed design method, supported by mapping between the RMT configuration and the configuration tree. In Step A1, we map the RMT configuration into the tree, and we define the configuration design variables by the node and edge sets. After selecting the most satisfactory configuration tree scheme, we determine the module and assembly relationships by the node in Step A2. Further, in Step A3, we form the edge sets and the RMT configuration design.
- ***Configuration Generation, Block B,*** Fig. 6.5: This step is the core of the design method supported by the constraint model of the required capability and functionality. These models support decision-making on node sets from the node library. We use the node set to satisfy the requirement of customized capability, afterward we use the proper node set to generate feasible configuration trees. At this stage, adding or (re)moving nodes in the configuration tree is used to represent the reconfiguration of modules in the RMT configuration.
- ***Configuration Evaluation, Block C,*** Fig. 6.5: We evaluate the performance of the feasible configuration with respect to cost and reconfigurability, which helps us to make decisions on the edge set with the fixed node set. In light of the trade-offs among multiple objectives, we choose the most satisfactory RMT configuration tree. Further, this enables the selected RMT configuration to satisfy the set of operational requirements with the trade-off between minimizing cost and maximizing reconfigurability.

In this monograph, the assumption is that the RMT is equipped with multiple-spindle heads due to the hard constraints specified in the design that limits the number of spindle-head to three. Similar to assumptions in the literature (Mpofu et al. 2008; Moon and Kota 2002), we assume that there are only basic modules in the RMT configuration. The module library has a limited number of modules where different modules are matched with different functions. We adopt the following assumptions to simplify the design process.

- ***Assumption* 1**: Processing requirements are given, and the task information includes the operation feature and the operation quantity;
- ***Assumption* 2**: Spindle heads in the same configuration are independent;
- ***Assumption* 3**: Modules are assembled via the standard interface, which is beneficial for the reconfiguration smoothness;
- ***Assumption* 4**: Errors in the cost of modules has little influence on the exploration process; and
- ***Assumption* 5:** Sequence among the modules yet to be assembled, the module related to the x-axis motion has priority to be installed, the second is y-axis motion, and the third is z-axis motion and the final is the rotary motion, a spindle head.

In summary, the RMT configuration design method includes the construction of a configuration tree; see Block A in Fig. 6.5; the construction of a constraint model; see Block B in Fig. 6.5; and the construction of an objective model; see Block C in Fig. 6.5.

6.3.2 Construction of the Configuration Tree

A configuration tree is a functional model that we use to describe the RMT configuration. In the RMT configuration we use various nodes to represent the modules, and the edges to represent the assembly relationship among the modules. We identify the steps for construction of the configuration tree based on the mapping relationship between the RMT configuration and the configuration tree, where we define the node library to express the function of the modules. To quantify the design problem, it is critical to propose a mathematical expression of the configuration tree. We represent the tree using sets due to the structure of its data.

Analysis of a Configuration Tree, Fig. 6.5, *Block A, Steps A*1 *and A*2

Given that the movement between the tools and the workpiece is independent, each other, each processing operation of the machine tool can be simplified to reflect the relative movement among several tools and the workpiece. Hence, the entire configuration design of the RMT is framed by tool-related and workpiece-related configurations. The tool-related configuration of the RMT consists of modules related to tool movement. The workpiece-related configuration consists of modules related to workpiece movement. The process of constructing a configuration tree has three steps due to two sub-configurations in the RMT configuration design.

***Step* 1**: Form the tool-side branch of the tree mapped in the modules of the tool-related configuration.
***Step* 2**: Form the workpiece-side branch of the tree mapped in the modules of the workpiece-related configuration.
***Step* 3**: Form the complete tree with the workpiece-side branch and the tool-side branch. Step 3 maps the entire RMT configuration.

The cornerstone of constructing the configuration tree is the node library. To map the modules and the nodes, each node represents the carrier of a single module function. Hence, one node corresponds to a single module in the module library, while each module is also expressed as a node based on its function. Accordingly, when the number of modules in the RMT configuration increases, the number of corresponding nodes and edges in tree increases. However, the computational complexity is increased as the size of tree increases. The quantity of nodes and edges in the tree structure is limited. Considering manufacturing performance, the functions we can achieve in the machine tool is also limited. Consequently, in general, the tree can represent any normal size of the RMT configuration.

In Table 6.2, we present seven different nodes to express the procedure in detail, as an example of the RMT configuration. The nodes are divided by function into (1) spindle head, (2) fixture, (3) slider, (4) cross-slider, (5) column, (6) rotary table, and (7) base. To identify the appropriate location of a module in the assembly relationship, we define the parent-child relationship of edges in the configuration tree model. If one module makes another module move, the former is the parent node of the edge, and the latter becomes the child node.

Table 6.2 Basic node domain of RMT configuration tree

Num.	Name	Node function
1	Spindle Head	Modules with a variety of tools
2	Fixture	Modules for positioning the workpiece
3	Slider	Modules for moving tool and workpiece left and right
4	Cross-slide	Modules for moving tool and workpiece forward and backward
5	Column	Modules for moving tool and workpiece up and down
6	Rotary table	Modules for allowing tool and work-piece rotary motion
7	Base	Modules for support and connection of the modules

Table 6.3 Variable i and Corresponding Node Type (Wang, et al. 2018)

i	u_{ijk}	Node type
1	u_{1jk}	Spindle head
2	u_{2jk}	Fixture
3	u_{3jk}	Slide
4	u_{4jk}	Cross-slide
5	u_{5jk}	Column
6	u_{6jk}	Rotary table
7	u_{7jk}	Base

Model of a Configuration Tree, Fig. 6.5, *Block A, Step A3*

In the process of the RMT configuration design, a configuration tree is not directly involved as a design variable. Hence, it is necessary to build a computational model of the configuration tree; see Eq. (6.1).

$$CT = \{u, v | u \in U, v \in V\} \tag{6.1}$$

In Eq. (6.2), we present the definition of the node set U.

$$U = \left\{u_{ijk} \middle| i \in (1, \cdots, 7), j \in N, k \in 0, 1\right\} \tag{6.2}$$

If $k = 0$, the node belongs to the tool-side branch and if $k = 1$ the node belongs to the workpiece-side branch.

An edge consists of two nodes in ordered pair. Every node in the configuration tree can be described by Eq. (6.2). In Eq. (6.3), we present the edge set V.

$$V = \left\{v = u_{ijk}, u_{i'j'k'} \middle| f_1(v) = 1, u_{ijk}, u_{i'j'k'} \in U, i \neq i'\right\} \tag{6.3}$$

$$f_1\left(u_{ijk}, u_{i'j'k'}\right) = \begin{cases} 1 \; existing\, the\, edge\, of\, u_{ijk} and\, u_{i'j'k'} \\ 0 \; non - existing\, the\, edge\, of\, u_{ijk}\, and\, u_{i'j'k'} \end{cases} \tag{6.4}$$

$f_I(<u_{ijk'}\ u_{i'j'k'}>)$ is a discriminant function of the edge; see Eq. (6.4). Further, when the function f_1 equals 1, the edge between u_{ijk} and $u_{i'j'k'}$ exists in the configuration tree. On the other hand, when the function f_I equal 0, there is no edge between u_{ijk} and $u_{i'j'kp}$.

To ensure manufacturing stability, we assume the following rules for the assembly relationship among modules:

***Rule* 1**: The modules for the same motion are always installed on different branches, and there is no assembly relationship among the modules on different branches. In Eq. (6.5), we present the mathematical expression of Rule 1.

$$\begin{gathered} u_{ijk}, u_{i'j'k'} = 0 \\ \sum_{k \neq k'} \sum_{i'} \sum_{j'} u_{ijk}, u_{i'j'k'} = 0 \end{gathered} \tag{6.5}$$

***Rule* 2**: There is at most one assembly relationship between any two modules. The module cannot be assembled with itself. In Eq. (6.6), we present the Rule 2.

$$\begin{gathered} \sum_{j} u_{ijk}, u_{i'j'k'} = 0 \\ \forall i' \neq i \sum_{j'} u_{ijk}, u_{i'j'k} \leq 1 \end{gathered} \tag{6.6}$$

***Rule* 3**: The module with the tools and the module for the workpiece are always installed on different branches. In Eq. (6.7) we present the Rule 3.

$$u_{1jk}, u_{2j'k}\ and\ u_{2jk}, u_{1j'k} = 0 \tag{6.7}$$

***Rule* 4**: A module for rotary motion is always installed on a module for linear motion. In Eq. (6.8), we present the Rule 4.

$$u_{6jk}, u_{3j'k}, u_{6jk}, u_{4j'k}\ and\ u_{6jk}, u_{5j'k} = 0 \tag{6.8}$$

***Rule* 5**: For linear motion, a module for up and down motion is always installed on a module for forward and backward motion, while the module for forward and backward motion is always installed on a module for left and right motion. In Eq. (6.9), we present the Rule 5.

$$u_{4jk}, u_{3j'k}, u_{5jk}, u_{4j'k}\ and\ u_{5jk}, u_{3j'k} = 0 \tag{6.9}$$

These rules are used in constructing the configuration tree.

6.3.3 *Construction of the Constrained Model*

We construct the constraint model in a way to generate the set of feasible configuration trees, which are designed to satisfy the requirements from the operation set. The operation set, corresponding to the RMT, includes the quantity and features of the required operations. To process the operation set, the RMT is required to provide suitable capability and functionality. The RMT capability refers to the number of operations which are finished in the RMT, measured as the number of tools in the RMT configuration. The RMT functionality refers to its operational features, measured as the relative motion between the tool and the workpiece. Therefore, this process is used to establish mathematical expressions for capability and functionality, and combine the processing parameters of the operational requirements, forming the constraint model.

Determination of Reconfiguration Capability from the Configuration Tree, Fig. 6.5, *Block B, Step B*1

There is a need for one-to-one relationship between the RMT tool and the required operations, that is, need for a tool to finish one operation at time. The greater the number of tools in the RMT configuration the greater the number of operations completed concurrently, therefore, the RMT capability is higher. In the node library, the spindle head is the module referring to the tool itself. Thus, in Eq. (6.10), we present the capability.

$$CCT = \sum_{j \in N} f_2(u_{1j0}) \tag{6.10}$$

If the configuration tree CT contains $u_{1j0}, f(u_{1j0})$ is equal to 1, otherwise $f(u_{1j0})$ is equal to 0.

Determination of System Functionality from the Configuration Tree, Fig. 6.5, *Block B, Step B*2

The process of determining RMT operation is specifying the relative motion between the tool and the workpiece in the machine tool. Hence, the functionality requirement is described by relative motions. In kinematics, the degrees of freedom (DOF) is a measure used to express the motion. There are 6 axes in the DOFs to define the motion directions X-, Y-, Z-, A-, B- and C- axes; see Fig. 6.6. The X-axis is used for left and right motion, the Y-axis is used for forward and backward motion, the Z-axis is used for up and down motion, the A-, B- and C- axes is for rotary motions on the X-, Y- and Z- axes respectively.

In view of the DOF, arbitrary RMT functionality is described as a set of six motion directions. In the node library, the slide, cross-slide, column, and rotary table are all directional nodes used to influence the motion of the tool and the workpiece. In Table 6.4, we present the mapping between the directional nodes and DOF.

In Eqs. (6.11) and (6.12), we present the functionality of the RMT.

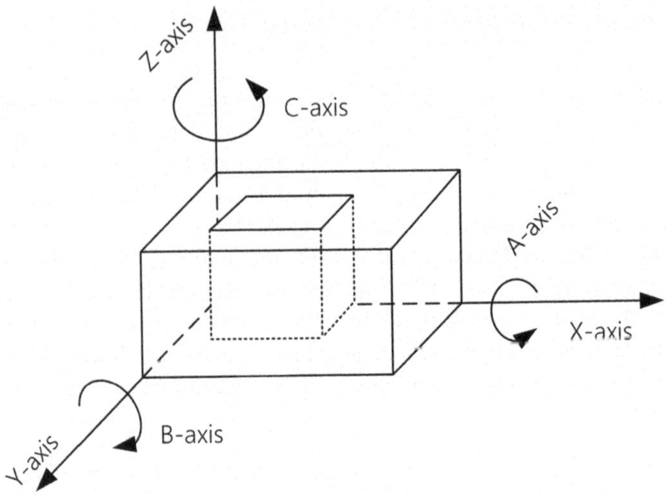

Fig. 6.6 The six axes of degrees of freedom (Wang et al. 2018)

Table 6.4 The mapping relationships among directional nodes and DOF (Wang et al. 2018)

Primitive	Linear (L) axes			Rotary (R) axes		
	X-axes	Y-axes	Z-axes	A-axes	B-axes	C-axes
Slide	1	0	0	0	0	0
Cross-slide	0	1	0	0	0	0
Column	0	0	1	0	0	0
Rotary table	0	0	0	1	1	1

$$T\left(u_{1j0}\right) = \left[t_{j1}t_{j2}t_{j3}t_{j4}t_{j5}t_{j6}\right] \tag{6.11}$$

$$H_{1x6} = [h_{11}h_{12}h_{13}h_{14}h_{15}h_{16}] \tag{6.12}$$

In the matrix T, the element t_{jr} is either 0 or 1. If the value of t_{jr} is 0, the related tool cannot realize motion in that direction. If the value of t_{jr} is 1, the motion can be realized in that direction. In matrix H, the element h_{1r} is either 0 or 1. If the value of h_{1r} is 0, the workpiece cannot realize motion in that direction. If the value of h_{1r} is 1, it can realize motion in that direction.

Constrains from the Operational Set, Fig. 6.5, *Block B, Step B3*

When the capability and functionality satisfy the requirements of the operational set, a feasible configuration tree is obtained. In terms of the operational set, we define a requirement by the number of operations, *n,* and the features of the operations. In Eq. (6.13), we present the operational requirement.

$$O_j = \left[o_{j1} o_{j2} o_{j3} o_{j4} o_{j5} o_{j6}\right] \tag{6.13}$$

If we need the motion in this direction for the operation, then the related element a_{jr} is equal to 1, otherwise it is equal to 0.

The constraint on the capability is the number of spindle heads in the configuration tree that must be equal to the number of operations performed. The constraint on functionality is that the motion realized by the direction nodes in the configuration tree must agrees with the motion required by the operational features. In Eqs. (6.14) and (6.15), we present the constraint model, which is used to design the feasible configuration tree.

$$CCT = n \tag{6.14}$$

$$FCT = t_{lr} \oplus h_{1r} = o_{lr} \tag{6.15}$$

If the RMT needs to be reconfigured but the operational requirement (for example, the features or quantity) are unknown, then the parameters on the right side of the constraint model are undefined. Because future reconfiguration is unpredictable, the RMT is required to be flexible enough to provide the required capability and functionality. Therefore, for incomplete operational requirement information, the constraint model must be modified. There are three different modes to address the incomplete information.

Mode 1: The feature O_j is unknown but the operation quantity n is given. In Eqs. (6.16) and (6.17), we present the corresponding modified constraint model.

$$CCT = n \tag{6.16}$$

$$\min FCT' = \sum_{l=n} \sum_{r=6} (1 - (t_{lr} \oplus h_{1r})) \tag{6.17}$$

The closer $t_{lr} \oplus h_{1r}$ is to 1, the smaller is the value of FCT'.

Mode 2: The feature O_j is given but the operation quantity n is unknown. In Eqs. (6.18) and (6.19), we present the corresponding modified constraint model.

$$CCT \leq n_{\max} \tag{6.18}$$

$$FCT = t_{lr} \oplus h_{1r} = o_{lr} \tag{6.19}$$

Mode 3: Both the feature O_j and the operation quantity n are unknown. In Eqs. (6.20) and (6.21), we present the corresponding modified constraint model.

$$CCT \leq n_{\max} \tag{6.20}$$

$$\min FCT' = \sum_{l=n} \sum_{r=6} (1 - (t_{lr} \oplus h_{1r})) \tag{6.21}$$

When the features and quantity of the required operations are known, then we use Eqs. (6.14) and (6.15) as the constraint model. When at least one of the features and quantities is unknown, the constraint model is modified into the corresponding model.

6.3.4 Construction of the Objective Model

Efficiency is the main prerequisite for ensuring the widespread use of configuration trees. After generating the set of feasible configuration trees, based on the performance, we select the most satisfactory configuration tree. Hence, in the selection process a decision making regarding the edge set of the configuration tree to solve trade-offs among multiple objectives is required. Compared with a conventional machine tool design process, the RMT is cost-effective with efficient reconfigurability. Cost-effective reconfiguration implies reduction in expenditure as much as possible in the reconfiguration process. Efficient reconfiguration requires that the modules in the RMT configuration are as reusable as possible in the reconfiguration process.

Cost of the Configuration Tree, Fig. 6.5, *Block C, Step C*1

In the reconfiguration process, the alteration of edges leads to the addition, deletion, or moving the nodes. Based on node movement, the cost is divided into fixed and assembly costs. Fixed cost refers to the cost due to adding a node to the RMT configuration tree and assembly cost refers to the cost of altering a node in the RMT configuration tree. Assembly cost is directly proportional to the number of node alterations. When the edge between two nodes is altered, the "number of node alterations" increases by 1. When a node is moved between different branches or deleted from the configuration, the labor loss and the installation time increases, so the "number of node alterations" increases by 2. Hence, in Eq. (6.22), we present the quantitative model of the cost for an RMT configuration design.

$$ECT = \sum_{u \in U_i'} g(u) + \sum_{u \in U_i'} q(u) \times s(u) \tag{6.22}$$

Reconfigurability of the Configuration Tree, Fig. 6.5, *Block C, Step C*2

Reconfigurability refers to the degree of difficulty to reconfigure the configuration. If more edges are altered from the previous to the next configuration tree it is more difficult to complete the reconfiguration. Reconfigurability is affected by the similarity and sustainability of configurations due to reconfiguration relationship between two consecutive RMT configurations. Configuration similarity refers to the utilization of the edges of the previous configuration tree in the current configuration tree. A higher utilization has a positive impact on the reconfigurability of the configuration tree. Configuration sustainability refers to the probability that the edges of the current configuration tree are the same to those of the next configuration tree. A higher probability has a positive impact on reconfigurability of the configuration tree. In Eqs. (6.23)–(6.25), we present the quantitative models for the reconfigurability of the RMT configuration tree.

$$RCT = UCT + PCT \tag{6.23}$$

$$UCT = \frac{card(V')}{card(V)} \tag{6.24}$$

$$PCT = \sum_{v \in V} p(v) \tag{6.25}$$

Higher values of RCT indicate a greater reconfigurability for the RMT configuration tree. Each branch of the RMT configuration tree allows up to four DOFs. Different DOF sets correspond to different node sets, which corresponds to different edge sets. Based on the combinations of six DOFs, a configuration tree has 15 candidate edge set schemes; see Table 6.5. The probability that edge v appears, $h(v)$, is equal to the frequency at which this edge appears in the 15 candidate edge sets.

Decision-Making for Multiple Performance Measures, Fig. 6.5, *Block C, Step C3*

Table 6.5 The Value of $p(v)$ (Wang et al. 2018)

#	Edge subset v	$p(v)$
1	$<u_{7jk}, u_{3jk}>$	8/15
2	$<u_{7jk}, u_{4jk}>$	4/15
3	$<u_{7jk}, u_{5jk}>$	2/15
4	$<u_{7jk}, u_{5jk}>$	1/15
5	$<u_{7jk}, u_{3jk}>, <u_{3jk}, u_{4jk}>$	4/15
6	$<u_{7jk}, u_{3jk}>, <u_{3jk}, u_{5jk}>$	2/15
7	$<u_{7jk}, u_{3jk}>, <u_{3jk}, u_{5jk}>$	1/15
8	$<u_{7jk}, u_{4jk}>, <u_{4jk}, u_{5jk}>$	4/15
9	$<u_{7jk}, u_{4jk}>, <u_{4jk}, u_{5jk}>$	2/15
10	$<u_{7jk}, u_{5jk}>, <u_{5jk}, u_{5jk}>$	4/15
11	$<u_{7jk}, u_{3jk}>, <u_{3jk}, u_{4jk}>, <u_{4jk}, u_{5jk}>$	2/15
12	$<u_{7jk}, u_{3jk}>, <u_{3jk}, u_{4jk}>, <u_{4jk}, u_{5jk}>$	1/15
13	$<u_{7jk}, u_{3jk}>, <u_{3jk}, u_{5jk}>, <u_{5jk}, u_{5jk}>$	1/15
14	$<u_{7jk}, u_{4jk}>, <u_{4jk}, u_{5jk}>, <u_{5jk}, u_{5jk}>$	2/15
15	$<u_{7jk}, u_{3jk}>, <u_{3jk}, u_{4jk}>, <u_{4jk}, u_{5jk}>, <u_{5jk}, u_{5jk}>$	1/15

The decision-making process for the edges of the RMT configuration tree uses weighted objectives, minimize cost while maximizing reconfigurability. To give a uniform format for the objective functions, the expressions for the capability and reconfigurability are normalized. Further, we introduce a deviation variable to describe how much the proposed scheme deviates from the most satisfactory value of the objective. In Eqs. (6.26) and (6.27), we present the method of developing expressions for cost (ECT), and reconfigurability (RCT).

$$\frac{\min(ECT)}{ECT} + d_{m1} = 1 \tag{6.26}$$

$$\frac{RCT}{\max(RCT)} + d_{m2} = 1 \tag{6.27}$$

The most satisfactory configuration tree has the smallest sum of deviations. Furthermore, based on the actual production processing, designer's focus on different aspects of performance. When we use RMT for processing various small batches of parts, then we emphasize reconfigurability. When we use the RMT for processing large batches of parts, then we emphasis cost. Thus, we introduce a weighting coefficient to signify priorities. In Eq. (6.28), we present the decision-making model in order to select the satisfactory configuration tree.

$$minI = w_1 d_{m1} + w_2 d_{m2} \tag{6.28}$$

The value of index I represents the standard for making the decision, when this value is close to 0, the corresponding scheme has a smaller deviation from the target value. Weights are determined by experts based on experience, where the sum of w_1 and w_2 is equal to one. Further, when it is preferred to maintain product diversity, the value of w_2 is greater than w_1, and when it is preferred to reduce cost, the value of w_1 is greater than w_2.

6.4 Designing a Reconfigurable Inspection System (RIS)—An Example

In this section, we illustrate the design method for a Reconfigurable Inspection System (RIS) by an example. In Sect. 6.4.1, we present the development of RIS configuration design. In Sect. 6.4.2, we present the development of specific model-based procedure for RIS configuration design.

6.4.1 Development of the RIS Configuration Design

The manufacturing process in the RMS consists of multiple stages; see Fig. 6.7. Workpieces are machined into products through multiple operations. In order to ensure excellent quality of the product there are four types of components (1) operational stations, (2) inspection stations, (3) delivery system, and (4) return system. The operational stations (i.e., Stations 1, 2, 4, …, n + k − 1) perform one or more operations via RMT, while the inspection stations (i.e., Stations 3, …, n + k) measure the product quality during the process via RIS. The workpiece is transferred from one station to the next using the delivery system. When errors are detected by the inspection stations, the return system is used to return the workpiece to related upstream operational stations. Therefore, the RIS in the kth inspection station focuses on detecting its upstream operational stations, which are behind the k − 1th inspection station.

In conventional manufacturing systems, the inspection station is usually at the end of the line. End-of-line detection is a popular and cost-effective approach in the industry. However, with this approach we can only detect some of the errors and due to delayed detection, the quality of product is reduced. On the other hand, in saturated detection each operational station is followed by an inspection station that results in high quality of product. However, this approach is expensive. Hence, we need to select minimum but sufficient number of inspection stations to monitor the manufacturing process. Hence, configuration design of the RIS is an attempt to make a trade-off between cost and diagnosability in the RMS's manufacturing process.

Detection Mechanism via RIS

In the RMS, the manufacturing process includes multiple operations, and the quality of the product is affected by the processing parameters of the operational stations. During the RMS's lifecycle, deviations in the processing parameters leads to variations in the product quality. Variations can propagate and accumulate among the stations. The influences of the different stations on the product quality is interrelated, and this interrelatedness creates a complex flow of information through the manufacturing process. To describe the information flow, Stream of Variation (SoV)

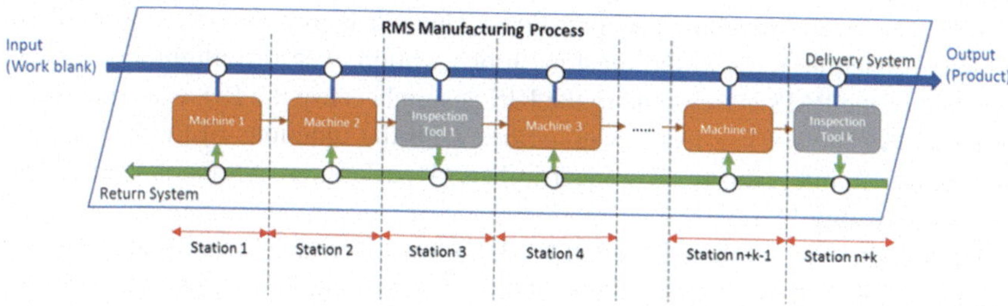

Fig. 6.7 Manufacturing process in the RMS

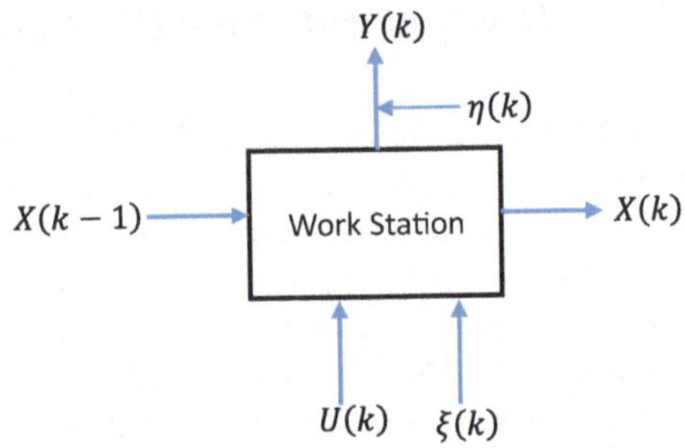

Fig. 6.8 Information flow at one station based on SoV theory, (Shang et al. under review)

theory is used (Shang et al. 2018). This theory clearly models the accumulation and propagation of errors in the production process; see Fig. 6.8.

Identifying the root cause of errors using SoV theory relies on the RIS configuration design as a basis for error diagnosis.

The essence of identifying root causes via SoV theory is to simulate the manufacturing process; see Eq. (2.1). A product quality error refers to a deviation of the actual position from the ideal position, and the main types of error include fixturing error caused by imperfect locators, datum errors caused by errors in the datum feature, and machine tool errors caused by loops in the machine's structure. Fixture and machine tool errors are part of the current station, whereas datum errors are influenced by other stations. Therefore, a virtual station is simulated using data from datum measurements. If the virtual station is in the same state as the corresponding real station, the real station is of high quality, otherwise, the station should be maintained or updated. When the SoV is used to identify the root cause of errors, it is paramount that the quality features of the product and the datum of the processing feature are detected.

Key Features of the RIS for Reconfiguration

In dynamic manufacturing process, the main function of the RIS is to ensure the quality of the final product. In order to do so, the RIS should be capable of quickly adjusting its detection functionality and capability. Detection functionality refers to the product features of the RIS that can be detected, and detection capability refers to the RIS's sensitivity to product errors. The RIS divides the manufacturing process into several machine tool partitions. The RIS is more sensitive when there are fewer machine tools in the machine tool partition. The reconfiguration design principle of the RMS is available to the RIS design because the RIS is a component of the RMS. The six key features combined with the functions of the RIS are (1) modularity, (2) integrability, (3) convertibility, (4) scalability, (5) customization, and (6) diagnosability.

Modularity: We combine different RIMs to form diverse RIS configurations and reconfigure the RIS by changing the RIM layout. For example, we can achieve different detection functionality and capability of the RIS by changing the number of RIM and their positions, and different detection tasks in the RIM by changing sensors. However, an excessive number of RIM in the RIS can result in high management costs and increase the system's ramp-up time. Similarly, having too many sensors in the RIM increases the structural complexity and decreases the equipment's reliability. Furthermore, the modularity of the RIS requires determination of a reasonable assignment of detection tasks and selection of the minimum sufficient number of RIM and sensors. We assume that one processing feature can be detected by each sensor. Based on the above description of the detection mechanism, the number of products is related to each process and datum feature. The number of sensors in the RIM is proportional to the processing characteristics measured by the RIM.

Integrability: Standardization of interfaces in the RIM is beneficial for the introduction of the modern technologies or new devices. When the number of sensors in the RIM is small and the number of ideal interfaces is large, the RIS has a greater configuration flexibility. The higher flexibility of the RIS, the lower is the reconstruction time and cost. Hence, in order to achieve integrability we need to determine a reasonable sensor density and select the corresponding number of sensors in the RIS. In addition, the ease of RIM reconfiguration is governed by the degree of sensor change. Therefore, the reconfiguration time and cost are proportional to the number of sensor changes.

Convertibility: It is the ***c***apability of the RIS to quickly adjust its detection functionality based on the process route. We can change the RIS's detection functionality in order to satisfy the requirements of different parts or part families. The required detection functionality in the RIS we provide by adding, deleting, or replacing sensors. The product quality is ensured if the RIS can detect all the product features. Accordingly, different product features alter the RIS's detection functionality. Furthermore, convertibility enables the RIS to detect the manufacturing process by selecting a sufficient number of RIM and sensors. Further, the number of sensors should be greater than the number of detected product features, which ensures that dimensional errors in each machine tool are detected. At the same time, when the product features change, we consider changing the number of sensors to change the RIS's detection functionality.

Scalability: It is the capability of the RIS to rapidly adjust its detection according to the process route. That is, we adjust the RIM's position according to the processing route to detect each operation in a timely manner. Timely detection of the operation refers to direct monitoring of the process as much as possible. Conversely, if the distance between the RIM and the operational station is greater, the RIM cannot detect the quality error in time, which results in unnecessary waste. Therefore, scalability requires the RIS to locate the RIM close to the operation. As the operation changes, the position of the RIM changes, minimizing the distance between the RIM and the operations.

Customization: The design of the RIS configuration is based on the quality detection required by a given process route while minimizing the redundant detection functionality and detection capabilities. Customizability is measured by its device utilization. The RIS is more customizable for higher device utilization rates. Therefore, the customizability of the RIS requires that it maximize its utilization of equipment to select the appropriate number of RIM for the required number of sensors. As the process route changes, the RIS adds the required equipment while removing unnecessary equipment to avoid redundancy.

Diagnosability: Validity and timeliness of the data collected by the RIS determines whether the data collected meets the needs of the root cause diagnosis process. The validity of data refers to the degree of influence of the device reliability on acquired data. The timeliness of data refers to the duration of the fault from the moment we generate data for detection. Therefore, diagnosability of the RIS requires to select enough RIM and sensors based on device reliability and the detection range while installing each RIM in a reasonable location. If the complexity of the RIS's device structure is lower, the device reliability is higher. As a result, the impact of the device on data acquisition is smaller, which makes the data more efficient. If the RIS's equipment distribution density is larger, the detection range is smaller. As a result, the data are available sooner as the device collects them faster.

Analysis of the key features shows that diagnosability is the core of the six key RIS's features, and the remaining five are designed to support diagnosability. In summary, the essence of the RIS's configuration design is to provide quality detection in the reconfiguration process through the number of RIM, their locations, and the number of sensors, where a satisfactory configuration has all the key features, especially diagnosability.

Overview of RIS Configuration Design

In RMS, the use of RIS and RIM to collect data provides an avenue for implementing quality assurance, process testing, and fault diagnosis. Based on the analysis in the Sect. 6.3, the capability of the RIS to collect data depends on its configuration design. Currently, other authors mainly focus on the detection of faults or the architecture of the control system. There is a lack of information on the physical configuration design of the detection system. Hence, we present a key-feature-based method for RIS configuration design. The method has three steps.

Step* 1**: ***Building RIS Configuration Design Model. In Sect. 6.3.2, we present the basis of the proposed method. Due to the basic components of the RIS, the design variables are defined as the number of RIM, their positions, and the number of sensors. The design variables are mathematically described using vectors. Further, the process route is analyzed, a computational model is constructed, and the configuration's design parameters are extracted.

Step* 2**: ***Exploring Feasible RIS Configuration Schemes. In Sect. 6.4.2, we present the core of the proposed method. The fundamental goal of the RIS is to detect the processing route. Further, in the process of RIS configuration design, the constraints of the configuration design are modeled according to the RIS's detection capability and detection functionality for the process route. Feasible RIS configuration schemes are generated.

Step* 3**: ***Decision-making based on Satisfactory RMT Configuration Scheme. In Sect. 6.5, we present the key to the proposed method. The final goal of the RIS is to increase the efficiency of the detection process. Therefore, we use cost and diagnosability to evaluate each RIS configuration, and we select the most satisfactory configuration design with the goal of minimizing cost and maximizing diagnosability.

6.4.2 A Method for Designing RIS Configuration

The RIS is composed of multiple RIM, as we indicated in the proceeding sections. Different numbers or different positions of RIMs generate different RIS configurations, where each RIM is composed of multiple sensors. With different sensors we can achieve different detection functionalities and capabilities, which affects the RIS's detection process. Hence, the RIS configuration design variables include the number and positions of the RIM, and the number of sensors. In Eqs. (6.29) and (6.30), we present a computational model to describe the RIS configuration.

$$CRIS = \{cr_1, cr_2, L, cr_M\} \tag{6.29}$$

$$cr_i = (y_i, x_i) \tag{6.30}$$

In the RIS configuration design in view of the importance of product features to product quality it is necessary to define a processing route. In Eq. (6.31), we present a better description of the RIS design process, based on a set of mathematical descriptions of the process.

$$C_p = \{cp_1, cp_2, L, cp_N\} \tag{6.31}$$

In the following section, we present constraints and goals of the model based on the model of the RMS configuration and the processing route.

Modeling RIS Constraint Design

We use the RIS configuration constraint model to explore the feasible domain of the RIS configuration design, in order to detect product quality based on the processing route. Quality detection is related to the ability to inspect errors and identify their root cause, which is a hard constraint for RIS configuration design. With the RIS we can successfully achieve a quality detection process if and only if the RIS's detection capabilities satisfy the processing route's quality requirements. The process of designing a feasible RIS configuration is as follows (1) establish a computational model with the detection functionality and capabilities necessary to create a feasible model of the domain design constraints, and (2) extract the routing parameters of the design process within the constraint model to obtain all feasible RIS configuration designs.

Construction of Detection Functionality Model

Detection tools in the RIS are sensors. Each sensor detects the operational feature. We can detect more operational features when there are more sensors, which improves the detection functionality. Detection functionality is expressed by the number of sensors; see Eq. (6.32).

$$FCRIS = \sum_{i=1}^{M} x_i \tag{6.32}$$

Construction of Detection Capability Model

We present detection capability of the RIS by RIS's sensitivity to quality errors. The sensitivity to errors is essentially the time at which the RIS detects a quality error after it occurs. We assume that the delivery time for the product in the system is constant. The length of time before a fault is detected is expressed as the distance from the fault station to its nearest detection machine. In Eq. (6.33), we present the detection capability of the RIS.

$$CCRIS = \frac{\sum_{i=1}^{m} \sum_{j=y_{i-1}+1}^{j=y_i} (y_i - j)}{\frac{(N-1)\times N}{2}} \tag{6.33}$$

CCRIS value is between 0 and 1. When *CCRIS* is closer to 0, the detection capability of the RIS is higher.

Modeling Constraints with Detection Functionality and Capability

RIS configuration design is feasible when the detection functionality and capability satisfy the product quality requirements. The requirements for detecting the processing route are (1) the number of processes N, (2) the characteristic surface of the product $FCP = (f_1, f_2, \cdots, f_{CM})$, (3) the number of characteristic surfaces of the product CM, and (4) the fault diagnosis distance T. The fault diagnosis distance T is the least upper bound on the distance between the fault station and its corresponding inspection station. The constraint model of the RIS configuration design is discussed based on the two aspects of detection functionality and detection capability. On the one hand, the number of processes and the surface of the product features in the processing route are constrained by the RIS's detection functionality. Furthermore, this indicates that the number of sensors in the RIS configuration is not less than the number of product features involved in the process. On the other hand, the number of processes in the processing route and the fault diagnosis distance constrain the RIS's detection capability. That is, with the position of the RIM we determine the distance available for detecting the error. In Eq. (6.34), we present the constraint model that we use to design feasible RIS configurations.

$$\begin{cases} FCRIS \geq \sum_{i=1}^{N} \sum_{j=1}^{CM} H(cp_i, f_j) \times \left(\sum_{k=1, k\neq j}^{CM} G(f_k) + 1 \right) \\ CCRIS \leq \frac{D(T)\times INT\left(\frac{N}{T}\right) + D\left(N - T\times INT\left(\frac{N}{T}\right)\right)}{D(N)} \\ 1 \leq M \leq N \end{cases} \tag{6.34}$$

If we machine the surface f_j operation cp_i, then, $H(cp_i, f_j)$ is equal to 1, otherwise, it is equal to 0. If surface f_k is the datum on surface f_j, then, $G(f_k)$ is equal to 1, otherwise, it is equal to 0.

Modeling the RIS Design Goal

The low cost and high diagnosability of the RIS configuration is the motivation for improving the RMS's efficiency and reducing ramp-up time. We can reduce the cost of the RIS design by minimizing design costs and resource waste, such as management costs, scrap costs, etc. The high diagnosability of the RIS design refers to maximized data collection performance, including timeliness, comprehensiveness, and effectiveness. Hence, we consider two goals, cost and diagnosability, in evaluating configurations. In addition, each decision about the RIS configuration design is weighed against the multi-goal conflict process. The goal is to analyze and establish the cost and diagnosability models for the RIS configuration design. Further, a design decision model is established based on the two mentioned quantitative models. We identify a satisfactory RIS configuration by solving these models.

Construction of the Cost Model

Different RIS configurations have different numbers of RIM and sensors, resulting in different design costs. The design cost is the cost of adding RIM and sensors to the RIS configuration, related to the number of new additions to the configuration design. As the RIS's product inspection becomes increasingly accurate, more RIM and sensors are needed in the RIS design, and the cost of the RIS design increases. Increases in the number of sensors in the RIS design the design flexibility increases and decreases the flexibility of the RIM design. In Eq. (6.35), we present a quantitative model of the costs of RIM configuration.

$$ECRIS = \sum_{i=1}^{M} (c_1 + x_i \times c_2) \tag{6.35}$$

Construction of the Diagnosability Model

Diagnosability is crucial to RIS configuration design, and its goal is to ensure the validity and timeliness of collecting process data. In the RIS, both the RIM and the sensors have certain levels of reliability. Only when a sensor is in its normal state can data be collected and used in fault diagnosis, otherwise, the data are invalid. On the other hand, when an error cannot be identified in time, the system produces products of poor quality. The later the error is detected the more serious the damage to the quality of product. The degree of damage to the system is defined as the number of workstations traversed from the failed process to the test equipment. The timeliness of the data is defined as the average of the reciprocal of the maximum degree of damage to the system for each line segment. When this value is close to 1 than the data is well-timed and there is less damage to the system. In Eqs. (6.36)–(6.38), we present the RIS constructs that are the quantitative models of the diagnosability.

$$RCRIS = XCRIS \times TCRIS \tag{6.36}$$

$$XCRIS = \frac{\sum_{i=1}^{M} p(x_1 + x_2 + \cdots + x_M)}{M} \tag{6.37}$$

$$TCRIS = \frac{\sum_{i=1}^{M} \frac{1}{(y_i - y_{i-1})}}{M} \tag{6.38}$$

Modeling Goals with Cost and Diagnosability

The decision-making process for the RIS configuration design is based on weighted goals in order to minimize cost and maximize diagnosability. When we weigh cost to be low then we compromise the RIS diagnosability, and RMT cost increase in terms of diagnosability. Hence, we can conclude that cost and diagnosability are conflicting goals. The decision-making process for the optimal design involves a trade-off between the two goals of cost and diagnosability. To give the goal functions uniform format, the data for the cost and diagnosability are normalized, and a deviation variable is introduced to describe how much the proposed scheme deviates from the most satisfactory value of the goal. In Eqs. (6.39) and (6.40), we present the cost and diagnosability.

$$\frac{\min(ECRIS)}{ECRIS} + d_{g1} = 1 \tag{6.39}$$

$$\frac{RCRIS}{max(RCRIS)} + d_{g2} = 1 \tag{6.40}$$

A feasible configuration design of the RIS tradeoff between cost and diagnosability. That is, we minimize the sum of the cost and diagnosability deviations. However, in actual production situations, designers focus on different levels of cost and diagnosability. Therefore, we introduce the weight coefficients to represent the priority of the target. In Eq. (6.41), we present the RIS configuration that uses the weight coefficient to select the goal.

$$minI = a_1 d_{g1} + a_2 d_{g2} \tag{6.41}$$

The sum of a_1 and a_2 is one. When we prefer maintaining product quality, a_2 is greater than a_1. When we prefer cost reduction, a_1 is greater than a_2.

6.5 A Method for Designing a Reconfigurable Manufacturing System (RMS)

In Sect. 6.5.1, we present the multi-granularity configuration in the RMS. In Sect. 6.5.2, we study the reconfiguration process using the proposed method. In Sect. 6.5.3, we present the RMT and RIS compromise Decision Support Problem (cDSP) models for decision-making in the RMS configuration design. In Sect. 6.5.4, we present the RMT and RIS models for studying the interactions in the RMS configuration design. In Sect. 6.5, we present a decision network for exploring the RMS reconfiguration method.

6.5.1 Multi-granular Configurations in RMS

The RMS is a multi-stage manufacturing process consisting of several stations in series, each station performing a set of operations. Manufacturing enterprises require these RMS to be cost-effective, scalable, and adaptable to meet changing requirements. The main elements of the RMS are RMT and RIM. RMT plays the role of manufacturing the blank or the intermediate product, while RIM performs the quality inspection on the intermediate product or final product. RMT and RIM consist of modular components, such as mechanical and sensing modules. On the other hand, the entire RMT in RMS is considered to be a subsystem that performs customized production, and is called a Reconfigurable Production Subsystem (RPSU). The RIMs are formed into another subsystem, called a Reconfigurable Inspection Systems (RIS), and is used for providing full real-time fault detection. Since the RMS configuration is multi-granular, the reconfiguration takes place on two different scales, namely, time and space. Along the lines of the discussion in Sect. 6.1, RMS configuration design can be divided into (1) module granularity, (2) equipment granularity, and (3) system granularity.

To meet the dynamic requirements, we propose reconfiguration on different scale and the consideration of configuration designs of different granularities. Smaller granularity of the configuration indicates greater emphasize on exploration for the ability to grasp details. In the bottom-up design principle, the manufacturing system is usually assembled from MGD to SGD followed by device assembly, subsystem assembly, and then integrating all the subsystem assemblies into a whole. In design of the manufacturing process, the goal is for each device to complete its corresponding operation task. The goal for the entire system is to complete the production and inspection of the manufactured product for diverse part families. In addition, the production or detection of subsystems depends on the functions and capabilities of the corresponding devices. The production and detection functions of a system depend on the functionality and capabilities of its subsystems. It can be seen that there is an interactive relationship between different granular configurations, especially the match between production and detection. The goal of multi-granularity configuration analysis is to integrate the different granularity configuration designs on the basis of interaction relationship to realize the overall design.

The RMS's single granular configuration design includes three steps (1) determination of the design goal, (2) analysis of the performance, and (3) exploration of the solution space. The design goal of the microscopic granularity depends on the macroscopic granular configuration design scheme. Further, in the microscopic granular configuration design we can adjust the design variables of the macroscopic granular configuration design and support the macroscopic granular configuration analysis. When the system receives a new order, the designer formulates the product design parameters according to the customer's requirements, and analyzes the functions and capabilities required for the manufacturing system. Thus, we explore the preliminary RMS configuration for SGD. According to the RMS configuration scheme, we analyze the specific production process and the requirements of the inspection process, where we explore the design of the production subsystem and the detection subsystem. The production subsystem configuration determines the layout of the RMT and their specific operation tasks. Based on the operation tasks, we can identify the configuration of each RMT. Similarly, we can determine the RIM layout and its detection tasks based on the configuration of the detection subsystem. We determine the RIM configuration according to the detection tasks. However, when we reconfigure a certain granular configuration its functions and capabilities change, which affects the performance of the macroscopic granularity. Hence, we need a new optimized design of the macroscopic particle size configuration. Furthermore, to meet the reconstruction process of the RMS there is a need to not only support single-granularity configuration design, but to also support interactions between configurations of different granularities. In this monograph, we define the reconfiguration method of the RMS multi-granularity configuration as the integration of the RMT, related to the production performance, and the RIS configuration design, related to the detection performance.

6.5.2 The Workflow of the Proposed Method

The RMS reconstruction process is a complex problem due to the multi-granular characteristics. The design process requires expert knowledge in manufacturing to support and adjust mechanical designs. The process is designed to meet the production and inspection requirements at different granularities. However, repeated iterations and changes may result in reduced work efficiency. The key to the RMS reconfiguration is to consider the configuration, design process of cooperation, and cooperation between different granularities. Implementing a concurrent design of multi-granular configurations requires fulfilling the following three main requirements.

Step A*: *Information Collection. During the design process, we need to exchange and integrate information from different design activities, and complete the decision-making based on the comprehensive design information. However, the RMS refactoring involves different granularity or design of production and inspection, leading to diversified knowledge information. Hence, in system design there is a need for a standardized model to promote parallel design among multiple activities.

Step B*: *Activity Interaction. In a design process there is interaction between different design activities. Furthermore, design decision of each activity may affect the decisions of other activities resulting in different design overall performance. In system design, simulation is needed to study the interaction between different granular configurations of the RMS. The simulations can help reduce unnecessary steps in the optimization process.

Step C: Explore the Overall Design Plan. To integrate the collection of information and the activity interaction model we propose to complete them based on their interaction. Hence, in system design there is a need to construct a decision network description of the RMS configuration design and realize the integration of the activity model, Steps 1 and 2 in the reconstruction process.

The above problem can be addressed by the proposed method for exploration of the RMS reconfiguration method. The method consists of three steps:

Build the Decision Models, Step A, Fig. 6.9: In Step A, we discuss the key issues of the RMT and the RIS configuration decision-making configuration. We present the decision-making models as cDSPs in order to manage the decision models.

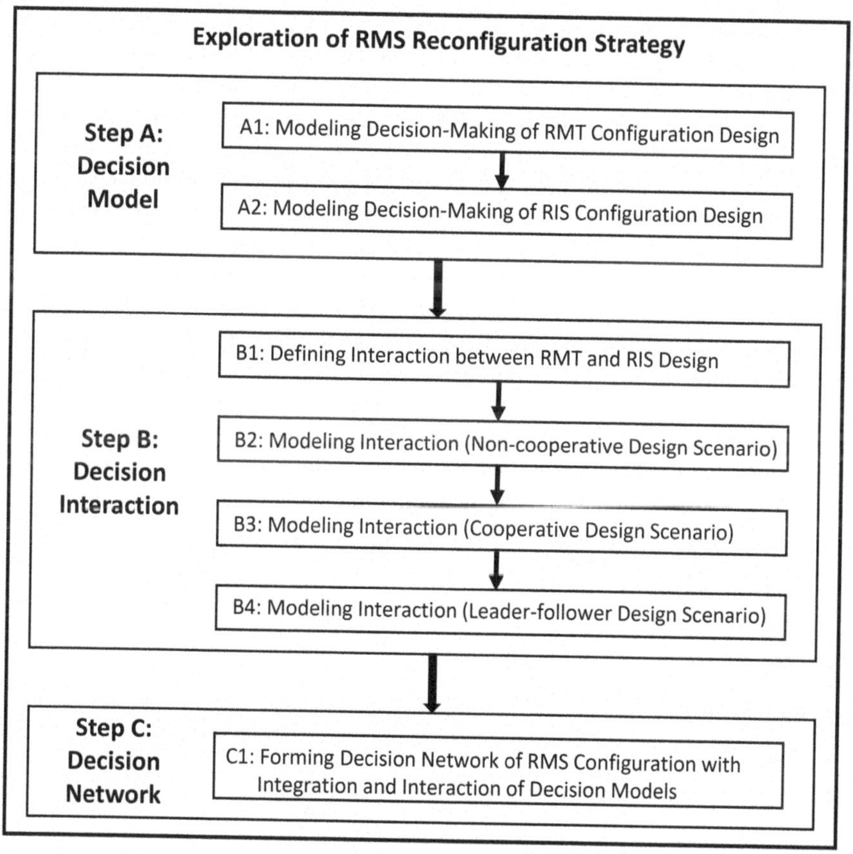

Fig. 6.9 The proposed specific process to explore the RMS reconfiguration method

Describe the Decision Interaction, Step B, Fig. 6.9: In Step A we present three types of interactions between the decision-making models (1) weak-weak relationships, (2) strong-strong relationships, and (3) strong-weak relationships. In Step B, we use the corresponding models in the game theory to describe different interactions.

Form the Decision Network, Step C, Fig. 6.9: In Step C, we present the decision network for exploring the RMS reconfiguration method according to the decision-making models and their interactions. The core of the network is the integration of decision-making and the consideration of interactions.

6.5.3 *The cDSP for Decision-Making in the RMS Configuration Design*

In this section, we focus on the RMT and the RIM configuration design in the exploration of the RMS reconfiguration method. The procedures of the RMT and the RIS configuration design are discussed by Wang and Shang (Wang et al. 2018; Shang et al. 2018a, b). The main tasks in the RMT configuration design is the selection of the appropriate module and its assembly relationship. Feasible solution should have available capability and function to satisfy the given process requirements while minimizing the cost and maximizing the reconfigurability. Thus, the variables in the RMT configuration design are the module and assembly relationships. The feasible design space is bounded by capability and functionality constraints. The requirements are the RMT's capability needs to match the number of operations, and the RMT's functionality needs to manufacture all the operating features. In the end, we select the final design from the feasible design space, which can be reconfigured via adding, or (re)moving modules in a cost-effective and utilization-effective manner. In addition, in view of the systemic exploration for the RMS reconfiguration, in the design process of the RMT we need to consider issues such as routing arrangements and quality inspections. Therefore, the decision-making of the RMT configuration we need to take RIM's design information into account. In Table 6.6, we present the decision-making model for the RMT.

Table 6.6 The cDSP construct of the RMT configuration design

Given		
Processing route	$C_p = \{C_{p1}, \cdots, C_{pN_{cp}}\}$	Fixed
Operation features	$O_{cpi} = \{O_{i1}, \cdots, O_{i6}\}$	Fixed
Limited number of workstations	N_s	Fixed
Module library	u_{ijr}	Fixed
Economic investment	E	Fixed

(continued)

In weak-weak interactions between the RMT and the RIS configuration design, the information exchange between two decisions does not exist. In order to maximize the comprehensive performance, a non-cooperative model in game theory is chosen to explore the solution space. In strong-strong interaction between the RMT and the RIS, the two teams cooperate and share information. At this point, the cooperation model in game theory is selected to explore the solution space. In strong-weak interaction between the RMT and the RIS, the design of the RMT configuration has precedence over the design of the RIS configuration. In this case, we choose the leader-follower model in game theory to explore the solution space.

The specific procedures of the solution algorithm of the above three scenarios we present in the following sections.

Scenario 1: Solution Space Exploration in Weak-Weak Interaction

In Fig. 6.10, we present Scenario 1, which is the solution space exploration for weak-weak interaction. Solution space exploration in Scenario 1 consists of four steps.

Step 1: Identify the design information, such as variables, constraints, and targets of the RMT and the RIS configuration design.
Step 2: Establish the cDSP models for the RMT and the RIS configuration design according to the design information.
Step 3: Take each other's feasible design space as the starting parameters in the decision model. The specific parameter value we define to solve the decision model and finally obtain the corresponding solution space.
Step 4: The point selected in the intersection of the two solution spaces in Step 3 is the satisfactory solution to the RMS systemic reconfiguration method.

Scenario 2: Solution Space Exploration in Strong-Strong Interaction

In Fig. 6.11, we present the Scenario 2, which is the solution space exploration for the strong-strong interaction. Solution space exploration in Scenario 2 consists of three steps.

Step 1: Identify the design information such as variables, constraints, and targets of the RMT and the RIS configuration design.
Step 2: Establish the cDSP models for the RMT and the RIS configuration design according to the design information.
Step 3: Combine the cDSP models for the RMT and the RIS configuration design into the cDSP model for the RMS reconfiguration design and explore the solution for the reconfiguration method.

Scenario 3: Solution Space Exploration in Strong-Weak Interaction

In Fig. 6.12, we present the Scenario 3, which is the solution space exploration in strong-weak interaction. Solution space exploration in Scenario 3 consists of four steps.

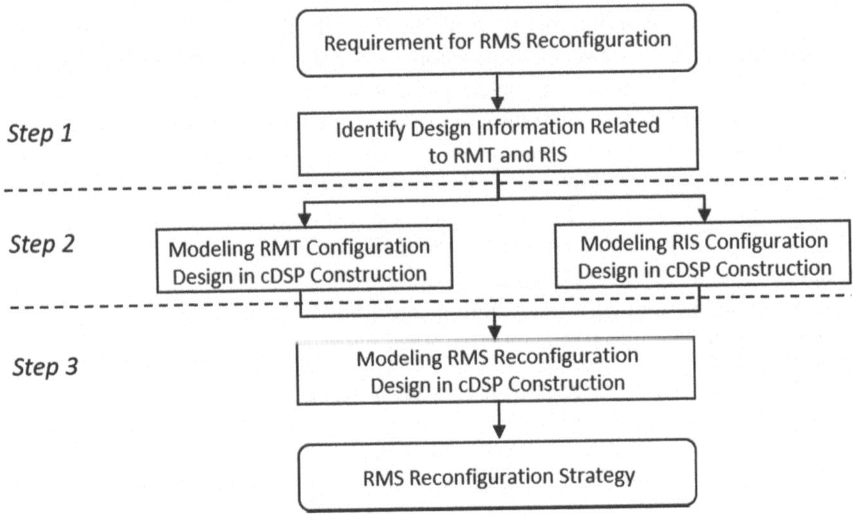

Fig. 6.11 Solution space exploration in strong-strong interaction (Shang et al. under review)

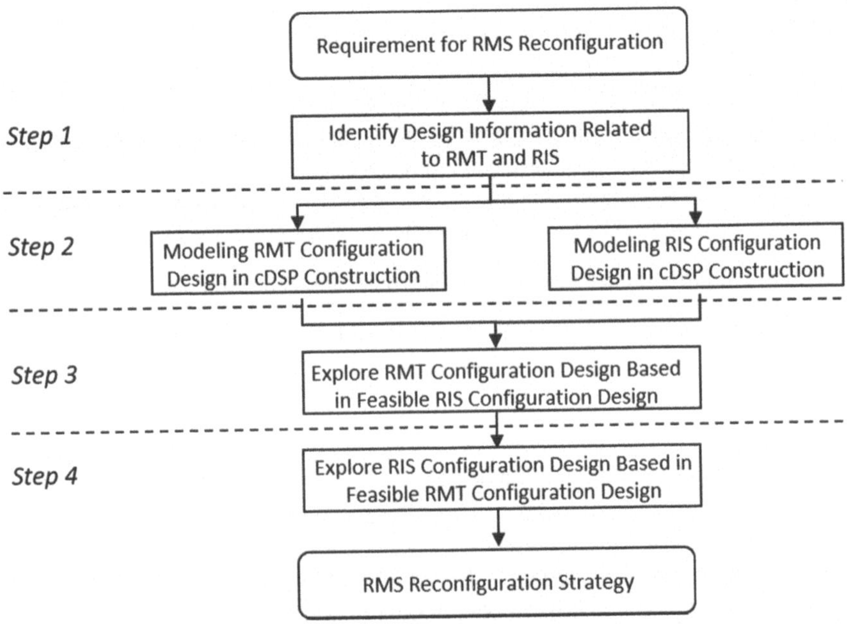

Fig. 6.12 Solution space exploration in strong-weak interaction (Shang et al. under review)

Step 1: Identify the design information such as variables, constraints, and targets of the RMT and the RIS configuration design.
Step 2: Establish the cDSP model for the RMT and the RIS configuration design according to the design information.
Step 3: Bring the feasible domain of the RIS configuration design as a hypothesis parameter to the decision model of the RMT configuration design. Further, select the specific parameter values to solve the decision model. Finally, obtain the corresponding solution space of the RMT configuration design.
Step 4: Bring the RMT solution space of Step 3 into the decision model of the RIS configuration design to obtain the corresponding solution space of the RIS configuration design.

6.5.5 *Decision Network for Exploring the RMS Reconfiguration Method*

In the reconfiguration process of the RMS there is a need for support from the single granularity configuration design, with emphasis on the interactive relationships between different granularity configurations. In this section, we present the combination the decision-making models (cDSPs) and interaction models (game theory models) in order to construct the decision network as a support to the RMS reconfiguration method exploration; see Fig. 6.13. This network lays the foundation for the dynamic management of the RMS reconfiguration.

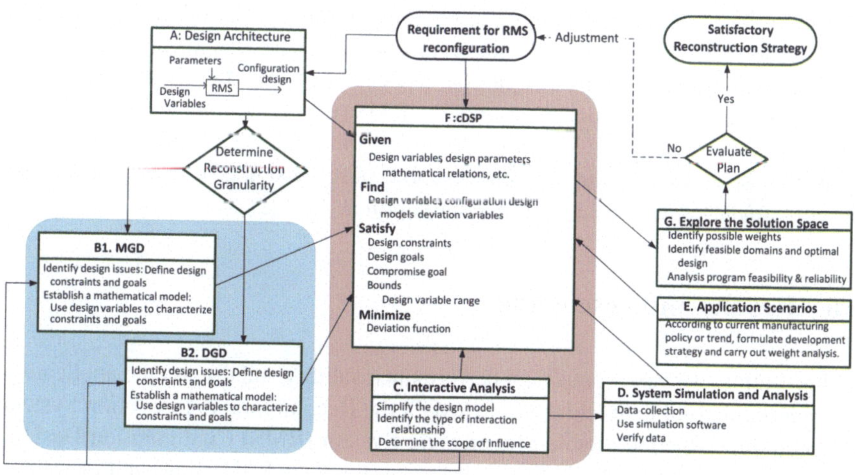

Fig. 6.13 Decision network for exploring the RMS reconfiguration method

The input of the decision network is the RMS reconfiguration requirement for the designer to formulate a systemic configuration design. Furthermore, design requirement determines the static problem analysis and dynamic data input for decision networks, including the design parameters, design variables, etc. The final output of the decision model is a satisfying RMS reconfiguration method that enables a designer to make rational decisions and improve the efficiency of the manufacturing process.

The first step is to preprocess the information due to diversity and complexity of the input information; see Step A in Fig. 6.13. By constructing a common design architecture, the relevant information of different granular configurations can be efficiently collected and analyzed. At the same time, design information is divided into design variables, parameters, and goals. The second step is to analyze and explore the single granularity configuration design as the initial solution; see Steps B and F in Fig. 6.13. With the increase in the number of the RMS devices, the complexity of the RMS's reconfiguration increases, which increase the uncertainty of the design solution. According to the key features of different RMS granularity configuration and the corresponding performance indicators, the relationship between design variables and design parameters is defined, and a single-grain configuration design scheme is explored. When performance indicators or design goals change, it requires a (re)exploration of the space exploration. A single-granularity configuration design scheme is used as the initial solution to explore the systematic reconfiguration method. The third step is to consider the interactions between different granularity configuration designs and establish exploration of the RMS reconfiguration method; see Steps C and F in Fig. 6.13. The solution of the cDSP model is the most satisfactory scheme. Its evaluation index mainly focuses on satisfying customer preferences and solving the trade-off of multiple objectives through weights setting. The fourth step is to derive the weights from system simulation; see Step D in Fig. 6.13; or application scenario analysis; see Step E in Fig. 6.13. The weights are also used to analyze the sensitivity of the design plan to dynamic demand changes, and further promote the designer's understanding of the RMS reconfiguration behavior. Finally, the designers evaluate the satisfactory design schemes that we use as the initial value of the next stage to carry out multiple iterations of the design process to promote the dynamic management of the RMS reconfiguration.

6.6 Results and Discussion

In this section, we verify empirical performance validity and discuss the application of the Reconfigurable Machine Tool (RMT), the Reconfigurable Inspection System (RIS), and the Reconfigurable Manufacture System (RMS) Configuration Design; see Quadrant 3 in Fig. 1.17. In Sect. 6.6.1, we present the RMT configuration design that we verify with a transmission box as Illustrative Example 3. In Sect. 6.6.2, we present the RIS configuration design that we verify with headstock as Illustrative Example 3. In Sect. 6.6.3, we present the RMS configuration design that we verify with an engine cylinder block as Illustrative Example 4.

6.6.1 *Application of the Reconfigurable Machine Tool (RMT) Configuration Design*

A transmission box in a vehicle plays a supportive and connective role in the entire reducer assembly. The quality of the transmission box directly affects not only the accuracy of the location of parts (for example, shafts and gears) but also the life and performance of the reducer. This box is a typical box-type part and the main operations of manufacturing this part include a high-precision plane, bearing holes, screw holes, etc. Due to application demands in the field and unstable demand/supply relationship, the multi-transmission boxes manufacturing processes face product variance or fluctuating order demand. Global market forces the manufacturing enterprises for reconfigurable manufacturing system with the capabilities of converting functionality and scaling the capability.

Three Design Scenarios for Transmission Box Production. In Table 6.8, we present simplified module library for the RMT design in the example according to the functions of modules, motion and support. The details of module division are discussed in the references (Moon and Kota 2002; Chen et al. 2005). To build the configuration tree, the premise is to define the mappings between the basic modules and the nodes; see Table 6.3. In Fig. 6.14, we present the one-to-one relationship between modules and nodes when the function is used as the mapping standard, where we also show the number of the node and the module.

The motivation for reconfiguration is to maintain and update the manufacturing performance of the RMT. In case the accuracy of the RMT deteriorates or if the RMT is damaged, then some of the modules in the RMT may have to be replaced. Furthermore, the product variants may require previously unavailable functions. The

Table 6.8 RMT basic modules and the corresponding costs (Wang et al. 2018)

#	Module name	Simplified module	Fixed cost [$]	Assembly cost [$]
1	Spindle head		2000	1000
2	Fixture	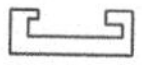	1500	800
3	Slider	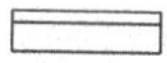	200	100
4	Cross-slider	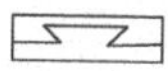	200	100
5	Column		600	200
6	Rotary table	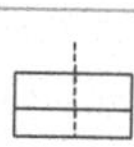	450	150
7	Base	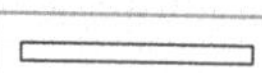	800	200

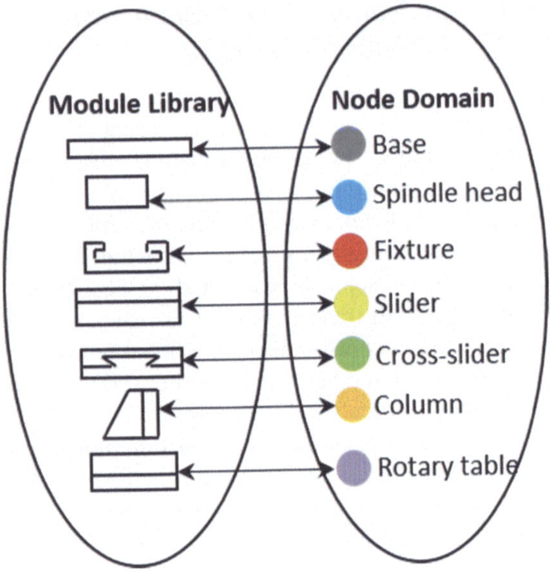

Fig. 6.14 Mappings between the module library and the node domain (Wang et al. 2018)

RMT may also have to add modules (for example, slides or columns) to enable new functionality or remove some existing modules to reduce redundancy.

In other instances, fluctuations in the demand can result in the need to scale the capability. If we need to increase the capability of the RMT we add spindle heads in the RMT to realize additional operations in the same configuration, or we remove some spindle heads in the RMT to reduce redundancy. In order to address these challenges in RMT configuration design we present three different scenarios in the remaining of the section.

Scenario 1: In the pre-planning phase, we design the RMT configuration only around a given set of operational requirements;
Scenario 2: The features of the operations changes but the quantity remains the same. We design a new RMT configuration on basis of the configuration scheme in the first case; and
Scenario 3: The quantity of the operations changes but the features remains the same. We design another RMT configuration based of the configuration scheme in the second case.

In all scenarios, there are two phases in the RMT configuration design process. The first is to generate feasible configuration trees with design constraints of capability and functionality. The second is to select the most satisfactory configuration tree with the design goals of cost and reconfigurability. According to the selected design scheme of configuration trees, the corresponding three-dimensional figures of the RMT configuration are printed via the relationship of modules and nodes; see Fig. 6.14.

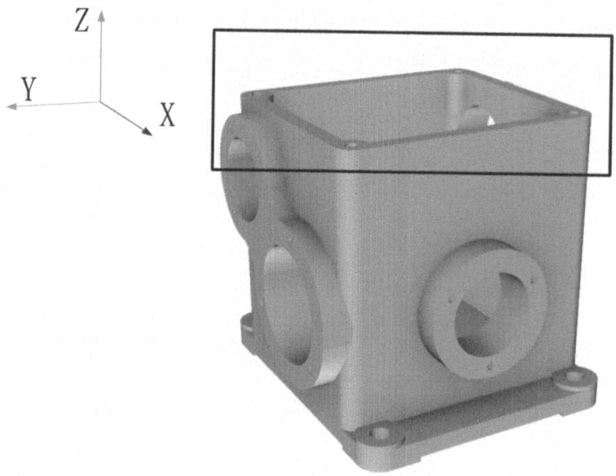

Fig. 6.15 Three-dimensional model of level-2 transmission box in scenario 1 (Wang et al. 2018)

Scenario 1: *The RMT Configuration Design and Decision-Making for Fixed Operation Requirements*

In Fig. 6.15, we present the transmission box type (Level 2) is in the pre-planning phase. The operational requirement for a specified RMT is to process the upper surface; see the rectangle in Fig. 6.15. As the RMT processes the upper surface the operational requirement contains one operation (n = 1). This operation requires relative motions between tools and workpieces along the X-, Y-, and Z- axes.

Generation of Feasible RMT Configuration Trees for Scenario 1. In Eq. (6.42), we present the constraint model of capability and functionality consistent with the operational requirement.

$$\begin{cases} \sum\limits_{j \in N} f_2(x_{1j1}) = 1 \\ \begin{cases} t_{1r} \oplus h_{1r} = 1(r = 1, 2, 3) \\ t_{1r} \oplus h_{1r} = 0(r = 4, 5, 6) \end{cases} \end{cases} \quad (6.42)$$

We determine the node set that meets the operation requirements by solving the constraint model; see Eq. (6.42). The set involves six node types: (1) the spindle head node, (2) slider node, (3) cross-slider node, (4) column node, (5) fixture node, and (6) base node. Except for the base node, the spindle head node and the fixture node, are the directional nodes are randomly assigned to the tool-side and workpiece-side branches in the assembly process; see Sect. 6.3.3. In Table 6.9, we present formed eight feasible RMT configuration trees.

Decision-making for the Most Satisfactory RMT Configuration Tree in Scenario 1. After generating the feasible design space, the decision indicators such as fixed cost $g(u)$, assembly cost $q(u)$, utilization rate UCT and similarity rate PCT are calculated; see Eqs. (6.23)–(6.25). In Table 6.10, we present the evaluation results. The configuration design process occurs in the pre-production phase in which there are no previous configurations for comparison, therefore, the utilization rate of all

Table 6.9 The feasible configuration trees designed for scenario 1 (Wang et al. 2018)

Scheme	Degree of freedom		Configuration tree	Mathematical sescription
	Tool-side branch	Workpiece-side branch		
1	None	X, Y, Z	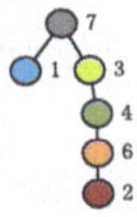	$C_1 = \left\{ \begin{array}{c} x_{710}, x_{110}, x_{311}, x_{411}, x_{511}, x_{211}, \\ \langle x_{710}, x_{110}\rangle, \langle x_{710}, x_{311}\rangle, \langle x_{311}, x_{411}\rangle, \\ x_{411}, x_{511}, x_{511}, x_{211} \end{array} \right\}$
2	X	Y, Z	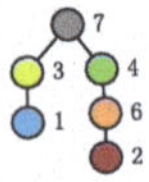	$C_2 = \left\{ \begin{array}{c} x_{710}, x_{110}, x_{310}, x_{211}, x_{411}, x_{511}, \\ \langle x_{710}, x_{310}\rangle, \langle x_{310}, x_{110}\rangle, \langle x_{710}, x_{411}\rangle, \\ x_{411}, x_{511}, x_{511}, x_{211} \end{array} \right\}$
3	Y	X, Z	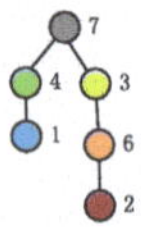	$C_3 = \left\{ \begin{array}{c} x_{710}, x_{410}, x_{110}, x_{311}, x_{511}, x_{211}, \\ \langle x_{710}, x_{410}\rangle, \langle x_{410}, x_{110}\rangle, \langle x_{710}, x_{311} rlangle, \\ x_{311}, x_{511}, x_{511}, x_{211} \end{array} \right\}$
4	Z	X, Y	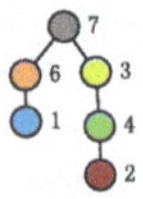	$C_4 = \left\{ \begin{array}{c} x_{710}, x_{510}, x_{110}, x_{311}, x_{411}, x_{211}, \\ \langle x_{710}, x_{510}\rangle, \langle x_{510}, x_{110}\rangle, \langle x_{710}, x_{311}\rangle, \\ x_{311}, x_{411}, x_{411}, x_{211} \end{array} \right\}$
5	X, Y	Z	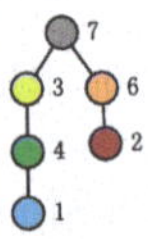	$C_5 = \left\{ \begin{array}{c} x_{710}, x_{310}, x_{410}, x_{110}, x_{511}, x_{211}, \\ \langle x_{710}, x_{310}\rangle, \langle x_{310}, x_{410}\rangle, \langle x_{410}, x_{110}\rangle, \\ x_{710}, x_{511}, x_{511}, x_{211} \end{array} \right\}$
6	X, Z	Y	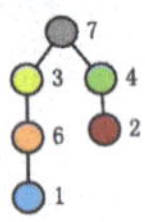	$C_6 = \left\{ \begin{array}{c} x_{710}, x_{310}, x_{510}, x_{110}, x_{411}, x_{211}, \\ \langle x_{710}, x_{310}\rangle, \langle x_{310}, x_{510}\rangle, \langle x_{510}, x_{110}\rangle, \\ x_{710}, x_{411}, x_{411}, x_{211} \end{array} \right\}$
7	Y, Z	X	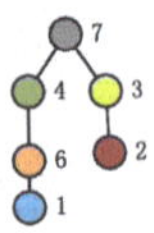	$C_7 = \left\{ \begin{array}{c} x_{710}, x_{410}, x_{510}, x_{110}, x_{311}, x_{211}, \\ \langle x_{710}, x_{410}\rangle, \langle x_{410}, x_{510}\rangle, \langle x_{510}, x_{110}\rangle, \\ x_{710}, x_{311}, x_{311}, x_{211} \end{array} \right\}$
8	X, Y, Z	None	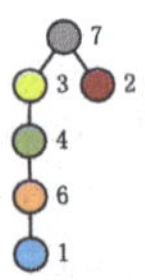	$C_8 = \left\{ \begin{array}{c} x_{710}, x_{310}, x_{410}, x_{510}, x_{110}, x_{211}, \\ \langle x_{710}, x_{310}\rangle, \langle x_{310}, x_{410}\rangle, \langle x_{410}, x_{510}\rangle, \\ x_{510}, x_{110}, x_{710}, x_{211} \end{array} \right\}$

1. Spindle Head; 2. Fixture; 3. Slider; 4. Cross-Slider; 5. Column; 6. Rotatory table; 7. Base

RMT configuration trees are 1. Further, we process the deviations of goals, such as cost and reconfigurability; see Eqs. (6.26) and (6.27). The weights of goals, reconfigurability and cost, are set to be 0.8 and 0.2 because the reconfigurability is more important than the cost in the initial RMT design phase. Finally, we substitute the deviation values and weight values in the Eq. (6.28) in order to obtain the objective function value. In Table 6.10, we present the deviation values and objective function

Table 6.10 **a** The RMT configuration tree decision indicators in scenario 1 (Wang et al. 2018); **b** selected objective function values of the rmt configurations tree in scenario 1 (Wang et al. 2018)

(a)

Index	$g(u)$	$q(u)$	ECT	UCT	PCT	RCT
Scheme	[$]	[$]	[$]	[−]	[−]	[−]
1	5300	3000	8300	1	0.133	1.133
2	5300	3000	8300	1	0.800	1.800
3	5300	3000	8300	1	0.400	1.400
4	5300	3000	8300	1	0.400	1.400
5	5300	3000	8300	1	0.400	1.400
6	5300	3000	8300	1	0.400	1.400
7	5300	3000	8300	1	0.800	1.800
8	5300	3000	300	1	0.133	1.133

(b)

Index	d_1	d_2	I
Scheme	[−]	[−]	[−]
1	0	0.3704	0.2963
2	0	0	0
3	0	0.2222	0.1778
4	0	0.2222	0.1778
5	0	0.2222	0.1778
6	0	0.2222	0.1778
7	0	0	0
8	0	0.3704	0.2963

values of all schemes. In Fig. 6.16, we present a trend chart illustrating the performance of the RMT configuration schemes based on the selective objective function values.

The decision objective function I has the smallest values for Schemes 2 and 7. Schemes 2 and 7 are relatively close to the ideal solution, which is expected to minimize the cost yet maximize the reconfigurability. Consequently, we select the two satisfactory RMT configuration trees. The result is the lack of reconfiguration of the process. Hence, some of indices related to reconfiguration, such as assembly cost and utilization rate, are useless for the design decision. Based on manufacturing expertise, when the workpiece is large and difficult to move, workpiece-related configurations must have as few DOFs as possible, thus, the number of DOFs in the tool-related configuration should be increased. Further, Scheme 2 is chosen as the most satisfactory RMT configuration tree in Scenario 1 due to the dimensions of the Level 2 transmission box; see Fig. 6.17.

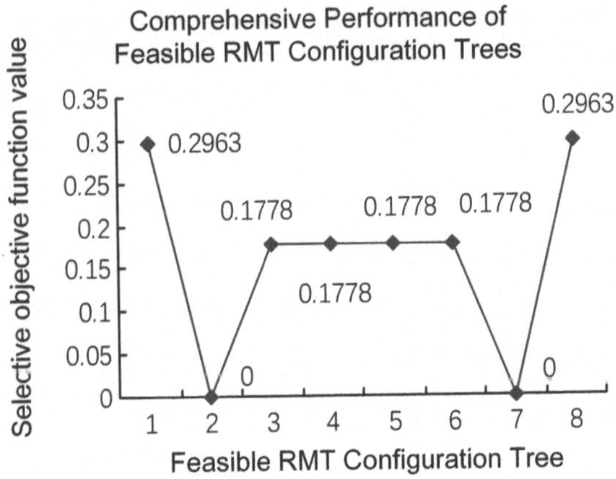

Fig. 6.16 Performance of the feasible RMT configuration (Wang et al. 2018)

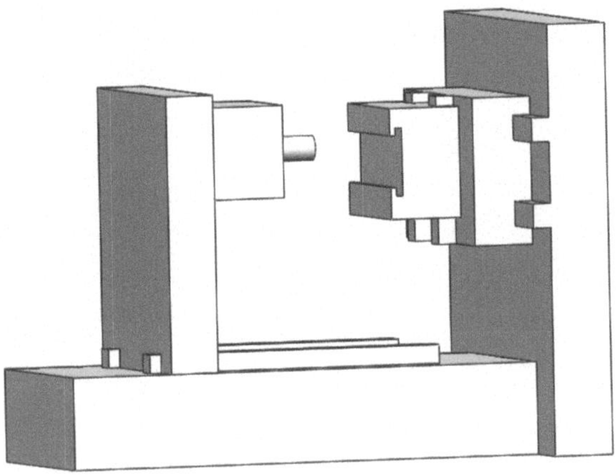

Fig. 6.17 The most satisfactory RMT configuration for scenario 1 (Wang et al. 2018)

Scenario 2: *The RMT Configuration Design and Decision-Making for Operational Feature Changes*

In a RMT, the type or batch size of the workpiece in new orders may change as a result of changing market conditions. Hence, there is a need to re-plan the processing route, which would result in reconfiguring some of the RMT. In this section, we apply the method we propose for redesign of the RMT configuration when there are changes in the operational features.

In Sect. 6.17, we present the desired type of transmission box for new orders. Hence, it is necessary for the RMT to process the whole side; see circle in Fig. 6.18. The operational set of the RMT contains only one operation ($n = 1$). Using the three-dimensional coordinate system, the operation feature is accomplished by the relative motions between tools and workpieces along the X- and A- axes. Compared with the operational requirements the number of machining operations required in the new process remains the same, but the operation itself is different; see Sect. 6.3.1. Hence, it is necessary for the RMT to be reconfigured from the original configuration.

Fig. 6.18 Three-dimensional model of level-2 transmission box in scenario 2 (Wang et al. 2018)

Generation of the Feasible RMT Configuration Trees for Scenario 2. In Eq. (6.43), we present the constraint model for the node set of decisions in this process.

$$\begin{cases} \sum_{j\in N} f_2(x_{1j1}) = 1 \\ \begin{cases} t_{1q} \oplus h_{1q} = 1(q = 1, 4) \\ t_{1q} \oplus h_{1q} = 0(q = 2, 3, 5, 6) \end{cases} \end{cases} \tag{6.43}$$

The node set for this process consists of (1) spindle head node, (2) base node, (3) fixture node, (4) slider node, and (5) rotary table node. These nodes are formed into four feasible RMT configuration trees; see Table 6.11.

Decision-making for the Most Satisfactory RMT Configuration Tree in Scenario 2. In Table 6.12, we present the values of decision indices of the feasible configuration trees in Scenario 2. In the development phase, the manufacturing enterprises tend to expand their market share. Hence, the manufacturing process is required to be both reconfigurable and cost-effective, where reconfigurability is equally important as is the cost in Scenario 2, therefore, the assigned weights are 0.5. In Table 6.13, we present the deviation values and decision function values of each scheme. In Fig. 6.19, we present the chart of the performance of RMT configuration schemes (Tables 6.14 and 6.15).

The value of function I for Scheme 2 is the smallest; see Fig. 6.19. Scheme 2 has smaller deviations from the ideal solution. Therefore, we select it as the most satisfactory RMT configuration tree. In Fig. 6.20, we present the RMT configuration based on the inverse mapping between the module library and the node library.

Scenario 3: *The RMT Configuration Design and Decision-Making for Machining Operation Quantity Changes*

In Scenario 3, we apply the method we propose to a situation in which the product quantity changes. When the batch of transmission boxes increases in a new production

Table 6.11 The feasible RMT configuration trees designed in scenario 2 (Wang et al. 2018)

Scheme	Degree of freedom		Configuration tree	Mathematical description
	Tool-side branch	Workpiece-side branch		
1	None	X, A	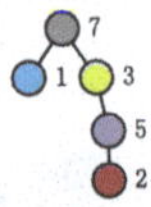	$C_1 = \left\{ \begin{array}{l} x_{710}, x_{110}, x_{211}, x_{311}, x_{611}, \\ \langle x_{710}, x_{110}\rangle, \langle x_{710}, x_{311}\langle, \langle x_{311}, x_{611}\langle, \\ x_{611}, x_{211} \end{array} \right\}$
2	X	A	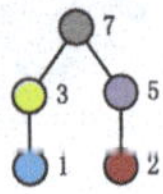	$C_2 = \left\{ \begin{array}{c} x_{710}, x_{110}, x_{310}, x_{211}, x_{711}, \\ \langle x_{710}, x_{310}\rangle, \langle x_{310}, x_{110}\rangle, \langle varvecx_{710}, x_{611}\rangle, \\ x_{611}, x_{211} \end{array} \right\}$
3	A	X	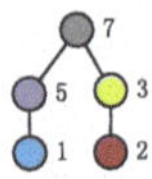	$C_3 = \left\{ \begin{array}{c} x_{710}, x_{610}, x_{110}, x_{311}, x_{211}, \\ \langle x_{710}, x_{610}\rangle, \langle x_{610}, x_{110}\rangle, \langle x_{710}, x_{311}\rangle, \\ x_{311}, x_{211} \end{array} \right\}$
4	X, A	None	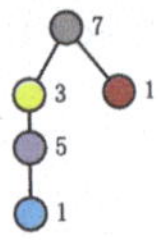	$C_3 = \left\{ \begin{array}{c} x_{710}, x_{110}, x_{310}, x_{610}, x_{211}, \\ \langle x_{710}, x_{310}\rangle, \langle x_{310}, x_{610}\rangle, \langle x_{610}, x_{110}\rangle, \\ x_{710}, x_{211} \end{array} \right\}$

1. Spindle Head; 2. Fixture; 3. Slider; 4. Cross-Slider; 5. Column; 6. Rotatory table; 7. Base

Table 6.12 The RMT configuration tree decision indicators in scenario 2 (Wang et al. 2018)

Index	$g(u)$	$q(u)$	ECT	UCT	PCT	RCT
Scheme	[$]	[$]	[$]	[−]	[−]	[−]
1	450	6100	11,850	0	0.0667	0.0667
2	450	3200	8950	0.5	0.600	1.1000
3	450	6100	11,850	0	0.600	0.6000
4	450	5400	11,150	0.25	0.0667	0.3167

Table 6.13 Selection objective function values of the RMT configuration trees in scenario 2 (Wang et al. 2018)

Index	d_1	d_2	I
Scheme	[−]	[−]	[−]
1	0.2447	0.9394	0.5921
2	0	0	0
3	0.2447	0.4545	0.7121
4	0.1973	0.7121	0.4547

order, multiple operations need to be combined in order to guarantee timely delivery. Consequently, the RMT is required to process both the side holes and the large upper-surface holes; see Fig. 6.21. The number of operations changes to 2 (n = 2). The features of the first operations are accomplished by the relative motions between the

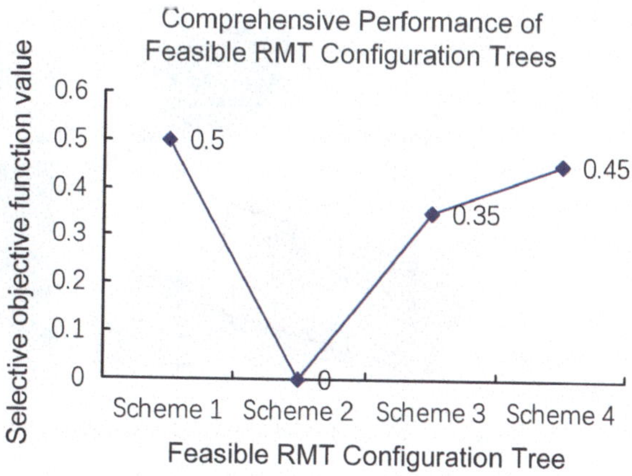

Fig. 6.19 Performance of the feasible RMT configuration trees for Scenario 2 (Wang et al. 2018)

Table 6.14 The RMT configuration decision indicators for scenario 3 (Wang et al. 2018)

Index	$g(u)$	$q(u)$	ECT	UCT	PCT	RCT
Scheme	[$]	[$]	[$]	[−]	[−]	[−]
1	3050	8050	11,100	0.25	0.1333	0.3833
2	2850	8750	11,600	0	0.7333	0.7333
3	3250	8150	11,400	0.25	0.4000	0.6500

Table 6.15 Selective objective function values of the RMT configuration trees for Scenario 3 (Wang et al. 2018)

Index	d_1	d_2	I
Scheme	[−]	[−]	[−]
1	0	0.4773	0.0954
2	0.0431	0	0.0345
3	0.0263	0.1136	0.0438

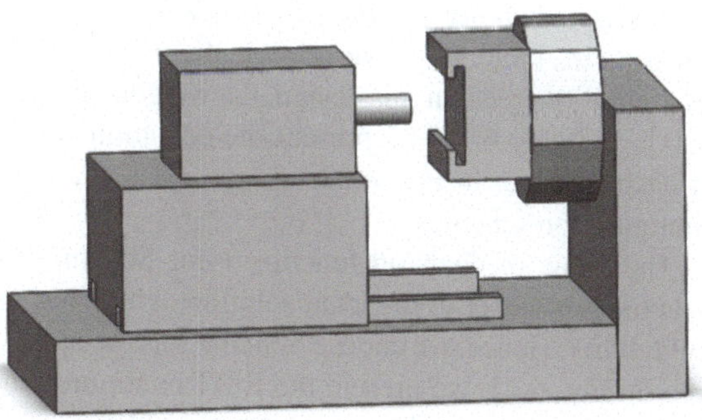

Fig. 6.20 The most satisfactory RMT configuration for Scenario 2 (Wang et al. 2018)

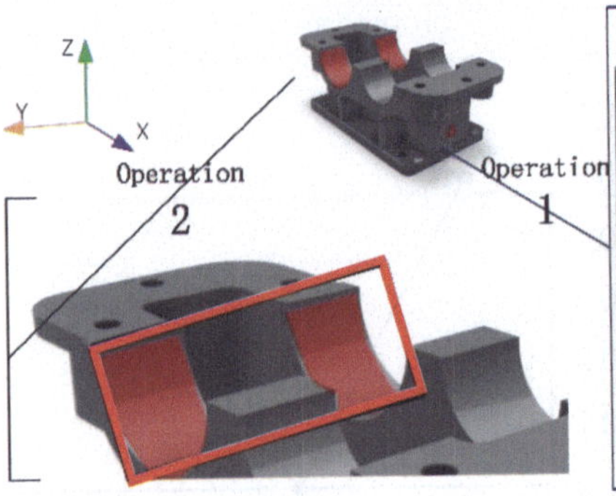

Fig. 6.21 Three-dimensional model of level-2 transmission box in Scenario 3 (Wang et al. 2018)

tools and workpieces along the X- and A- axes, and then along the X-, Y-, and B- axes. Compared with the process described in the Scenario 2, the quantity of operations increases. Consequently, the RMT cannot fulfill the operational requirements; see Fig. 6.20. It is necessary to reconfigure the RMT based on the second configuration.

Generation of the Feasible RMT Configuration Trees. In Eq. (6.44), we present the constraint model for the node set decision of this process. In Table 6.16, we present three feasible RMT configuration trees.

$$\begin{cases} \sum\limits_{j \in N} f_2(x_{1j1}) = 2 \\ \begin{cases} t_{1r} \oplus h_{1r} = 1(r = 1, 4) \\ t_{1r} \oplus h_{1r} = 0(q = 2, 3, 5, 6) \\ t_{2r} \oplus h_{2r} = 1(r = 1, 2, 3, 6) \\ t_{2r} \oplus h_{2r} = 0(r = 4, 5) \end{cases} \end{cases} \tag{6.44}$$

Decision-making for the Most Satisfactory RMT Configuration Tree in Scenario 3. In Table 6.14, we present the values of decision indices of the feasible configuration trees. In this design, the weights of cost and reconfigurability are equal to 0.5. In Table 6.15, we present the deviation values and decision function values of each scheme. In Fig. 6.22, we present a chart of the performance of the RMT configuration schemes.

The value of decision function I for Scheme 2 is the smallest; see Fig. 6.22. Scheme 3 is closer to the ideal solution, which has great balance of cost and reconfigurability. Hence, we choose Scheme 3 as the most satisfactory RMT configuration tree. In Fig. 6.23, we present the RMT configuration based on the inverse mapping between the module library and the node library.

Table 6.16 The feasible RMT configuration tree schemes in Scenario 3 (Wang et al. 2018)

Scheme	Degree of freedom			Configuration tree	Mathematical description
	Tool-side branch	Tool-side branch	Workpiece-side branch		
1	X, A	X, Y, B	None	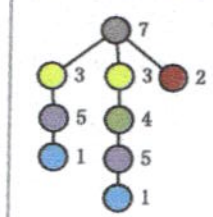	$C_1 = \left\{ \begin{array}{c} x_{710}, x_{310}, x_{320}, x_{410}, x_{610}, x_{620}, x_{110}, \\ x_{120}, x_{211}, \langle x_{710}, x_{310} \rangle, \langle x_{310}, x_{410} \rangle \\ \langle x_{410}, x_{610} \rangle, \langle x_{610}, x_{110} \rangle, \langle x_{710}, x_{320} \rangle, \\ \langle x_{320}, x_{620} \rangle, \langle x_{620}, x_{120} \rangle, \langle x_{710}, x_{211} \rangle \end{array} \right\}$
2	A	Y, B	X	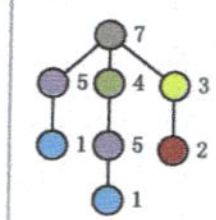	$C_2 = \left\{ \begin{array}{c} x_{710}, x_{410}, x_{610}, x_{620}, x_{110}, x_{120}, x_{311}, \\ x_{211}, \langle x_{710}, x_{410} \rangle, \langle x_{410}, x_{610} \rangle \\ \langle x_{610}, x_{110} \rangle, \langle x_{710}, x_{620} \rangle, \langle x_{620}, x_{120} \rangle, \\ \langle x_{710}, x_{311} \rangle, \langle x_{311}, x_{211} \rangle \end{array} \right\}$
3	Y, X, A	B	X, Y	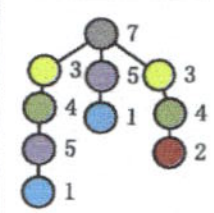	$C_3 = \left\{ \begin{array}{c} x_{710}, x_{310}, x_{410}, x_{610}, x_{620}, x_{110}, x_{120}, \\ x_{321}, x_{421}, x_{211} \langle x_{710}, x_{610} \rangle, \\ \langle x_{610}, x_{110} \rangle, \langle x_{710}, x_{310} \rangle, \langle x_{310}, x_{410}, \\ \langle x_{410}, x_{120} \rangle, \langle x_{710}, x_{321} \rangle, \langle x_{321}, x_{421} \rangle, \\ \langle x_{421}, x_{221} \rangle \end{array} \right\}$

1. Spindle Head; 2. Fixture; 3. Slider; 4. Cross-Slider; 5. Column; 6. Rotatory table; 7. Base

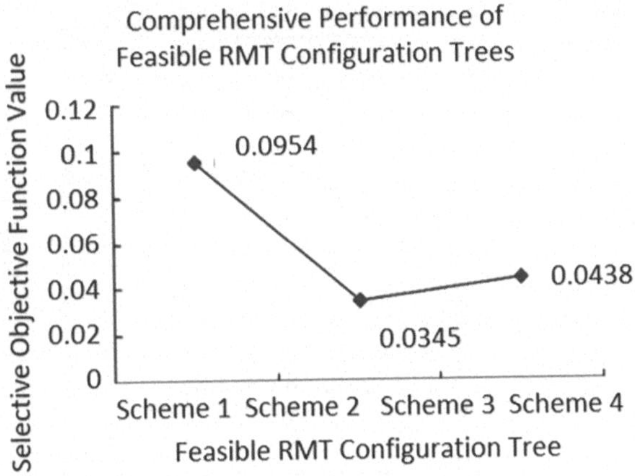

Fig. 6.22 Performance configuration schemes for changes in quantity (Wang et al. 2018)

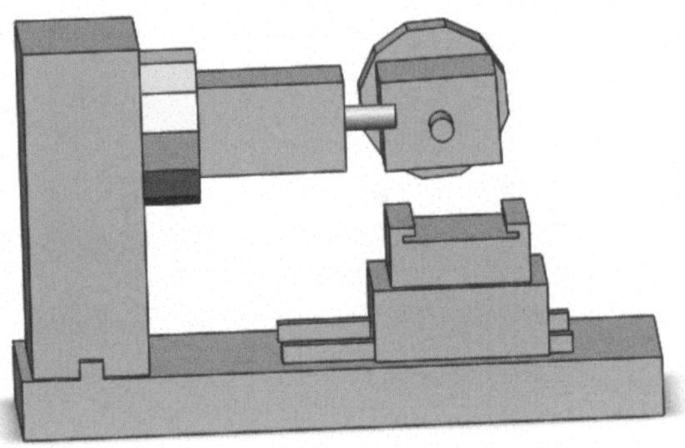

Fig. 6.23 Most satisfactory RMT configuration for Scenario 3 (Wang et al. 2018)

6.6.2 Application of the Reconfigurable Inspection System (RIS) Configuration Design

The headstock is an important part of the lathe and is usually used to lay out the machine tool spindle, its drive parts and the corresponding auxiliary mechanisms. The spindle box is the basic component of the headstock, which machines the shaft, sleeve, gear, etc., to ensure correct positioning of these components. Furthermore, the processing quality of the box directly affects the accuracy, performance, and life of the machine. The method is described and verified based on the lathe spindle, and the spindle housing; see Table 6.17.

Facing the production issues, the manufacturing enterprise would establish an efficient detecting system and ensure the product quality. Currently, there are 5 stations in the manufacturing process and the error is required to be detected up to 3 stations. According to constraint models of the RIS configuration design from the perspective of detection capability and functionality, it is calculated that there are at least 14 sensors in the RIS and the distance from all of the processes to their respective detectors is no more than 40 [mm]; see Eq. (6.45). According to the permutations

Table 6.17 The process requirement example

Number	Processing surface	Datum surface
1	M	R
2	N	R
3	Slide	R
4	P	M, N
5	Q	M, N

and combinations, a total of 13 feasible design solutions are formed; see Table 6.18.

$$\begin{cases} FCRIS \geq \sum_{i=1}^{5}\sum_{j=1}^{7} H(cp_i, f_j) \times \left(\sum_{k=1,k\neq j}^{5} G(f_k) + 1 \right) = 14 \\ CCRIS \leq \frac{D(3)\times INT(\frac{5}{3})+D(5-3\times INT(\frac{5}{3}))}{D(5)} = 0.4 \\ 1 \leq M \leq 5 \end{cases} \quad (6.45)$$

In Table 6.19, we present the cost and diagnosability decision index values for each of the RIS feasible configuration designs we describe in the proceeding. The reliability of sensor is 0.9 [−], the cost of installing one RIM averaged 10,000 [$] and the cost of one sensor averaged 1000 [$]. The amount of deviation of cost and diagnosability is obtained by processing the data; see Table 6.19. The diagnosability and the cost weight are 0.5, respectively. Finally, we obtain the target function value by combining the deviation value and the weight value, and the variation tendency diagram; see Fig. 6.24.

According to the changing tendency of the RIS feasible design performance level, Scheme 5 has the lowest objective function value and its cost and diagnosability reach a relative balance; see Fig. 6.24. Hence, we can conclude that Scheme 5 satisfies the RIS configuration solution. In the current process route, the satisfied quality detection relies on four RIM and focuses on the downstream area.

6.6.3 Application of the Reconfigurable Manufacturing System (RMS) Configuration Design

The replacement of automotive product designs is accelerating at a faster rate leading to an increasing degree of disparity of the car designs. Hence, the design of the engine is changing accordingly. Currently, the common engine structures include in-line four-cylinder engine (L4), V-type six-cylinder engine (V6) or V-type eight-cylinder engine (V8). Under the proceeding circumstances, different car manufacturer chooses different types of engines based on parameters such as displacement or noise reduction. For example, Audi S3 uses an L4 engine, Audi A6 uses a V6 engine, and Audi A8 uses a V-type eight-cylinder engine. In Fig. 6.25, we present the cylinder block of the L4 and V8 engines. In this context, the manufacturing enterprises

Table 6.18 The design scheme of RIS configuration

#	RIS Configuration	Model
1		$\boldsymbol{CRIS}_1 = \{(1,2),(2,2),(3,2),(4,3),(5,3)\}$
2		$\boldsymbol{CRIS}_2 = \{(1,2),(3,3),(4,3),(5,3)\}$
3		$\boldsymbol{CRIS}_3 = \{(1,2),(2,2),(4,5),(5,3)\}$
4		$\boldsymbol{CRIS}_4 = \{(1,2),(2,2),(3,2),(5,4)\}$
5		$\boldsymbol{CRIS}_5 = \{(2,2),(3,2),(4,3),(5,3)\}$
6		$\boldsymbol{CRIS}_6 = \{(1,2),(4,5),(5,3)\}$
7		$\boldsymbol{CRIS}_7 = \{(1,2),(3,3),(5,4)\}$
8		$\boldsymbol{CRIS}_8 = \{(1,2),(2,2),(5,6)\}$
9		$\boldsymbol{CRIS}_9 = \{(3,4),(4,3),(5,3)\}$
10		$\boldsymbol{CRIS}_{10} = \{(2,3),(4,5),(5,3)\}$
11		$\boldsymbol{CRIS}_{11} = \{(2,3),(3,2),(5,4)\}$
12		$\boldsymbol{CRIS}_{12} = \{(2,3),(5,6)\}$
13		$\boldsymbol{CRIS}_{13} = \{(3,4),(5,4)\}$

try to enhance adaptability of the manufacturing system in order to accommodate dynamic market demands. With the premise of maintaining low cost and high quality, the manufacturing enterprises produce more diverse parts to meet the requirements of different automotive assemblies and improve the competitiveness of the enterprise. Therefore, in the automotive industry, the RMS has unlimited development

Table 6.19 The index data of RIS design

Index	Solution 1	Solution 2	Solution 3	Solution 4	Solution 5
ECRIS [−]	$6.2*10^3$	$5.1*10^3$	$5.2*10^3$	$5.0*10^3$	$5.0*10^3$
XCRIS [−]	0.5335	0.5362	0.5340	0.5866	0.5733
TCRIS [−]	1	0.875	0.875	0.875	0.875
RCRIS [−]	0.5335	0.4692	0.4672	0.5132	0.5016
Index	Solution 6	Solution 7	Solution 8	Solution 9	Solution 10
ECRIS [−]	$4.0*10^3$	$3.9*10^3$	$4.0*10^3$	$4.0*10^3$	$4.1*10^3$
XCRIS [−]	0.5457	0.5960	0.6049	0.4944	0..4911
TCRIS [−]	0.7778	0.6667	0.7778	0.7778	0.6667
RCRIS [−]	0.4244	0.3973	0.4705	0.3845	0.3274
Index	Solution 11	Solution 12	Solution 13		
ECRIS [−]	$3.9*10^3$	$2.9*10^3$	$2.8*10^3$		
XCRIS [−]	0.5690	0.5582	0.5433		
TCRIS [−]	0.6667	0.4167	0.4167		
RCRIS [-]	0.3793	0.2326	0.2264		

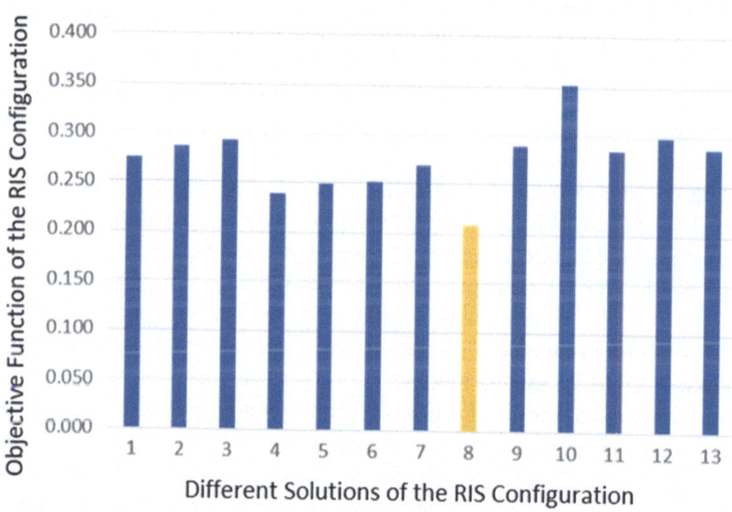

Fig. 6.24 The result of RIS configuration design objective function

and application space, and the RMS is an indispensable new manufacturing force for the future emerging enterprises. The proposed method for exploring the RMS reconfiguration method is illustrated using the manufacture of the engine cylinder block as an example; see Sect. 6.5.

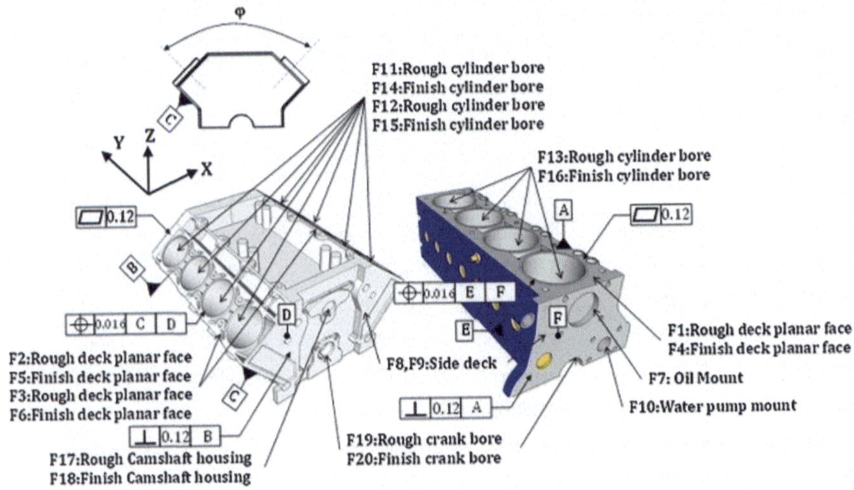

Fig. 6.25 The example of the L4 (right) and V8 (left) engine block (Abbas et al. 2016)

The machining characteristics of the engine cylinder block include multiple machining surfaces and holes, and the required machining accuracy is also different. Therefore, the premise of parts processing is the reasonable arrangement and division of the process route. In this example, we explore the RMS reconfiguration method in two scenarios. The first scenario is the transition between different part families that leads to the RMS reconfiguration. The second scenario is the transitions between different parts in one-part family that leads to RMS reconfiguration. According to the structure of the cylinder block, two families of parts are extracted, the L-type and V-type engine blocks. The linear engine cylinder block family includes one component, L4 engine block. The V-type engine block family includes two components, the V6 and V8 engine blocks. Hence, the first design requirement is reconfiguring the RMS to match L4 to V6 engine block. The second design requirement is reconfiguring the RMS to match V6 to V8 engine blocks.

Before the detailed design of the reconfiguration configuration, we analyze the processing routes of the proceeding three engine blocks. Due to the large number of machining features of the cylinder block, we consider the key operations in the text example. The L4, V6, and V8 engine blocks we consider in this case are sourced from a company; see Fig. 6.26. In Tables 6.20 and 6.21, we present the key operating information. During the reconstruction process, the machining process route of the L4 engine block is Op11 - Op12 - Op13 - Op14 - Op15 - Op16. The machining process route of the V6 engine block is Op21 - Op22 - Op23 - Op24 - Op25 - Op26 - Op27 - Op28. Further, in order to reduce the fixture error caused by the fixture change, the V8 engine block's process path change is Op21 - Op25 - Op22 - Op23 - Op24 - Op27 - Op26 - Op28.

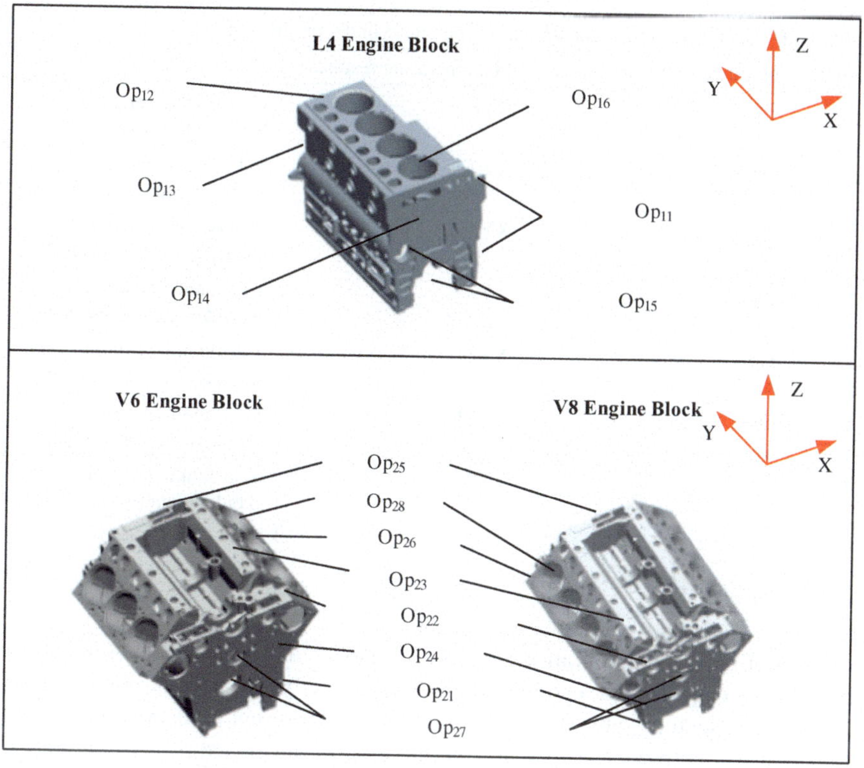

Fig. 6.26 The example of three types of the cylinder block

Table 6.20 The key operation for L4 engine cylinder production

Num	Manufacturing operations	Manufacturing features	Reference surface
1	Op_{11}	XYZ	Undersurface and two locating pin holes
2	Op_{12}	XYZ	Side datum, position shoulder and two pin holes
3	Op_{13}	XYZ	Side datum, position shoulder and two pin holes
4	Op_{14}	XYZ	Side datum, position shoulder and two pin holes
5	Op_{15}	YZB	Side datum, position shoulder and two pin holes
6	Op_{16}	YZC	Side datum, position shoulder and two pin holes

Table 6.21 The key operation for V6 and V8 engine cylinder production

Num	Manufacturing operations	Manufacturing features	Reference surface
1	Op_{21}	XYZ	Undersurface and two locating pin holes
2	Op_{22}	XYZ	Side datum, position shoulder and two pin holes
3	Op_{23}	XYZ	Side datum, position shoulder and two pin holes
4	Op_{24}	XYZ	Side datum, position shoulder and two pin holes
5	Op_{25}	XYZ	Side datum, position shoulder and two pin holes
6	Op_{26}	XYZB	Side datum, position shoulder and two pin holes
7	Op_{27}	YZB	Side datum, position shoulder and two pin holes
8	Op_{28}	XYZBC	Side datum, position shoulder and two pin holes

The RMS initial configuration is designed to meet the requirements of manufacturing the L4 engine block. In order to simply and directly represent the configuration design, the system or subsystem configurations we represent by using a block diagram, and we represent the specific configuration of the device by a configuration tree. In Fig. 6.27, we present the initial configuration of the RMS in the example. This configuration has 9 workstations in total, 6 are RMT and three are RIM. Each RMT configuration we present in the corresponding configuration tree. The first production partition contains RMT_1, that is, RIM_1 detects the state of RMT_1, and RIM_1

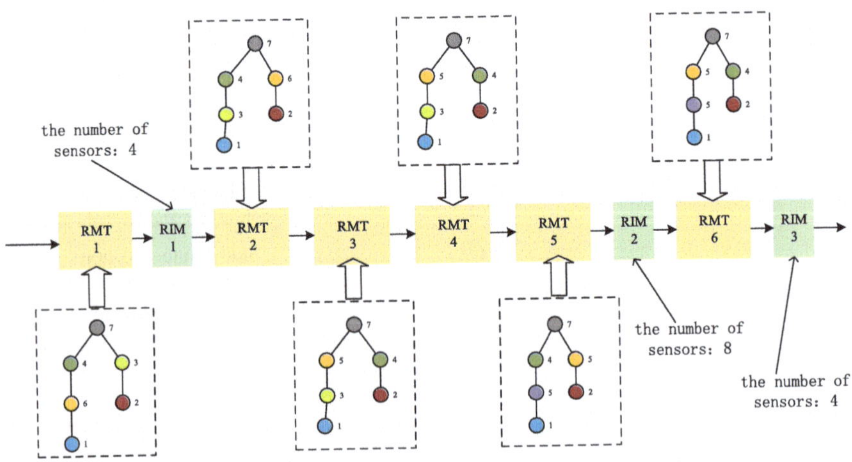

Fig. 6.27 The RMS initial configuration in text example

includes four sensors. The second production partition contains four machine tools RMT_2, RMT_3, RMT_4 and RMT_5, that is, RIM_2 is to inspect the states of these four machine tools and RIM_2 has eight sensors. The final production partition contains RMT_6, in which the RIM_3 detects the state of RMT_6, and the number of sensors is four.

In addition, in order to consider the manufacturing enterprise's values and preferences, we investigate their manufacturing needs and extract the required parameter data, including module fixing, assembly costs and the number of modules of the machine tool (one machining operation per machine tool). Sensor reliability (average is 0.9), sensor cost (average is \$146), detector installation cost (average is \$1460), quality inspection time (up to 3 units of work), and company's expected investment cost (less is better), the number of work units expected by the company (no more than 12), etc.

The reconfiguration process of this case consists of two stages, which are reconfiguration from the L4 engine cylinder block to the V6 engine block and from the V6 engine block to the V8 engine block. In order to simplify the description, we present the following two phases of reconfiguration, Phases A and B, respectively. The requirement in the Phase A is the conversion between different part families to the reconfiguration of the system. The designers need to (re)design and (re)configure the production process and detection process. In Fig. 6.10, we use a cooperative model to explore the solutions for the problem of multi-granularity interactions that exist at this stage. In Fig. 6.28, we present the satisfying solution for Phase A.

The design of Phase B belongs to the conversion of different parts between the same part families, leading to the reconfiguration of the system. Normally, there are many similar processes with the parts in the part family, so designers are more focused on optimizing the detection process. Since the inspection process may change as the production process changes, the production process is designed first, and then the inspection process is formed for a specific production process. Therefore, the leader-follower model is used to solve this stage, where the leader is the RMT configuration design. In Fig. 6.29, we present the satisfying solution for Phase B.

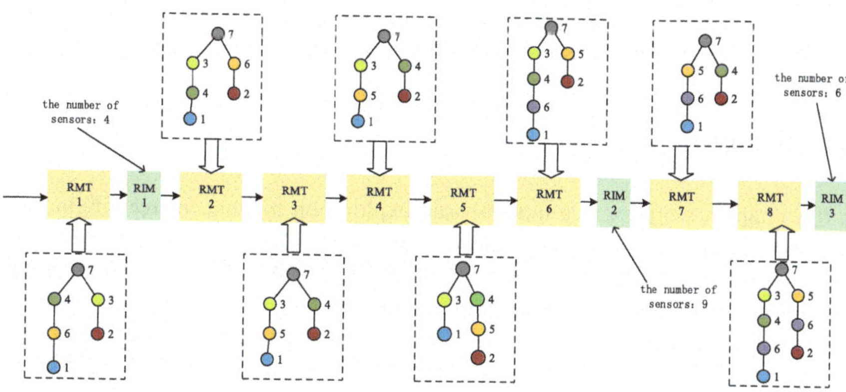

Fig. 6.28 The solution of phase A with the cooperative model

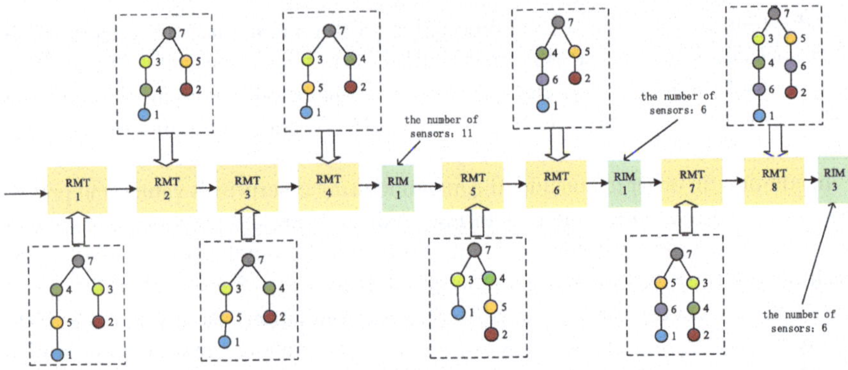

Fig. 6.29 The solution of phase B with the cooperative model

The reconfiguration process of the RMS becomes more complicated as the number of workstations increases. For example, there are 71 feasible design for the reconfiguration of detection process in the RMS with 12 workstations. The configuration design method based on the cDSP construction and game theory we propose for effective solution space exploration. According to the reconfiguration method, the results obtained by the proposed method can make the good use of the original resources. For example, the process path of the cylinder block of the L4 engine has the same processing characteristics as the first four steps of the process line of the cylinder of V6 engine. The corresponding configuration in the design is also the same, so as to effectively avoid unnecessary reconfiguration. At the same time, different interaction relationships are used to describe the design priorities and priorities among designers. For example, the reconfiguration of Phase A corresponds to different part families. Therefore, both the production and the detection process need to be reconfigured, so that the cooperation model is applied. The solution also maximizes the balance between reconfigurability and diagnosability. Phase B corresponds to different parts and has many similar processes. Therefore, the reconfiguration of the detection process is more affected by the production process, so the leader-follower model is applied. The resulting design solution improves the diagnosability of the specific production process during the processing of the part family.

6.7 Synopsis

In this chapter, we present a model-based exploration method for reconfiguration method of Reconfigurable Manufacturing System (RMS), where we establish connectivity between the Reconfigurable Machine Tool (RMT) and the Reconfigurable Inspection System (RIS). The reconfiguration method is a critical component of the computational framework, Design for Dynamic Management; see Quadrant 4 in Fig. 6.1. There are three main phases in the proposed method. In Sect. 6.3, we present Phase 1 of the method wherein we model the performance of RMT layout, such as capacity, and we further discuss the reconfiguration design of RMT layout in the cDSP construct. In Sect. 6.4, we present the Phase 2 of the method wherein we model the performance of RIS distribution, such as diagnosability, and we discuss the reconfiguration design of RIS distribution in the cDSP construct. In Sect. 6.5, we present Phase 3 of the method wherein we integrate multiple cDSPs to explore the RMS reconfiguration method to meet overall system goals, such as cost and flexibility. In Sect. 6.6, we present empirical performance validity and we illustrate the capability to produce useful results for the chosen example problems, reconfiguration of Transmission Box as Illustrative Example 3 and reconfiguration of Engine Cylinder Block as Illustrative Example 4 (Fig. 6.30).

In Chap. 7 we provide an overview of the monograph and some thoughts on what is needed to transition what is covered in this monograph to industry. Specifically, in Sect. 7.1, we provide a summary of the monograph. In Sect. 7.2, we revisit the questions we address in the monograph and we validate the hypotheses. In Sect. 7.3, we present the contributions and limitations of the proposed methods regarding adaptability, operability, and reconfigurability. In Sect. 7.4, we identify what is needed to transition the methods documented in this monograph to industry.

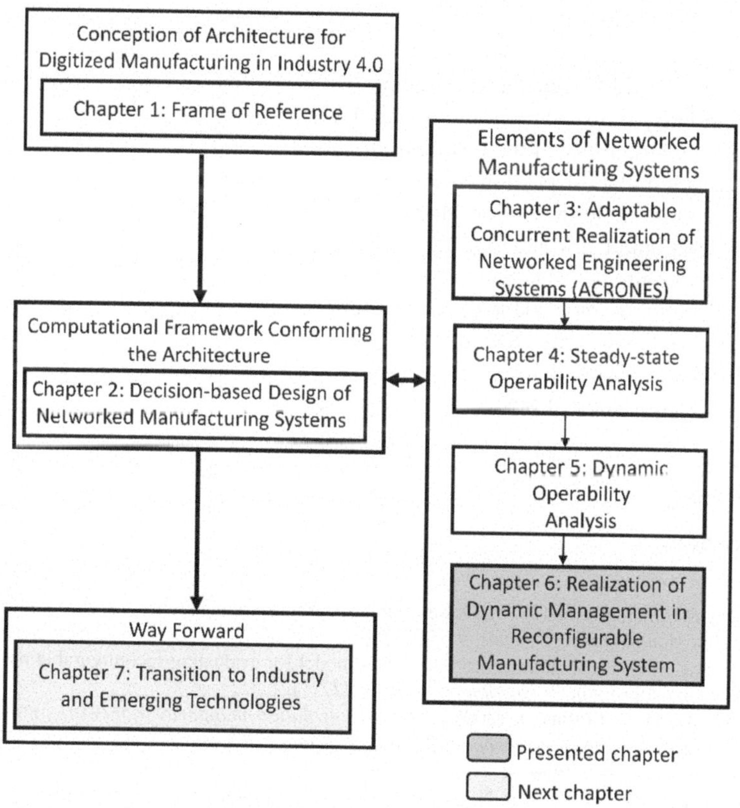

Fig. 6.30 Organization of the monograph—presented and next chapter

Glossary

cDSP compromise Decision Support Problem

Configuration Tree A functional model that we use to describe the RMT configuration

CPS Cyber-Physical System

DFDM Design for Dynamic Management

DGD Device Granularity Design

Digital Thread Refers to the communication framework that allows a connected data flow and integrated view of the asset's data throughout its lifecycle across traditionally siloed functional perspectives. Typically, the digital thread connects digital twins, digital models of physical assets, or groups of assets

Digital Twin Refers to computerized companions of physical assets that can be used for various purposes. Digital twins use data from sensors installed on physical objects to represent their near real-time status, working condition or position

DOF Degree of Freedom

ECT Cost of the Configuration Tree

IoT Internet of Things

MGD Module Granularity Design

NMS Networked Manufacturing Systems

PCT Similarity Rate of the Configuration Tree

RCT Reconfigurability of the Configuration Tree

RIS Reconfigurable Inspection System

RIM Reconfigurable Inspection Machine

RMS Reconfigurable Manufacturing System

RMT Reconfigurable Machine Tool

RPSU Reconfigurable Production Subsystem

SGD System Granularity Design

SoV Stream of Variation

UCT Utilization Rate of the Configuration Tree

References

Abbas, M., & Elmaraghy, H. (2016). Design synthesis of machining systems using co-platforming. *Journal of Manufacturing Systems, 41*, 299–313.

Abdi, M. R. (2009). Fuzzy multi-criteria decision model for evaluating reconfigurable machines. *International Journal of Production Economics, 117*(1), 1–15.

Ahuett, H., Aca, J., & Molina, A. (2005). *A directed evolution modularity framework for design of reconfigurable machine tools, cooperative design, visualization, and engineering* (pp. 243–252). Berlin Heidelberg: Springer.

Ai-Ping, L. I., Meng, Q. D., Xie, N., & Zeng, F. L. (2011). Method for configuration design of reconfigurable machine tools based on process requirement. *Machine Design and Research, 27*(5), 72–71.

Al-Zaher, A., Elmaraghy, W., Pasek, Z. J., & Elmaraghy, H. (2013). Design of reconfigurable automotive framing System. *CIRP Annals Manufacturing Technology, 62*(1), 491–494.

Andersen, A. L., Brunoe, T. D., Nielsen, K., & Rösiö, C. (2017). Towards a generic design method for reconfigurable manufacturing systems: Analysis and synthesis of current design methods and evaluation of supportive tools. *Journal of Manufacturing Systems, 42,* 179–195.

Attanaik, L., & Kumar, V. (2010). Multiple levels of reconfiguration for robust cells formed using modular machines. *International Journal of Industrial and Systems Engineering, 5*(4), 424–441.

Barhak, J., Djurdjanovic, D., Spicer, P., & Katz, R. (2005). Integration of reconfigurable inspection with stream of variations methodology. *International Journal of Machine Tools and Manufacture, 45*(4–5), 407–419.

Benderbal H. H., Dahane M., & Benyoucef L. (2017). Layout Evolution Effort for Product Family in Reconfigurable Manufacturing System Design. IFAC *(International Federation of Automatic Control) paper online, 50*(1), 10166–10171.

Bruccoleri, M., Pasek, Z. J., & Koren, Y. (2006). Operation management in reconfigurable manufacturing systems: Reconfiguration for error handling. *International Journal of Production Economics, 98*(1), 87–100.

Chen, L., & Macwan, A. (2005). Optimal module selection for preliminary design of reconfigurable machine tools. *Journal of Manufacturing Science and Engineering, 127*(1), 104–115.

Deif, A. M., & Elmaraghy, W. H. (2006). A Control approach to explore the dynamics of capacity scalability in reconfigurable manufacturing systems. *Journal of Manufacturing Systems, 25*(1), 12–24.

Dou, J. P., Dai, X. Z., & Meng, Z. D. (2009). Graph theory-based approach to optimize single product flow-line configurations of RMS. *The International Journal of Advanced Manufacturing Technology, 41*(9), 916–931.

Gabor, T., Belzner, L., Kiermeier, M., Beck, M. T., & Neitz, A. (2016). A Simulation-Based Architecture for Smart Cyber-Physical Systems. In *IEEE International Conference on Autonomic Computing* (pp. 374–379).

Gadalla, M., & Xue, D. (2016). Recent advances in research on reconfigurable machine tools: A literature review. *International Journal of Production Research, 55*(5), 1440–1454.

Gadalla, M., & Xue, D. (2017). An approach to identify the optimal configurations and reconfiguration processes for design of reconfigurable machine tools. *International Journal of Production Research, 4,* 1–21.

Goyal, K. K., Jain, P. K., & Jain, M. (2012). Optimal configuration selection for reconfigurable manufacturing system using NSGA II and TOPSIS. *International Journal of Production Research, 50*(15), 4175–4191.

Goyal, K. K., Jain, P. K., & Jain, M. (2013). A novel methodology to measure the responsiveness of RMTs in reconfigurable manufacturing system. *Journal of Manufacturing Systems, 32*(4), 724–730.7.

Heisel, U., & Meitzner, M (2006). Progress in reconfigurable manufacturing systems. In reconfigurable manufacturing systems and transformable factories (pp. 47–62). Springer.

Jackson, M. O. (2011). A brief introduction to the basics of game theory.

Koren, Y., Heisel, U., Jovane, F., Moriwaki, T., Pritschow, G., Ulsoy, A. G., et al. (1999). Reconfigurable Manufacturing Systems. *CIRP Annals Manufacturing Technology, 48*(2), 527–540.

Koren, Y., & Ulsoy, A. G. (2002). Reconfigurable manufacturing system having a production capacity, method for designing same, and method for changing its production capacity. US Patent 6,349,237.

Koren, Y. (2013). The rapid responsiveness of RMS. *International Journal of Production Research, 51*(23–24), 6817–6827.

Koren, Y., & Kota, S. (1999). Reconfigurable Machine Tool. U.S. Patent No. 5943750, Washington, DC, USA.

Koren, Y., & Shpitalni, M. (2010). Design of reconfigurable manufacturing systems. *Journal of Manufacturing Systems*, *29*(4), 130–141.

Kuzgunkaya, O., & ElMaraghy, H. (2007). Economic and strategic perspectives on investing in RMS and FMS. *International Journal of Flexible Manufacturing Systems, 19*(3), 217–246.

Lee, E, A. (2008). Cyber physical systems: design challenges. In *IEEE symposium on object oriented real-time distributed computing IEEE computer society* (pp. 363–369).

Lee, J., Bagheri, B., & Kao, H.-A. (2015). A cyber-physical systems architecture for industry 4.0-based manufacturing systems. *Manufacturing Letters, 3,* 18–23.

Liu, W., & Liang, M. (2008). Multi-objective design optimization of reconfigurable machine tools: A modified fuzzy-chebyshev programming approach. *International Journal of Production Research, 46*(6), 1587–1618.

Mehrabi, M. G., Ulsoy, A. G., & Koren, Y. (2000). Reconfigurable manufacturing systems: Key to future manufacturing. *Journal of Intelligent Manufacturing, 11*(4), 403–419.

Mittal, K. K., & Jain, P. K. (2014). An overview of performance measures in reconfigurable manufacturing system. *Procedia Engineering, 69*(1), 1125–1129.

Milisavljevic-Syed, J., Commuri, S., Allen, J. K., & Mistree, F. (2019b). Concurrent design exploration method for realizing networked manufacturing systems for industry 4.0. In *52nd CIRP conference on manufacturing systems*, PROC-D-18–00262.

Moon, Y.-M., & Kota, S. (2002). Generalized kinematic modeling of reconfigurable machine tools. *Transactions-American Society of Mechanical Engineers Journal of Mechanical Design, 124*(1), 47–51.

Mpofu, K., Kumile, C. M., & Tale. N. S. (2008). Adaption of commercial off the shelf modules for reconfigurable machine tool design. In *International conference on mechatronics and machine vision in practice IEEE* (pp. 144–150).

Nan, X., Aiping, L. I., & Wei, X. (2012). Cooperative optimization of reconfigurable machine tool configurations and production process plan. *Chinese Journal of Mechanical Engineering, 25*(5), 982–989.

Padayachee, J., & Bright, G. (2012). Modular machine tools: Design and barriers to industrial implementation. *Journal of Manufacturing Systems, 31*(2), 92–102.

Padayachee, J., Bright, G., & Masekamela, I. (2009). Modular reconfigurable machine tools: Design, control and evaluation. *South African Journal of Industrial Engineering*, *20*(2), 127–143.

Pattanaik, L. N., Jain, P. K., & Mehta, N. K. (2007). Cell Formation in The Presence of Reconfigurable Machines. *International Journal of Advanced Manufacturing Technology, 34*(3/4), 335–345.

Puik, E., Telgen, D., Moergestel, L. V., & Ceglarek, D. (2016). Assessment of reconfiguration schemes for reconfigurable manufacturing systems based on resources and lead time. *Robotics and Computer-Integrated Manufacturing, 43,* 30–38.

Roozenburg, N. F., & Eekels, J. (1995). *Product design: Fundamentals and methods*. Chichester: Wiley.

Shabaka, A. I., & Elmaraghy, H. A. (2007). Generation of machine configurations based on product features. *International Journal of Computer Integrated Manufacturing, 20*(4), 355–369.

Shang, X., Milisavljevic-Syed, J., Wang, G., Allen, J. K., & Mistree, F. (2018). A key feature-based method for configuration design of reconfigurable inspection system. *International Journal of Production Research*, TPRS-2019-IJPR-1260.R1.

Son, S. Y., Olsen, T. L., & Yip-Hoi, D. (2001). An approach to scalability and line balancing for reconfigurable manufacturing systems. *Integrated Manufacturing Systems, 12*(7), 500–511.

Son, H., Choi, H. J., & Park, H. W. (2010). Design and dynamic analysis of an arch-type desktop reconfigurable machine. *International Journal of Machine Tools and Manufacture, 50*(6), 575–584.

Spicer, P., Koren, Y., Shpitalni, M., & Yip-Hoi, D. (2002a). Design principles for machining system configurations. *Annals of the CIRP, 51*(1), 275–280.

Spicer, P., Koren, Y., Shpitalni, M., & Yip-Hoi, D. (2002b). Design principles for machining system configurations. *CIRP Annals Manufacturing Technology, 51*(1), 275–280.

Wang, G., Shang, X., Yan, Y., Allen, J. K., & Mistree, F. (2018). A tree-based decision method for the configuration design of reconfigurable machine tools. *Journal of Manufacturing Systems, 49*, 143–162.

Wang, W., & Koren, Y. (2012). Scalability planning for reconfigurable manufacturing systems. *Journal of Manufacturing Systems, 31*(2), 83–91.

Xia, T., Xi, L., Pan, E., & Ni, J. (2016). Reconfiguration oriented opportunistic maintenance policy for reconfigurable manufacturing systems. *Reliability Engineering and System Safety, 166*, 87–98.

Xu, Z., Xi, F., Liu, L., & Chen, L. (2017). A method for design of modular reconfigurable machine tools. *Machines, 5*(1), 5.

Youssef, A. M., & ElMaraghy, H. A. (2006). Modelling and optimization of multiple aspect RMS configurations. *International Journal of Production Research, 44*(22), 4929–4958.

Youssef, A. M. A., & Elmaraghy, H. A. (2007). Optimal configuration selection for reconfigurable manufacturing systems. *International Journal of Flexible Manufacturing Systems, 19*(2), 67–106.

Chapter 7
Closure

In this monograph we address the challenges in architecting networked engineered systems that are anchored in the Industry 4.0 construct. Global markets, rapidly evolving technologies, and changing customer preferences have all given rise to distributed manufacturing connected by internet technologies. While digitization helps in connecting erstwhile distributed systems and enables the use of digital models of physical processes in design and analysis, it does not automatically translate into the design of "smart" systems. In this monograph, we present a computational framework, Design for Dynamic Management (DFDM), conforming the architecture for realization of digitized manufacturing. The computational framework, DFDM, is designed to facilitate the harmonization of conflicting needs associated with the design of Networked Engineered Systems (NES), such as flexibility, robustness, resilience, operability, connectivity, and the incorporation of real-time analytics and smart sensing. The application of the computational framework, DFDM, is not limited to design of manufacturing systems but design of engineered systems in general, such as design of chemical processes, and health care systems.

In this chapter, we present a summary of the monograph, revisit the questions and hypotheses we address, and suggest a way forward for transitioning DFDM to industry; see Table 7.1. In Sect. 7.1, we present a short summary of the monograph. In Sect. 7.2, we revisit and evaluate the questions and the hypotheses introduced in Chap. 1 with special emphasis on the validity of the hypotheses. In Sect. 7.3, we present the contributions embodied in this monograph and the limitations of the proposed computational framework. Lastly, in Sect. 7.4, we identify guiding principles for Architecting Networked Engineering Systems to transition to industry.

7.1 A Summary of the Monograph

In this monograph we advocate that integration of adaptability, operability, and reconfigurability in the design of NES. Furthermore, current approaches are neither agile nor rapidly configurable and do not have built-in flexibility in selection and determination of values of design parameters, such as tools and/or sensor, and tolerances

J. Milisavljevic-Syed et al., *Architecting Networked Engineered Systems*,
https://doi.org/10.1007/978-3-030-38610-8_7

Table 7.1 Connection between questions, contributions, limitations, and future work

Chapter	Secondary questions	Contributions	Limitations	Future work
Chapter 3	What is the computational framework in the design method that facilitates adaptable design in the realization of networked engineered systems? Section 7.2.1	• Identifying drivers and relations in concurrent design • Build in flexibility in selection/determination of design parameters • Managing the high-complexity mathematical problem • Creating decision network structure as a decision support • Integration of process- and product-related decision models Section 7.3.1	• Capture the knowledge in design of NES • Multidisciplinary knowledge exchange between different domains • Interfacing domain ontologies, converting different analogies between different domains Section 7.3.1	Knowledge-based decision-based design ontology Section 7.4.1

(continued)

Table 7.1 (continued)

Chapter	Secondary questions	Contributions	Limitations	Future work
Chapters 4 and 5	What is the computational framework in the design method that facilitates dynamic change in the requirements in the realization of functional networked engineered systems? Section 7.2.2	• Framing operability/disturbance space without domain knowledge • Operability analysis of engineering systems • Managing dynamic performance of the system Section 7.3.2	• Decision-based design templates for integrating consumer and producer preferences Section 7.3.2	Decision-based design platform Section 7.4.2
Chapter 6	What is the reconfiguration strategy for reestablishing connection among elements in the system in the realization of networked engineered systems to remain competitive in the market? Section 7.2.3	• Reconfiguration strategy of manufacturing tool • Reconfiguration strategy of inspection system • Reconfiguration of manufacturing system Section 7.3.3	• Multiple configurations of elements from different systems • Reach sustainability by integrating cyber-physical-social systems in design • Cyber-social design decision network sensitive to dynamic changes of consumer's and producer's preferences in design Section 7.3.3	Integration of cyber-physical-social systems in the platform Section 7.4.3

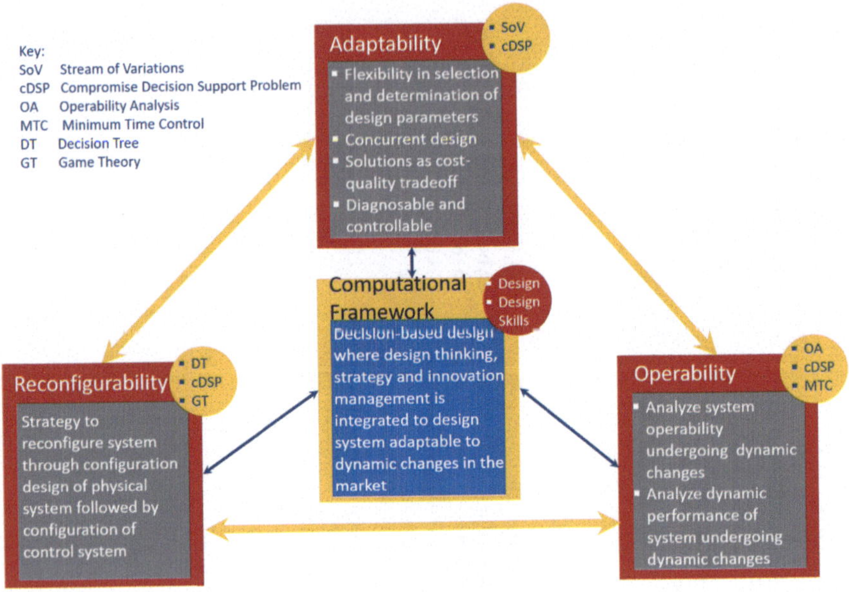

Fig. 7.1 Computational framework, design for dynamic management (Milisavljevic-Syed et al. 2019b)

for errors that are introduced, accumulated and propagated in the process. We suggest that the computational framework, DFDM, can be generalized and applied for product development and design of service systems; see Fig. 7.1.

7.1.1 *Method for Adaptive Concurrent Realization of Networked Engineering Systems (ACRONES)*

Networked engineering systems such as multistage manufacturing processes (MMPs) are complex processes consisting of multiple manufacturing stations and operations and are commonly encountered in applications such as automotive machining, assembly of electronic products, and semiconductor lithography processes (Shi and Zhou 2009). Ensuring quality of the produced goods in these MMPs requires the capability to detect defects, identify their likely causes and then adjust control parameters to eliminate or reduce these defects. In the past, given an implementation of MMP, state-space representation of the MMP is first derived and this mathematical representation was then used to analyze the root cause of failures (Ding et al. 2002). This representation is also used to determine whether the implementation is controllable, for example, if the cause of the defect could be eliminated through the proper adjustment of control parameters. While this approach is suited for analyzing the properties of MMP, it is not useful for selecting sensing and control components and their respective placements in the MMP. Further, the overall cost and performance of the MMP is dictated by this choice and is hard to ascertain at design time. Next generation manufacturing processes have to be adaptable, for example, able to handle many different product types, robust to fixturing and other forms of errors that can degrade product quality and offer cost-effective implementation of the entire process.

In this monograph, we propose a systematic method for the concurrent design of NES. We use the method to exploit the flexibility in selection and determination of the values of process/systems variables at design time to simultaneously address requirements such as controllability and diagnosability and lower the overall cost during the execution of NES cDSPs while ensuring that system constraints are satisfied; see Adaptability in Fig. 7.1.

In Fig. 7.2, we present the solution scheme of the ACRONES. The method is based on the compromise Decision Support Problem (cDSP) construct (Mistree et al. 1993) for an MMP, where an MMP is described by a Stream of Variation (SoV) model (Ding et al. 2000); see Fig. 7.3. The proposed method is illustrated using an example of an automotive panel stamping process. For more information about the method; see Chap. 3. The results are presented in several publications (Milisavljevic-Syed et al. 2019a; Milisavljevic et al. 2017; Milisavljevic 2015).

7.1.2 *Method for Operability Analysis in the Realization of Networked Engineering Systems*

One of the challenges faced in architecting networked engineering systems is big data analysis in real time. For example, the effect of dynamic market changes on an NES can only be ascertained by observing the data from all the processes in the NES over an extended period of time. In this monograph, we demonstrate one such approach using a technique called Operability Analysis. Operability Analysis provides the connection between the design and control of the process (Georgakis et al. 2003) and can be used to analyze the dynamic and steady-state performance of the NES as the system requirements change with dynamic market changes. Operability analysis is accomplished in two stages (1) steady-state operability analysis, and (2) dynamic operability analysis.

Steady-State System Operability

We use the steady-state operability to analyze how different requirements are changing system functionality that are driven by customer needs. System functionality is analyzed using Steady-State Operability Model (SSOM) based on operability analysis and the compromise Decision Support Problem (cDSP) construct. We use the cDSP construct to manage the structure and information of decision-making; see Operability in Fig. 7.1.

In Fig. 7.4, we present the solution scheme of the steady-state operability model. The proposed model is verified using an illustrative example of an automotive panel stamping process. For more information about the SSOM; see Chap. 4. The results are presented by Milisavljevic-Syed et al. (2019a).

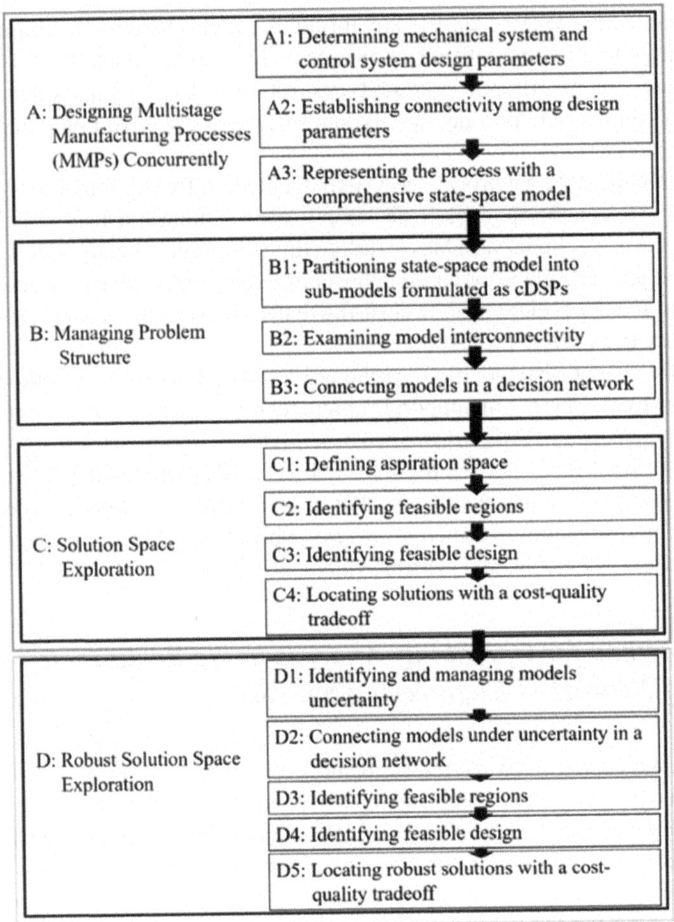

Fig. 7.2 Adaptive concurrent realization of engineering systems (ACRONES) framework (Milisavljevic et al. 2017)

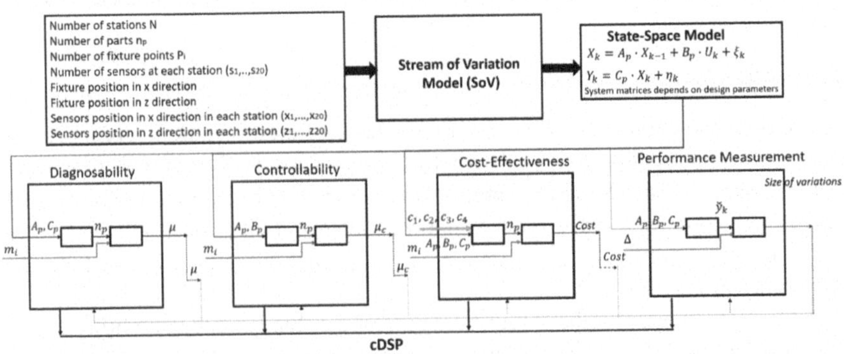

Fig. 7.3 Connection between SoV and cDSPs

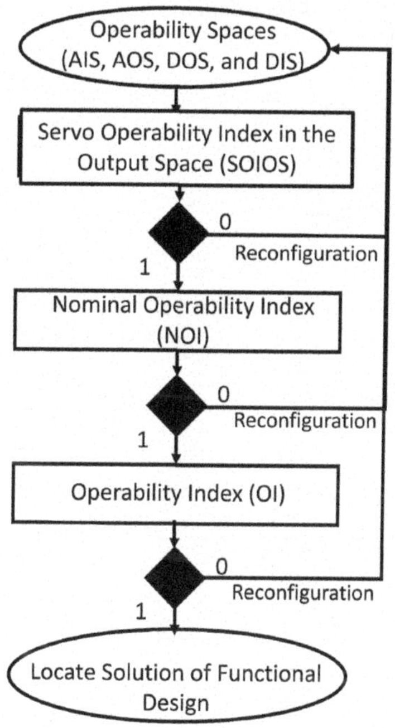

Fig. 7.4 SSOM solution scheme (Milisavljevic-Syed et al. 2018)

Dynamic System Operability

We use the dynamic operability to analyze dynamic performance of the NES and how that performance is changing over time due to changes in requirements. We analyze dynamic performance of a system with the Dynamic Operability Model (DOM) that is based on operability analysis and minimum-time control; see Operability in Fig. 7.1.

The solution scheme to ensure dynamic operability is presented in Fig. 7.5. The proposed method is illustrated using an example of a continuous stirred tank reactors. For more information about the DOM; see Chap. 5. For additional details; see Milisavljevic-Syed (2018).

7.1.3 *Reconfigurable Realization of Networked Engineering Systems*

Often, deterioration of product quality or small changes in the product specifications can be met by suitably modifying the associated sensors, tools, and control software of the NES. However, sometimes the desired changes necessitate large modifications in the functioning of the NES that cannot be met through simple changes. In such instances, it might be necessary to implement an altogether new design to

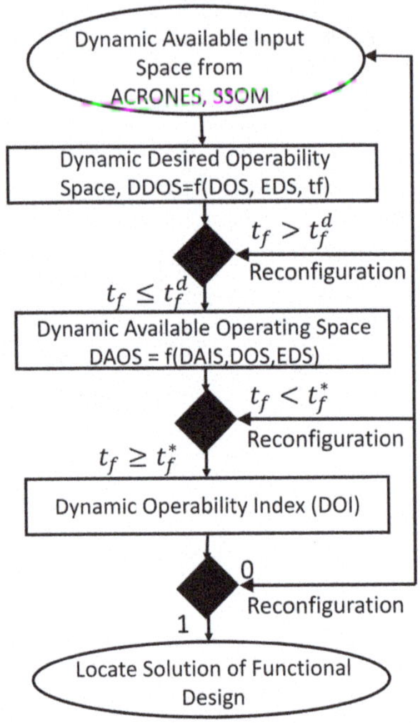

Fig. 7.5 DOM solution scheme

meet the new specifications. In this monograph, we advocate that, with design of Reconfigurable Manufacturing System (RMS) customized manufacturing processes that meet the changes in operational requirements or machine status is possible. If NES is designed to be a superset of all necessary designs to address all possible performance scenarios, then appropriate designs can be selected at reconfiguration time and thereby meet the operational requirements without the need to completely redesign the system from ground up. We show that effective realization of an RMS is supported by the computational framework, DFDM, that allows the detection of process errors and manages the reconfiguration strategy to meet requirements, see Reconfigurability in Fig. 7.1. As an example, we consider the reconfiguration of tools and sensors to accommodate process errors in a NES. Here, the reconfiguration is accomplished in two stages (or two parts) (1) reconfiguration of machine tools (RMT), and (2) reconfiguration of the inspection system (RIS).

A Reconfigurable Machine Tool (RMT) is repeatedly configured in order to respond to a wide range of requirements, such as the change in function or capacity. Module selection and the assembly relationships among modules in the end-assembly of the machine tool are central to any method for configuration design of reconfigurable machine tools. Typically, module models reside in different libraries and their descriptions vary. In this monograph, a tree-based method is proposed to tie information in different libraries (Wang et al. 2018). The method includes (1) defining a common model for reconfigurable machine tool configuration using a tree structure; (2) identifying concepts that combine the desired function and capability; and

(3) determining those concepts while minimizing cost and maximizing flexibility; see Fig. 7.6. For more information about the RMT; see Sect. 6.3. The results are presented by Wang et al. (2018).

Typically, research in reconfiguration revolves around production while ignoring the inspection of the system. In the RMS, Reconfigurable Inspection System (RIS) is designed to achieve the data-oriented detection of the product quality with the minimum but sufficient inspection machines. Thus, a key feature-based method is

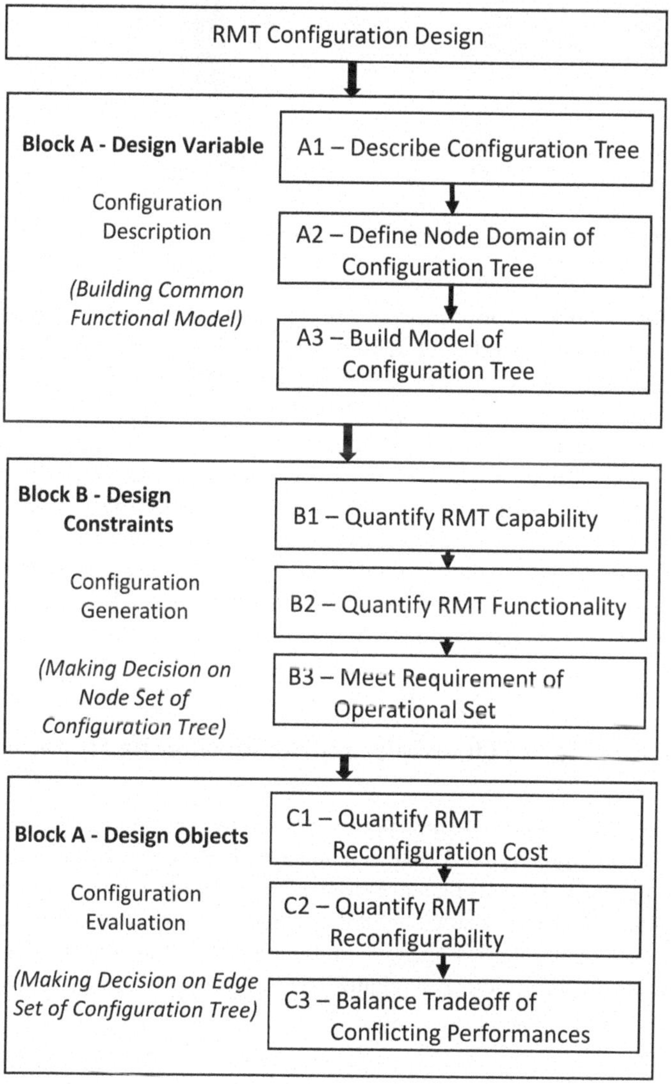

Fig. 7.6 Design method for the RMT configuration (Wang et al. 2018)

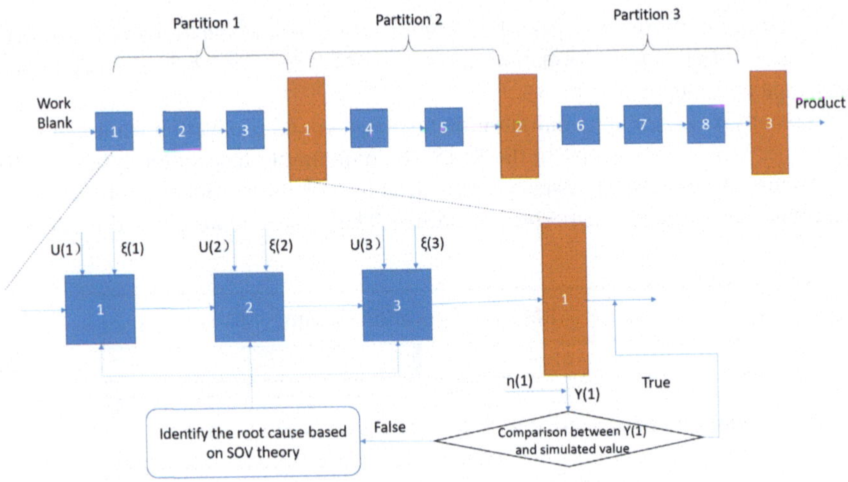

Fig. 7.7 Detection Mechanism via Integration of RIS and SoV

proposed for RIS configuration design to determine the satisfied RIS design (number of inspection stations, number of sensors and position of sensors) that detect the different processes, thereby meets the inspectional requirement for each phase of RMS lifecycle. First, the key features of RIS (for example, Modularity, Integrability, Customization, Scalability, Convertibility, and Diagnosability) are identified based on the detection mechanism of the RMS. Second, the model-based specific procedure to explore the RIS configuration design is introduced to ensure the RIS satisfies the multiple constraints (for example, inspectional functionality and capability) and goals (for example, cost and diagnosability). In Fig. 7.7, we present the detection mechanism of the RIS. For more information about the RMT; see Sect. 6.4. The results are presented in publications (Milisavljevic-Syed 2018).

7.2 Answering the Questions and Validating the Hypotheses

In this monograph we propose a computational framework suitable for the design of low-cost and high-quality NES adaptable to ambitious market demands as a support to further digitization (smart manufacturing). The primary motivation is to establish a computational framework suitable for the design of low-cost and high-quality NES adaptable to dynamic market changes in the context of Industry 4.0. The concept of computational framework is exploited in the context of addressing the primary question; see Sect. 1.4.2.

Primary Question (PQ). *In the context of Industry 4.0 what is the computational framework that facilitates the decision-based design of networked engineered systems adaptable to ambitious market demands as a support to further digitization?*

Primary Hypothesis (PH). *We hypothesize, that by applying the computational framework, we can support flexible, operable and rapidly configurable design of the manufacturing processes. Furthermore, by introducing flexibility in selection of design parameters, concurrent design of a mechanical and a control system, analyzing dynamic performance of the systems, and applying different reconfiguration strategies further digitization of NES can be ensured.*

The secondary questions are unified in the primary question; see Sect. 1.4.2. With SQ1 we address adaptable concurrent design of NES. With SQ2 we address operability analysis of NES. With SQ3 we address reconfiguration strategy of NES. The knowledge we obtain by answering the SQ 1 we further use to answer the SQ2, and knowledge we obtain by answering the SQ2 we further use to answer the SQ3.

Secondary Question 1 (SQ1). *What is the computational framework in the design method that facilitates adaptable design in the realization of networked engineered systems?*

Secondary Question 2 (SQ2). *What is the computational framework in the design method that facilitates dynamic change in the requirements in the realization of functional networked engineered systems?*

Secondary Question 3 (SQ3). *What is the reconfiguration strategy for reestablishing connection among elements in the system in the realization of networked engineering systems to remain competitive in the market?*

We propose the hypotheses in order to answer secondary questions as a support to the principal goal of the monograph. The end result is a new knowledge, a computational framework, DFDM, suitable for the design of low-cost and high-quality NES adaptable to dynamic market changes. Furthermore, we use gained insights to identify research gaps for way forward and transition to industry. We validate hypotheses in each chapter according to verification and validation roadmap presented in Sect. 1.5. Further, in Tables 7.2, 7.3 and 7.4, we revisit and summarize validation of hypotheses

7.2.1 Secondary Hypothesis 1 (SH1): Adaptable Concurrent Design of Networked Engineering Systems

In this section, we discuss SH1 in order to answer the SQ1. We introduce SH1 in Chap. 1 and further addressed in Chap. 3; see Table 7.2.

Secondary Hypothesis 1 (SH1). *Design of mechanical and control system concurrently while accounting for different types of uncertainty and extensive (robust) solution space exploration facilitates the adaptable design of networked engineering systems.*

In the design of NES there are subjective elements involved in decision making that reflects the initial design stage of the computational framework implementation,

Table 7.2 Validation of SH1—summary

Hypothesis	Validation	Details
SH1 Facilitating flexible design of NES with concurrent design while accounting for uncertainty and solution space exploration	*Theoretical structural validation*	
	Motivation for adaptable concurrent design of NES	Section 1.1
	Introduction to concurrent design of NES	Sections 1.2, 1.3
	Mathematical background of adaptable concurrent design model	Section 2.3.1
	Literature review	Section 3.1
	Empirical structural validation	
	Illustrative example, panel assembling process	Section 2.1.3, Fig. 2.3
	Designing NES concurrently	Section 3.3.1, Fig. 3.3
	Managing problem structure	Section 3.3.2, Fig. 3.4
	Solution space exploration	Section 3.3.3
	Robust solution space exploration	Section 3.3.4, Fig. 3.6
	Empirical performance validation	
	Adaptable concurrent realization of networked engineering systems (ACRONES)	Section 3.3, Fig. 3.3
	The comprehensive computational model	Section 3.4
	The cost-quality relationship analysis	Section 3.5, Figs. 3.13–3.15

therefore, there is a need to undertake the verification and validation of a computational framework/result due to the fact that a development of a computational framework includes many abstract elements and there is no unique answer.

The "Validation Square" is a method where designers build confidence in the utility of methods and examples that are used to verify the method (Seepersad et al. 2006), see Sect. 1.3.4. Further, the validation square is used to determine whether the method provides correct design solutions regarding structural validity and regarding performance validity (Choi 2005).

The validation square is the process of validation and the validation quadrants are:

- ***Quadrant 1****. Theoretical Structural Validity.* Examining the structural/logical validity and overall consistency of the proposed computational framework.
- ***Quadrant 2****. Empirical Structural Validity.* Includes building the confidence of the example problems chosen to verify a suggested design computational framework.

Table 7.3 Validation of SH2—summary

Hypothesis	Validation	Details
SH2 Determine functionality of the system at steady-state and dynamic state allowing a system to adjust and stabilize in presence of change	*Theoretical structural validation*	
	Motivation for operability analysis in design of NES	Section 1.1
	Mathematical background of operability analysis model	Section 2.3.2
	Literature review	Sections 4.1, 5.1
	Empirical structural validation	
	Illustrative example, continuous stirred tank reactors	Section 2.1.3, Fig. 2.4
	Steady-state operability model	Section 4.3, Fig. 4.4
	Dynamic operability model	Section 5.2, Fig. 5.3
	Empirical performance validation	
	Operability analysis model	Sections 4.3, 5.2
	Steady-state operability model	Section 4.5
	Dynamic operability model	Section 5.1

- ***Quadrant** 3. Empirical Performance Validity*. Used to build confidence in the applicability of a computational framework for the example problems that are chosen.
- ***Quadrant** 4. Theoretical Performance Validity*. Building confidence in the general use of the computational framework and determining is it useful for other problems beyond the example problems.

In this monograph, the validation square is adopted to illustrate the layout and the connections between the chapters, see Fig. 7.8.

Theoretical Structural Validation

Integrating flexibility in the selection of design parameters, such as tool and sensor attributes is necessary in order to better address the issue of dimensional errors. Furthermore, concurrent design of mechanical and control system is necessary in order to design a process that is both diagnosable and controllable. Hence, SH11 is the important link in the computational framework, Design for Dynamic Management, due to dynamic market changes that influence the design of NES. We present a motivation for the decision-based design of NES; see Sect. 1.1. In Sect. 3.1, we present existing design approaches. We present a problem of Decision-Based Design (DBD) of NES; see Sect. 2.1. In Sect. 2.2, we present a generic cDSP formulation. In Sect. 2.3, we present the mathematical background of the computational framework. Further, in Sect. 2.4, we present the overall computational framework, Design for Dynamic Management. Lastly, in Chap. 3, we present a method, Adaptable Concurrent Realization of Networked Engineering Systems (ACRONES).

Table 7.4 Validation of SH3—summary

Hypothesis	Validation	Details
SH3 Machine tool and inspection system reconfiguration allows reconfiguration of manufacturing system	*Theoretical structural validation*	
	Motivation for reconfiguration strategy in design of NES	Section 1.1
	Mathematical background of reconfiguration strategy	Section 2.3.3
	Literature review	Section 6.1
	Empirical structural validation	
	Illustrative example, transmission box and engine cylinder block	Section 2.1.3, Figs. 2.5 and 2.6
	Reconfigurable machine tool (RMT)	Section 6.3
	Reconfigurable inspection system (RIS)	Section 6.4
	Application of RMT configuration design	Section 6.6.1
	Application of RIS configuration design	Section 6.6.2
	Empirical performance validation	
	Reconfigurable manufacturing system (RMS)	Section 6.5
	Application of RMS configuration design	Section 6.6.3

Empirical Structural Validation

Adaptable Concurrent Realization of Networked Engineering Systems (ACRONES) is a comprehensive model that consists of (1) model for concurrent design of NES, (2) managing problem structure by integrating process decision models (diagnosability, controllability, and cost-effectiveness) with product quality models (performance measurement) in a comprehensive adaptability model, (3) solution space exploration, and (4) robust solution space exploration. In Sect. 3.3.1, we present the concurrent design of NES for identifying flexible design parameters, establishing their connectivity, and representing the process with a comprehensive state space model. In Sect. 3.3.2, we manage problem structure for determining interconnections between networked engineering system modules and the mathematical representation of the complete system. In Sect. 3.3.3, we explore the solution space for appropriate solutions. In Sect. 3.3.4, we explore the robust solution space for identifying and managing different types of uncertainty, exploring solution space, and identifying robust solutions of the design problem. We include drivers from both the mechanical and

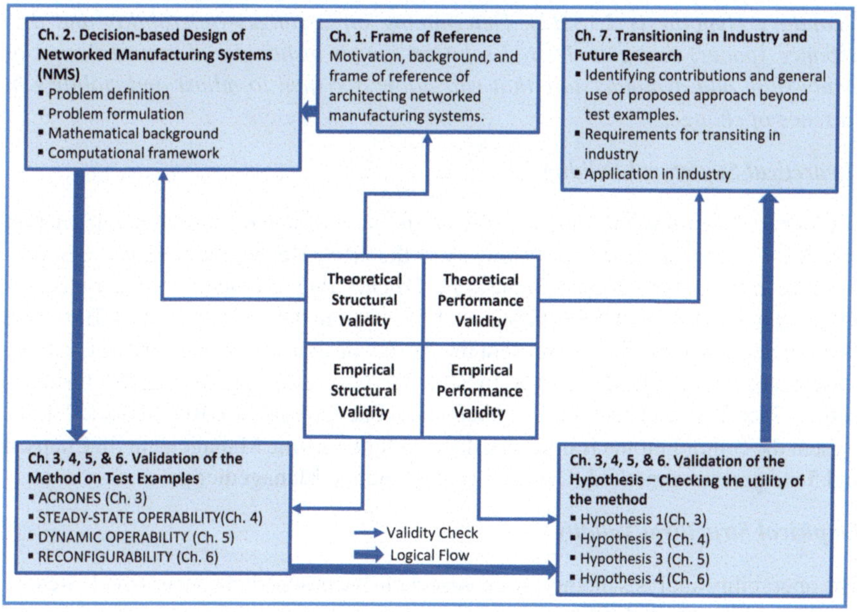

Fig. 7.8 Validation strategy of monograph

control systems in the models we for adaptable concurrent design in order to unify simulation parameters and represent the appropriate example for validating the SH1.

Empirical Performance Validation

Adaptable Concurrent Realization of Networked Engineering Systems (ACRONES) is a design method, a part of the computational framework, Design for Dynamic Management. Empirical performance validity of the ACRONES is performed in four steps (1) concurrent design of NES, (2) managing problem structure, (3) solution space exploration and locating solutions as a cost-quality tradeoff, and (4) robust solution space exploration and locating robust solutions as a cost-quality tradeoff. In Sect. 3.3, we present the ACRONES. In Sect. 3.5, we present the empirical performance validation of the ACRONES.

7.2.2 Secondary Hypothesis 2 (SH2). Operability Analysis of Networked Engineering Systems

In this section, we discuss SH2 in order to answer the SQ2. We introduce the SH2 in Chap. 1 and further address in Chaps. 4 and 5; see Table 7.4. SH1 is a support to operability analysis presented in SH2.

Secondary Hypothesis 2 (SH2). *Determining input ranges (operability and disturbance spaces) that give desired solution ranges of functional system design at steady-state and dynamic state that will allow a system to adjust and stabilize in presence of change.*

Theoretical Structural Validity

In order to determine the functionality of the system undergoing dynamic market changes we propose operability analysis of the NES. Hence, the SH2 is the important link in the computational framework, DFDM, due to dynamic market changes influencing the design of NES. In Sect. 1.1, we present the motivation for DBD of the NES. In Sects. 4.1 and 5.1, we present the current design approaches. In Sect. 2.1, we present a problem of DBD of NES. In Sect. 2.2, we present a general cDSP formulation. In Sect. 2.3, we present the mathematical background of DBD. In Sect. 2.4, we present the computational framework, Design for Dynamic Management. In Chaps. 4 and 5 we present a method, Operability in Dynamic Management.

Empirical Structural Validity

The operability analysis method is a comprehensive model that consists of (1) steady-state operability model (SSOM), and (2) a dynamic operability model (DOM). The steady-state operability model (SSOM) analyses the operability spaces obtained from ACRONES and disturbance spaces formed by customer needs in order to identify possible solutions of the functional system. The dynamic operability model (DOM) examines dynamic performance of the system undergoing dynamic changes. Operability analysis method is foundation for analyzing the functionality of the system undergoing dynamic changes, therefore, appropriate for validating SH2.

Empirical Performance Validity

Operability analysis is a design method, a part of the computational framework, Design for Dynamic Management. Empirical performance validity of the operability analysis is performed in three steps (1) obtaining information from ACRONES to frame operability spaces and information from customer to frame disturbance spaces, (2) steady-state operability analysis of the system, and (3) dynamic performance of the system and locating operable (functional) solutions. In Sects. 4.3 and 5.2, we present the operability analysis method. In Sects. 4.5 and 5.4, we present the empirical performance validation of the operability analysis.

7.2.3 Secondary Hypothesis 3 (SH3). Reconfiguration Strategy of Networked Engineering Systems

In this section, we discuss the SH3 in order to answer the SQ3. We introduce SH3 in Chap. 1 and further address in Chap. 6; see Table 7.3.

Secondary Hypothesis 3 (SH3). *Machine tool reconfiguration followed by inspection system reconfiguration will allow reconfiguration of the manufacturing system and re-establishing connection among elements in the system.*

Theoretical Structural Validation

In order to establish and maintain connectivity we propose reconfiguration strategies we use to accomplish multiple configurations of elements in the system if required. Hence, SH3 is the important link in the computational framework, Design for Dynamic Management, due to dynamic market changes in the design of NES. In Sect. 1.1, we present the motivation for the DBD of NES. In Sect. 6.1, we present the current reconfiguration methods. In Sect. 2.1, we present a problem of DBD of NES. In Sect. 2.2, we present the general cDSP formulation. In Sect. 2.3, we present the mathematical background of the computational framework. In Sect. 2.4, we present the computational framework, Design for Dynamic Management. In Sect. 6.5, we present a method, Reconfigurable Manufacturing System.

Empirical Structural Validation

The reconfigurable Manufacturing System (RMS) discussed in this monograph consists of (1) Reconfigurable Machine Tool (RMT), and (2) Reconfigurable Inspection System (RIS). The Reconfigurable Machine Tool (RMT) consists of multiple configurations of the tools and reconfiguration is accomplished by the dynamic selection of tools during the operation of RMS. The Reconfigurable Inspection System (RIS) consists of multiple inspection tools and reconfiguration of RIS is accomplished by the selection of appropriate inspection tools during the operation of RMS. Reconfiguration methods is the foundation for reconfiguration of manufacturing system and allows multiple simultaneous configurations of elements in the system in same time, therefore, are appropriate for validating SH3.

Empirical Performance Validation

Dynamic reconfiguration of manufacturing system is a part of the computational framework, Design for Dynamic Management. Validity and empirical performance of the RMS is performed in two steps (1) reconfiguration of a machine tool, and (2) reconfiguration of an inspection system. In Sect. 6.5, we present the RMS. In Sect. 6.6.3, we present the empirical performance validation of the RMS.

7.3 Contributions and Limitations

The contributions in the monograph are divided into three categories. The first contribution is adaptable concurrent design method, ACRONES, directly related to SH1; see Sect. 1.4.2. Further, the adaptive concurrent design has a direct influence on the creation of the decision-based design platform, particularly of the knowledge-based decision-based design ontology for multidisciplinary knowledge exchange, which we identify as a way forward; see Sect. 7.4.1. The second contribution is method for operability analysis in the steady-state and dynamic-state, SSOM and DOM, in the design of networked engineering systems directly related to SH2; see Sect. 1.4.2. Further, operability analysis has a direct influence on a decision-based design platform, particularly integrating producer and consumer needs in design, which we identify as a way forward; see Sect. 7.4.2. The third contribution is the reconfiguration method of networked engineering systems related to SH3; see Sect. 1.4.2. Further, reconfiguration method takes decision-based design platform step forward in order to fit any engineered system in the platform, such as cyber-physical-social system, which we identify as a way forward; see Sect. 7.4.3.

7.3.1 Adaptive Concurrent Design of Networked Engineering Systems

Expected contributions related to SH1 are the following:

- Identifying the mechanical and control system drivers and their relations in concurrent design;
- Building in flexibility in selection and determination of values of design parameters in both systems;
- Manage the structure of the high-complexity mathematical problem;
- Create effective and efficient decision network structure as a decision support; and
- Integration of process and product-related decision models in the comprehensive model.

Identified limitations related to SH1 are the following:

- Unable to capture the information/knowledge in the design of networked engineering systems;
- Unable to achieve multidisciplinary knowledge exchange between different domains, beyond mechanical and control engineering; and
- In concurrent design of a mechanical and a control system there is need for interfacing domain ontologies. The identified limitation is inability to interface domain ontologies, and converting different analogies between different domains in order to exchange knowledge between different users and make effective and efficient design decisions.

7.3.2 Operability Analysis of Networked Engineering Systems

The contributions related to SH2 are as follows:

- Frame operability and disturbance space with limited domain knowledge;

- Analyze the operability of engineered systems; and
- Manage the dynamic performance of the system when requirements of the product change.

The limitation related to SH2 follows:

- Unable to integrate consumer and producer preferences in system design, product and service system design requires the use of templates for decision-based design.

7.3.3 *Reconfiguration Strategy of Networked Engineering Systems*

The contributions related to SH3 are the following:

- Strategy for Reconfiguration of Manufacturing System (RMS);
- Strategy for Reconfiguration of Manufacturing Tool (RMT); and
- Strategy for Reconfiguration of Inspection System (RIS).

The limitations related to SH3 are the following:

- Unable to manage multiple configurations of elements from different systems from different domains, such as social, economy, etc.;
- Unable to reach sustainability plateau without integrating cyber-physical-social systems in the design of systems, product-service systems; and
- Unable to integrate cyber-social design decision network in order to be sensitive to dynamic changes of consumer's and producer's preferences. However, this limitation we can overcome with integration of Cyber-Physical-Social Systems which we identified as a way forward; see Sect. 7.4.3.

7.4 Future Work—Architecting Networked Engineering Systems Transition to Industry

According to Schaefer (2017) the evolution of cyber-physical systems and the integration of Internet of Things (IoT) and Internet of Services (IoS) to address the needs of the industries of future is anchored in the Industry 4.0 construct that was first presented at the Hanover Trade Fair in Germany in 2011. Political changes in the 90's paved the way for globalization of manufacturing industries. As a result, collaborations that were previously limited to engagements between countries (G1 in Fig. 7.9) now shifts to collaborations between companies (G2), and lastly to collaborative efforts between individual collaborators (see G3), thereby creating the environment for new business opportunities (see the first row in Fig. 7.9). These direct interactions with collaborators in the society require tools and processes to

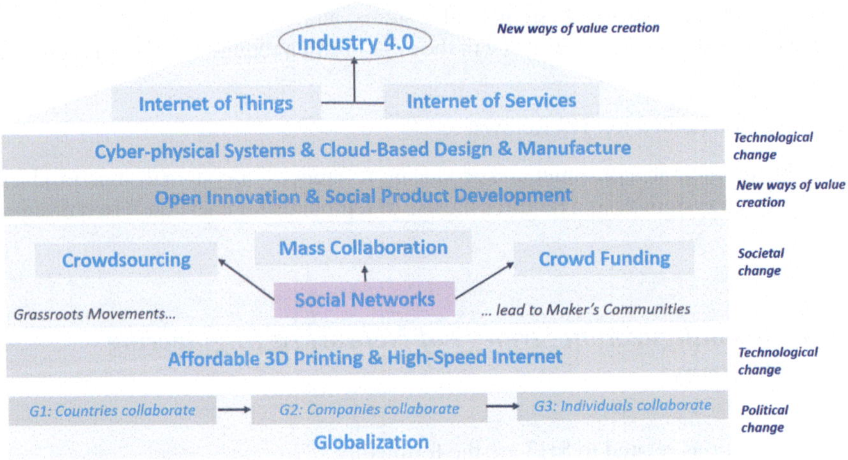

Fig. 7.9 The evolution of industry 4.0 (Schaefer 2017)

support Open Innovation and Product Development. Further, these interactions can only be possible when there are virtual representations of every entity in the manufacturing process. The new ways of value creation that emerged through this social partnership, spurred technological changes that resulted in the integration of cyber and physical systems as well as development of tools for cloud-based design and manufacture (see Fig. 7.9).

Trend of Product-Service-Systems and Cloud-Based Design and Manufacturing

As the cyber and physical worlds converge, many of the 'decision making' tasks hitherto performed in the physical world are likely to be performed in the cyber realm. This is especially advantageous as large data analytic methods can now be used to mine low-level sensor data for enterprise-wide solutions (Intelligroup 2019). The convergence of cyber and physical worlds has also indirectly spurred the convergence of cyber-physical and social systems (CPSS). Researchers have realized that the Internet infrastructure that is central to the development of CPS is not conducive to human interactions that are central to social networks. As a result, there has been increased interest in modeling human interactions both in CPS and CPSS. Therefore, a new paradigm called 'Internet of People' was proposed to capture the human-centric nature of the next generation of systems*(Conti et al. 2017). Since Industry 4.0 embodies collaborative relationships at all levels, implementation of manufacturing enterprises based on the Industry 4.0 construct must account for the interactions between the IoT and the IoP with the IoP (individual and social behaviors).

The trend of Cloud-based Design and Manufacturing (CBDM) is a service-orientated product development construct where consumers are afforded opportunity to design and customize products (Schaefer 2017). IoT and IoS play a major role and make possible realization from conceptualization to production just based on a consumer's idea of a new product. For customized design and manufacture to fully

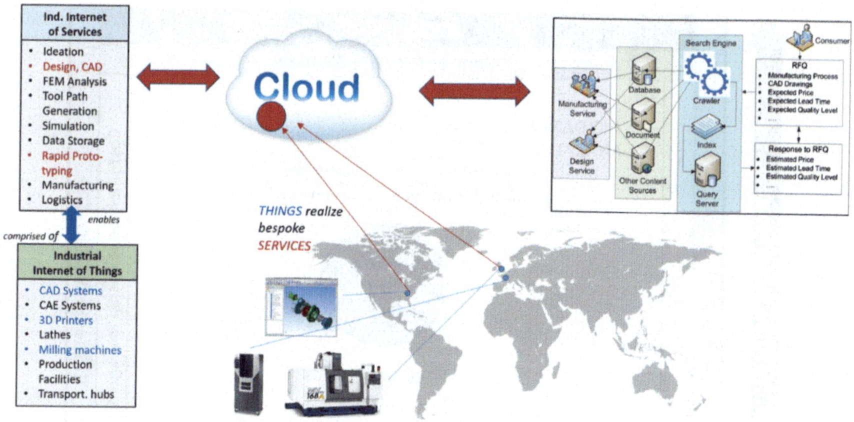

Fig. 7.10 A service-oriented cloud-based design and manufacturing scenario (Schaefer 2017)

succeed, one has to first model the process, represent the entire production stream by its equivalent digital thread, and allow access to information at every stage of design and manufacturing. Typical service-based cyber-physical product creation scenario is best explained by Schaefer (2017); see Fig. 7.10.

Transition to Industry through Platformization

Digitization of manufacturing systems that conforms to the Industry 4.0 construct is not a trivial task and can only be realized through the adoption of an appropriate computational architecture; see Fig. 7.11. The architecture must facilitate the integration of digital threads in two ways. The *vertical integration* of a digital thread facilitates the flow of information between the enterprise and resource planning entities to manufacturing services and planning and control entities within a manufacturing facility, see Fig. 7.11. In the vertical integration low-level components such as IoT sensors, machine controllers and their digital twins across the manufacturing enterprise are networked with big data analytics to monitor the manufacturing process and improve system performance and product quality. Vertical integration also allows for the creation of interoperable 'systems of systems' that can integrate seemingly diverse set of machines, sensors and controllers to produce a resilient manufacturing process that is capable of withstanding disruptions in the supply network, environment and in the market. *Horizontal integration* deals with data and material flows across the end-to-end value chain, see Fig. 7.11. The horizontal integration is represented by several digital twins from supply chain and the manufacturing processes, data flows and IT systems in the product development, to logistics, distribution and ultimately the customer. In Fig. 7.12, we present the elements of the architecture that can be expanded on in the context of digitized manufacturing processes.

In the following section we address the following questions (1) what are the Requirements for Transition to Industry?; and (2) what is the Way Forward?.

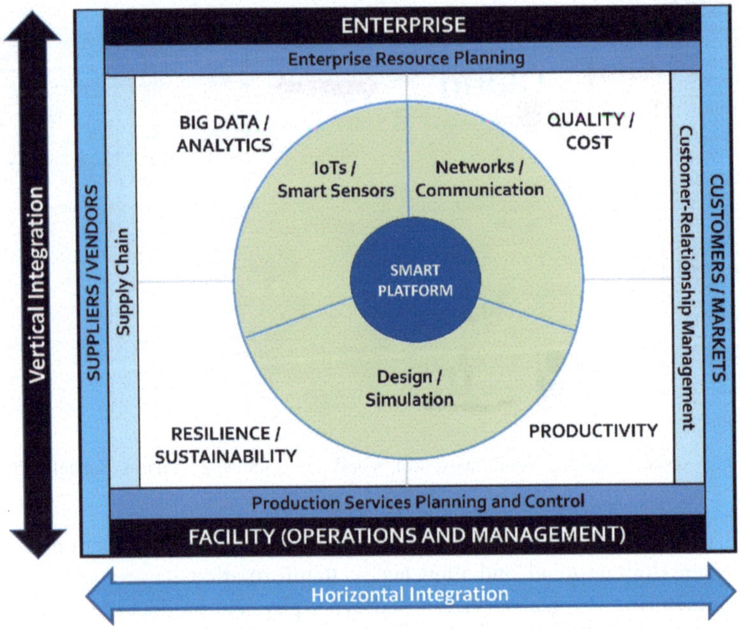

Fig. 7.11 Architecture for realization of digitized manufacturing in Industry 4.0 (Milisavljevic-Syed et al. 2019b)

Fig. 7.12 Elements of a networked engineered system

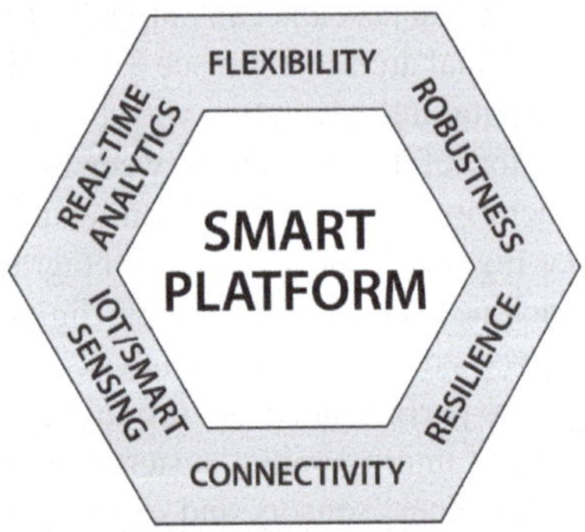

What Are the Requirements for Transition to Industry?

This monograph represents a first step in the realization of benefits of Industry 4.0 construct. The selection of appropriate IoT sensors and actuators is necessary for detecting the source and location of faults and accommodating them during the operation of the system. However, this is no different than Fault Detection, Identification, and Accommodation (FDIA) techniques that have been in use for several decades. The connection between low-level controls and 'enterprise level decision making' that is possible through IoTs and internet connectivity in the DFDM framework enables the analysis of high-level decision making on the performance of the manufacturing system. Further, Big Data Analytics can be used to mine large amounts of data collected from sensors and devices across the enterprise to anticipate the effect of disruptions on the performance of the NES and implement proactive strategies instead of hitherto reactive responses to disruptive events. The integration of Customer Relations Management (CRM) and Supply Chain functions with ERP systems can help the organization be responsive to customer demands and market changes while insulating it from disruptions in the supply chain. Ultimately, it helps the manufacturers develop and produce high-quality products that are tuned to customer preferences in a timely and cost-efficient manner. It also helps the manufacturers to gauge changing customer preferences and develop innovative designs, design and manufacturing methodologies, and novel products for the market.

In this monograph, a framework DFDM for designing smart platforms that are central to one instantiation of the Industry 4.0 construct is proposed, see Fig. 7.13. Incorporation of flexibility, robustness, resilience, and smart sensing into the design is analyzed using computational models and the performance is demonstrated using illustrative examples. However, the issues in connecting different elements of the system and performing analytics on large volumes of data is an open question. While the concept of a smart platform is explored from the perspective of a manufacturing system, in reality, The Industry 4.0 framework can be applied to a whole range of industries, both in the manufacturing as well as service sectors.

Implementation of Industry 4.0

Implementation of the Industry 4.0 construct has the potential to unify the relationships between manufacturing, customers and suppliers. Several industries are leveraging IoTs, Data Analytics, Cloud Computing, Artificial Intelligence, Robotics and

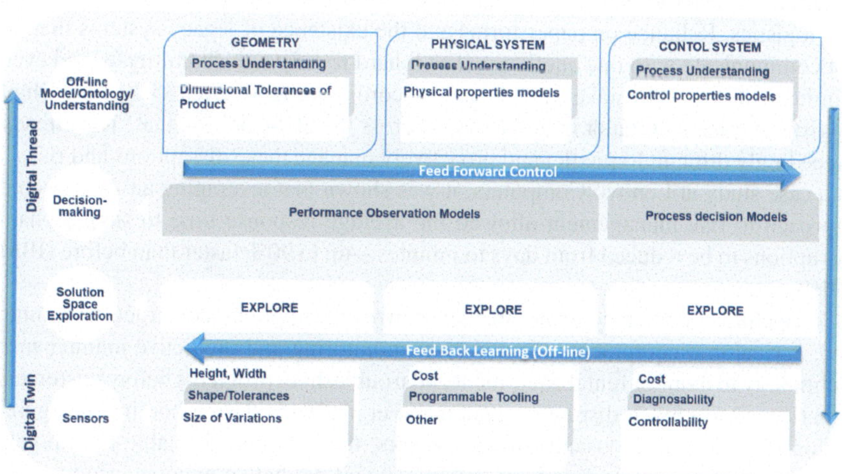

Fig. 7.13 Decision-based design platform for manufacturing and services

Automation to transform their global business processes and services. Manufacturers are embodying four key characteristics of modern manufacturing, namely, Creativity, Smartness, Value, and Humans in the Process to develop new cost-effective manufacturing processes that are adaptive and agile; increase value to stakeholders; improve efficiency and overall cost; offload repetitive jobs from human operators and leverage human creativity (Dassault Systems 2019). Fujitsu, one the world leaders in automation, has put forward the concept of a 'Hyper Connected Business' with tight value-network integration within the production line and with suppliers via de-centralized control of production processes. The ultimate goal in adopting Hyper Connectivity is to achieve individual production down to lot-size one at costs close to mass production (Fujitsu 2019).

Joby Aviation, a Santa Cruz, California-based company is using 3-D printing and additive manufacturing to design the world's first "flying car." The parts are first designed and tested in a virtual environment to lower design and testing costs and reduce the time to market. The prototypes for specific parts are then produced through rapid prototyping tools such as 3-D printers. "3-D printing gives Joby designers the freedom to feed a highly iterative loop that constantly improves their designs, enabling them to build structures that tightly couple the simulation during the concept phase and empower the designer to optimize for weight." (Dassault Systems 2019).

Similarly, Swedish truck and bus manufacturer Scania AB has transformed its business to quickly deliver highly configured and customized products. Using the concept of the 'Digital Twin', Scania was able to "enable full traceability and digital continuity from design through manufacturing" and produce fully customized trucks for its users in 8 weeks' time at a cost comparable to that of a standard truck.

IBM is developing services that enable manufacturers to connect erstwhile siloed data and use analytics to automate decision-making, optimize supply chains, and improve work environment and worker safety (IBM 2019). The global manufacturing industry is greatly dependent on manufacturing supply chain and related transport and logistics. Reliance on paper forms and the existence of legacy systems that do not communicate with one another makes it hard to help reduce, mitigate, and even eliminate, transport mistakes and delays. According to IBM, over 85% of the Chief Logistics Officers of major manufacturers across the globe believe that "it is already exceedingly difficult to predict and proactively manage these disruptions and risks". In a case study at Lenovo Computers, it was shown that leveraging an AI-powered approach to risk management allowed the average response time to supply chain disruptions to be reduced from days to minutes—up to 90% faster than before (IBM 2019).

Zeppelin, a German company that specializes in supplying construction, mining and, agricultural equipment, adopted asset monitoring and predictive maintenance technology to monitor rental equipment and troubleshoot problems before customers experienced too much disruption (Rüßmann et al. 2015). Using continuous aggregation of data from thousands of IoT sensors, the company was able to remotely monitor all of its rental equipment and implement predictive maintenance to avoid costly shutdown of the equipment. This resulted in better customer service, reduced equipment downtime, and an overall improvement in system efficiency.

Remote chiller plant is used to provide cooling for large commercial customers. In order to improve operational efficiency of these plants, Trane, the world's largest air conditioning company uses hundreds of IoT sensors to monitor the performance of chiller plants, implement forecast-based disaster risk management, and improve operational efficiency. By implementing processes that provided end-to-end visibility into the operations, Trane was able to reduce reporting times for outages from 2 to 3 hours to real-time.

Implementation of the Industry 4.0 construct in the United States, Europe, and Asia is already at a fast pace. It is anticipated that over the next five years, the implementation of the Industry 4.0 construct will result in increased productivity across all German manufacturing sectors by €90 billion to €150 billion (Splunk 2019). Overall productivity gains of 5–8% are expected while automotive manufacturers could see increases between 10 and 20%. Demand for new products and services are also expected to add €30 billion a year, or roughly 1% of Germany's GDP over the same period. Employment in Germany is also expected to increase by 6% over the next decade. Data driven decision making is one of the benefits of Industry 4.0 and it is expected to reduce machine downtime by 30–50% and increases machine life by 20–40% (Magic Software 2019). In addition to this, manufacturers expect to see increased output, increased error detection rates (up to 90%), improved product quality, and lower energy consumption. In the following paragraphs we will discuss the impact of Industry 4.0 on one facet of the economy, namely, Health Care.

Impact of Industry 4.0 on Health Care

Industry 4.0 can be instrumental in addressing many societal challenges such as equal access to resources, energy, urban migration and demographic changes. Industry 4.0 is fueled by rapid proliferation of internet technologies, mobile connectivity, and big data analytics. The effect of this has been felt in many areas other than just manufacturing, but none more so than in Health Care (Washington 2018; Deutsche Bank Research 2015).

Health Care worldwide is often plagued by lack of adequate resources and diagnostic capabilities. The problem is further compounded by poverty, poor hygienic and sanitary conditions, lack of transportation and critical infrastructure, inadequate/infrequent power and water supply. Technological advances in biology and computer science are having profound effects on healthcare worldwide. The development of low cost, portable diagnostic devices coupled with mobile networks has opened a whole new industry in telemedicine. Fundamental discoveries in the human genome and immune system has resulted in new approaches to early diagnosis and treatment of various illness. Wearables, digital apps and m-health programs are helping in early detection, real-time patient tracking and treatment adherence among patients. New biometric devices will also help connect remote patients with their physicians and improve the quality and timeliness of care. Over time, data analytics on large volumes of data collected from patients will help identify disease mechanisms and lead to development of new diagnostic methods and response methods to control the spread of communicable diseases (Ernst and Young 2017).

Today, medical practice is evolving at a rapid pace due to advances in medicine and technology. Conventional medicine and its empirical approach to treatment is gradually giving way to a preventive and ultra-personalized approach to medicine (Guedim 2017). Google is the world leader in Data Analytics and with its ability to collect and cross-check data from millions of individuals, Google can easily monitor the health of its customers and is working on developing early diagnostic procedures. Multiple Google services, such as Google Cloud, Google Fit and Search are already making their presence felt in healthcare and fitness. Not to be left behind, Apple has competing services through the iPhone market. Through its Health apps, Apple is addressing the digitization of health records including prescriptions, and the collection of individual's health and diet data such as body temperature, heart rate, activity levels, and alcohol consumption. As a result, the future of medicine could very likely be dictated by large tech companies like Apple and Google.

HealthCare 4.0 (Monteiro et al.* 2019*)

The evolution of component technologies spurred the growth in implementations of the Industry 4.0 construct. With growth came complexity and the need for a framework for implementation of Industry 4.0. Similarly, the rapid proliferation of wearable biometric devices, computer and mobile networks, and digitization of medical records has led to the implementation of complex and unmanageable healthcare solutions. The difficulty in providing quality health care in a timely and cost-effective manner has led researchers into investigating Frameworks for the efficient implementations of healthcare solutions influenced by the Industry 4.0 construct. HealthCare 4.0 is a tactical deployment and a managerial model for healthcare inspired by the Industry 4.0 construct (Monteiro et al. 2019). HealthCare 4.0 framework allows the development of digital models for virtualization of healthcare solutions in real-time for patients, workers, and both formal and informal caretakers. Such models can be personalized for each patient and create cyber physical systems encompassing cloud computing, extension of IoTs termed as Internet of Everything (IoE) including appliances, services, people, and mobile communication networks. HealthCare 4.0 allows for theragnostics in personalized and precision medicine.

7.4.1 What Is the Way Forward?

The following are some of the key issues to be addressed as industries move to implement the Industry 4.0 construct. These include:

1. the integration of digital twin and digital thread;
2. the integration of models and simulation tools spanning processes and length scales (the different domains in axiomatic design);
3. the definition of computational workflows that support decision making, and span multiple activities and users;
4. the definition of modular, reusable sub-workflows for specific processes;

5. the design of a cyber-social decision network;
6. the implementation of reliable and stable connections to external databases on materials, products, processes, and customer surveys;
7. the development of knowledge-based assistance mechanisms to enable different types of users to participate in design-related decision making;
8. the measures for security of data and privacy control (cyber security);
9. the capability to explore the design and solution space for implementation of systems that conform to Industry 4.0 construct; and
10. the robust implementation of Industry 4.0 construct through dynamic and cost-efficient reconfiguration of manufacturing processes during their operation.

In addition, the development of a skilled workforce to run the manufacturing enterprises of the future must be addressed. While Industry 4.0 will usher in economic growth and profitability, there will be loss of blue-collar jobs to the process of automation. If unaddressed, this could result in enormous social strife and discontent.

We identify three steps as a way forward:

- Step 1: Knowledge-Based Decision-Based Design Ontology,
- Step 2: Decision-Based Design Platform, and
- Step 3: Integration of Cyber-Physical-Social Systems in the Platform.

The issues involved in the implementation of these steps, the gaps in knowledge pertaining to these issues, and some suggestions on addressing these gaps are discussed next.

7.4.2 Knowledge-Based Decision-Based Design Ontology

Dynamic requirements of a global market are forcing enterprises to pay closer attention to the design process underlying the process development. Developing a process to translate a set of requirements that addresses current market needs and accommodates impending changes enhances the agility of the enterprise to respond to dynamic market changes. It is efficacious to capture and reuse the knowledge embedded in decisions made during the execution of a design process. Processes built on Ontology can be used to model engineering knowledge for sharing and reuse.

The goal is to learn how knowledge can be captured and reused in the design of networked engineering systems. NES, multistage manufacturing systems, have both characteristics of the mechanical and control systems that are typically designed separately. In order to exchange information between different domains we propose a taxonomy for both domains, interfacing the domain ontologies, and converting mechanical engineering into an electrical engineering analogy in order to make effective and efficient decisions that can be utilized in the design of various multistage engineered systems. In future work, we consider multidisciplinary knowledge exchange between mechanical, industrial, and control engineering in order to improve the design for dynamic management of NES.

The expected outcomes are the scientific and educational foundations of multidisciplinary knowledge exchange in the design of NES that will transform the way systems, product-service systems are designed and form a critical component of an enterprise's intellectual capital. These activities are expected to contribute to managing design processes among different disciplines by providing an ontology for capturing and reusing process-related knowledge associated with decision-based workflows using decision support problems (Ming et al. 2018).

7.4.3 Decision-Based Design Platform

We propose a digital platform to facilitate the automation, reuse, and integration of knowledge between users from different domains. In this section, we identify the gaps and address possible strategies for constructing such a platform.

What are the Identified Gaps?

We identify three gaps that will be further addressed as the Step 2 of the way forward:

1. Decision-based design templates for integrating consumer and producer preferences in system design, product and service system design;
2. Managing dynamic performance of any system with changes in the requirements; and
3. Decision-based design decision-network for managing dynamic performance of any system with changes in the requirements.

What is the Digital Platform Construct?

We anticipate that the decision-based design digital platform construct will consist of:

1. Decision Support Problem constructs;
2. Decision Templates; and
3. Knowledge Management through Ontology.

7.4.4 Integration of Cyber-Physical-Social Systems in the Platform

Integration of Cyber-Physical-Social Systems into the digital platform is necessary to foster societal and technological innovations and reach symbiotic and sustainability design. Further, this gives innovators a chance for collaborative design in large-scale (social) networks where communication and collaboration data can be analysed to provide new insights into Design Thinking, and help develop opportunities for Social Network Analysis and Big Data Analytics.

In this section we present the gaps we identified that are relevant to integration of cyber-physical-social systems in the platform. At this point we do not have enough information to predict the structure of the CPSS in the platform.

What are the Identified Gaps?

We identify three gaps are identified in this monograph that will be further addressed as the Step 3 of a way forward:

1. Allow multiple configurations of elements from different systems in sustainable design of systems, product-service systems;
2. Reach a sustainability plateau by integrating cyber-physical-social systems in design of systems, product-service systems; and
3. Cyber-social design of a decision network in order to be sensitive to dynamic changes of consumer's and producer's preferences in design of systems, product-service systems.

7.4.5 Evolution of Industry 4.0 and Its Impact on Factories and Societies

Digitization that is powering the transformation of the industries is disrupting our world and shaping the challenges that engineers in the workforce and newly graduated engineers will need to address in their professional careers. While a 'Digitized Workplace' has its own challenges, it is possible to adjust to these changes through a combination of new processes and training. 'Digitally Transformed Enterprises (DTEs)' on the other hand are constantly evolving and changing with technological advances. To successfully and sustainably manage change in DTEs, we suggest that people in the workforce need to continually hone non-technical, career-sustaining competencies. DTEs must foster an atmosphere of inclusiveness and innovation with special attention given to the management of workforce. It is not possible or economical to hire people with new skill sets, especially when the required skills are evolving at a rapid pace. Organizations must recruit and train people who are adaptive and able to acquire learn new skills as technology evolves. It is in this context that we pose the question:

a. What are the non-technical, career-sustaining competencies for success of people in enterprises characterized by the forces released by the adoption of Industry 4.0?

We suggest that the following non-technical competencies are necessary for generative learning that is foundational to innovation in a DTE:

1. *The ability to learn, unlearn and relearn through reflection on doing and the associated creation and articulation of knowledge.*

2. *The ability to speculate, articulate a vision and to identify gaps in state of the art or state of practice that may lead to innovation.* One cannot innovate without being able to identify gaps that are foundational to generative learning.
3. *The ability to ask questions, actively listen, reflect and identify opportunities worthy of further investigation.* Questions embody gaps in a person's knowledge and the answers may lead to generative learning.
4. *The ability to make decisions to move forward using incomplete information.* Innovation involves risk. Generative learning has utility only when a decision is made to move forward.
5. *The ability to assess and think critically (deductive reasoning and inductive speculation) and identify a way forward.* The ability to think critically is foundational to moving forward on a technical project and in identifying what one needs to learn next. For learning to start a person must know what he/she does not know and is motivated to find out. There are two foci, one that is external to the individual and the other that is internal. When focusing internally the individual through critical introspection identifies what competency he/she needs to develop further and why, and then identifies a way forward. The external focus takes the form of diagnosing what needs to be fixed in a system (product), the root cause, and then a way forward.

b. What role does an organization play in fostering a climate for continuous learning and adaptation for all its employees?

Organizations play a vital role in enabling their employees to continuously 'upskill' their talents. Peter Senge, an American System Scientist and the founder of the Society for Organized Learning, postulates that *Learning Organizations* are those organizations where "people continually expand their capacity to create the results they truly desire, where new and expansive patterns of thinking are nurtured, where collective aspiration is set free, and where people are continually learning to see the whole together." (Senge 2006). The basic rationale for such organizations is that in situations of rapid change only those that are flexible, adaptive and productive will survive. For this to happen, it is argued, organizations need to 'discover how to tap people's commitment and capacity to learn at all levels.'

Peter Senge identifies five basic disciplines or components of a learning organization: (1) systems thinking; (2) personal mastery; (3) mental models; (4) shared vision; and (5) team learning. People need structures and systems that are conducive to learning, reflection, and engagement. These disciplines can be approached through (i) Principles, that is, guiding ideas and insights; (ii) Practice, that is, application of Principles; or (iii) Essences, that is, achieving a high level of mastery in the selected discipline. The four traits of such an organization are: constant readiness, continuous planning, improvised implementation, and action learning. These four traits define a learning organization and are indicative of how they adapt to change.

Learning organizations are by nature localized as business responsibility lies with every contributing member of the organization. Contrary to Traditional Organizations where the organizational structure is hierarchical and power flows vertically upward with decisions being concentrated in a few individuals, learning organizations seek to empower their members while retaining coordination and decentralized control. Therefore, Leaders in a learning organization have to develop mechanisms to incorporate the five disciplines and create shared leadership models that will communicate shared vision, empower employees, and enable them to adapt to changes.

c. What is the responsibility of the Organization towards Society?

The digitization associated with the Industry 4.0 construct has resulted in dramatic transformation in the society. While the influence of digitization on engineering the 'factories of the future' was expected, the influence on the society has been extremely profound! The desire to manufacture products of 'lot size of one' at a cost associated with mass manufacturing has brought the customer closer to the design of the product. This customer-centric focus of manufacturing has revolutionized the way we think about manufacturing. In the past decade, we have transitioned from "manufacturers shaping customer behavior" to "customers influencing manufacturer's behavior"! Further, customers have come to expect responsive and timely service. While the customer's perception is that manufacturing is global and integrated, the reality for the manufacturers is that the customer base is fragmented and distributed. This has led to mismatch between the expectations of the manufacturers and customers. To survive in this new era, industries must anticipate customer preferences and produce quality products at competitive prices. Companies must embrace digitization and implement end-to-end visibility of their processes in order to maximize efficiency and control costs. These changes also require overhauling conventional methods of design and implementation of these factories, tighter integration with supply and distribution chains, and analytics driven decision making. These changes also require a workforce with new and complex skills that cannot be achieved through retraining of existing workforce. Such changes are likely to cause significant upheaval in the society.

One of the criticisms of Senge's approach is the perception of disconnect between 'Learning Organizations' and 'Learning Society' (Leadbeater 2000). While organizations previously focused on profitability and sustainability, the rapid growth of manufacturing power houses in the 1970's and 80's contributed to increased focus on profits and the creation of 'Shareholder Value'! As a result, one of the leading drivers for organizations became maximizing the 'earnings per share' for the shareholders of the company. The relentless focus on profitability had the downside of disregarding employee well-being and the impact of organizational decisions on society and environment. In recent years, there has been renewed focus on the societal responsibilities of corporations in addition to their responsibilities to employees and stockholders. 'Corporate Social Responsibility' has become an increasingly important part of the business environment and has altered the relationship of businesses with the state and society. Leading factors for this change have been the globalization of trade, the increased size and influence of companies, demand of people world-over for a slice of prosperity, and the impact of corporate brand and public perception on stock valuations.

Corporate Social Responsibility (CSR)

There is a growing realization that the stake holders of companies are not just the shareholders and employees and that the corporation is a part of the wider society (Chandler 2016). Therefore, increasing attention is been given to the concept of Corporate Social Responsibility (CSR), defined in terms of the responsiveness of businesses to stakeholders' legal, ethical, social and environmental expectations. According to the United Nations, CSR is a management concept used to help companies create processes that integrate social, environmental, and economic concerns into their values and operations in a transparent and accountable manner (UNIDO 2019). CSR encompasses issues as diverse as corporate governance, supply chain management, environmental pollution, corruption, employee pay and compensation. To achieve favorable results, such as a high-quality talent pool and positive affective, cognitive and behavioral employee outcomes, companies have to develop and implement consistent policies and procedures for (a) embeddedness of CSR values, and (b) levels of internal CSR (Carlini et al. 2019). Crane et al. (2019) state that Corporate Accountability and Corporate Citizenship are both important elements of CSR. CSR implementations must include processes to (i) manage business ethics, (ii) safeguard rights and responsibilities of stakeholders, (iii) safeguard employee well-being, (iv) safeguard customer rights, (v) maintain ethical supply chain and industrial ecosystems, and (vi) comply with governmental laws and regulations. Schwartz and Carroll (2003) propose a three-domain approach that is centered on social, environmental, and economic components for the sustainability of organizations.

Growth in industries and trade over the past three centuries has had the effect of reducing poverty and increasing the prosperity of nations. However, this prosperity has come at the expense of several nations across the globe and there remain considerable differences between and within regions, countries and societies (UNIDO 2013). Even in affluent societies, inequities exist among racial minorities and women. Another consequence of the industrial growth is the environmental impact and the issues of effective waste management and proper control of soil, water and air pollution. Given that the current industrial society is not fully inclusive or properly sustainable, the Member States of UNIDO, at their General Conference in Peru in December 2013, adopted the Lima Declaration in which they agreed that inclusive and sustainable industrial development must become an important part of the world's long-term development agenda.

The member nations agreed that the following precepts are necessary to implement inclusive and sustainable industrial development (UNIDO 2013):

- Every country achieves a higher level of industrialization, and benefits from the globalization of markets for industrial goods and services.
- No one is left behind in benefiting from industrial growth, and prosperity is shared among women and men in all countries.
- Broader economic and social growth is supported within an environmentally sustainable framework.
- New approaches are introduced to harness and share available knowledge, technology and innovation.

– All of the above are supported through multi-stakeholder partnerships.

Role of the Government in Regulating CSR

Small and Medium Enterprises (SMEs) constitute more than 90% of all businesses that exist around the globe and are found at the core of an economy (UNIDO 2007). While it is crucial that these enterprises integrate socially and environmentally responsible concepts into their business strategies, it is unreasonable to expect that these small companies drive this change. Multinational companies (MNCs) with huge annual revenues can (and do) exert enormous pressure on their suppliers and local governments. Traditionally, MNCs embraced social responsibility and their approach has been primarily geared towards philanthropy and charitable contributions. SME's to a large extent, ignored or were hostile towards CSR. While philanthropic contributions such as those by Warren Buffet, Bill Gates Foundation, Rockefeller and Carnegie Foundations, have had an enormous beneficial impact on the society, these activities were out of the scope of a company's core business, with the purpose of enhancing its reputation or mitigating negative impacts caused by the operation. CSR on the other hand, encourages businesses to do what they have always been doing, but in a way that benefits the business, and its stakeholders alike. Given the impact on the local and global economy, it is necessary for national and local governments to take the lead in regulating both MNCs and SMEs and ensuring that equitable growth and prosperity is achieved in an environmentally sustainable manner.

There are several companies that recognized the need for CSR and developed processes for implementing both philanthropy and socially responsible policies. Conscious that the task of social progress in any country, cannot be undertaken by the Government alone, J. R. D. Tata—the Chairman of the Tata Group from 1938 to 1991, believed that, "to create good working conditions, to pay the best wages to its employees and provide decent housing to its employees are not enough for the industry, the aim of an industry should be to discharge its overall social responsibilities to the community and the society at large, where industry is located." (Srivastava et al. 2012). In general, about 66% of the profits of Tata Group was spent on charity (VBCSD 2018). The TATA group is using this systematic plough back of wealth to the society to help establish and finance numerous quality research, educational and cultural institutes in India.

In its attempt to build on the success of the TATA group, the Government of India launched a CSR initiative in 2013 mandating that all enterprises above a certain size, both public or private, spend 2% of their profits on corporate social responsibility projects. However, in 2017–18, only a little over half of those required to spend on CSR have filed reports on their activity to the government. The other half either did not comply or simply failed to file. Even those that did, spent less than the minimum stipulated amount.

The need for government participation, however, in fostering sustainable and equitable growth has not been lost. United Nations Industrial Development Organization held an Expert Group Meeting on "Corporate Social Responsibility and Public Policy: The role of Governments in facilitating the uptake of CSR among SMEs in developing countries." (UNIDO 2007). The following questions were framed to highlight the importance of international competitiveness as an important driver of public sector engagement in CSR.

- What role does CSR play in strengthening national economic competitiveness?
- Could CSR policies help developing countries, given that the competitive advantage of many is primarily based on low costs?
- How might CSR be included in the design of competitiveness strategies and policies of national and local governments?
- Can we expect from businesses located in developing countries to adhere to social and environmental standards?

Ethics and Morality of Organizations in the Era of Industry 4.0

Globally connected businesses form the backbone on an interconnected economy. The rapid proliferation of internet technologies brings with it the risk of privacy and security. Companies in their rush to be responsive to the needs of their customers should not sacrifice security and ensure that every aspect of their organization from enterprise level functions such as logistics, operational planning, and supply chains to low-level manufacturing controls are secure. As more and more system functions are migrated into the cloud, cybersecurity takes on an even greater importance.

In recent years, internet-enabled technologies have increased the ability of companies to acquire large amounts of data and mine it for information. This has led to the realization that data is valuable and provides both monetary and competitive benefits. This has incentivized several unscrupulous companies to trade the personal data of their employees and customers. In other instances, lack of attention to data privacy has led to serious breaches with serious ramifications both to customers as well as the economy. In this age of Industry 4.0, it is necessary for organizations to reflect on their role and responsibility to the society at large. It is in this context that we pose the question:

a. What constitutes ethical and moral code of conduct for organizations in the era of Industry 4.0?

Morality deals with what is right or acceptable from an individual or societal point of view in relation to norms, values and beliefs embedded in social processes. Ethics is the codification of morally acceptable courses of actions (Crane et al. 2019). While morality is subjective and may differ from society to society, companies operating in the global landscape must be respectful and attentive to the values and morals of the societies they operate in. Organizations must be founded on Integrity, Objectivity, Professional Competence, Confidentiality, and Professional Behavior and these ethical values must manifest themselves in every facet of the organization.

Closing Remarks

Technological changes that are sweeping across the industrial landscape will have a transformative effect on the manufacturing industry and societies across the world. While these transformations will likely result in prosperity and wealth, it will also be a period of enormous turbulence in the industry. Critical to the survival of the industries during this period is their ability to adapt and change. In the words of Charles Darwin, "*It is not the strongest of the species that survive, nor the most intelligent but ones most responsive to change.*"

At What Price Change?

While it is undeniable that industries and companies will be the generators of wealth and progress on a global scale, it is imperative that there be a discussion on where companies fit within the social fabric. Responsible governance of these companies encompass issues such as corporate governance, supply chain management, environmental pollution, corruption, employee pay and compensation. Fundamental questions that arise are:

- Why does a business exist?
- What does profit represent?
- What obligations do businesses have to the societies in which they operate?
- Are these obligations voluntary or should they be mandated by law?
- To whom are the companies accountable?

Similarly, society has to address ways to harness natural resources in a sustainable manner and meet the aspirations of its population. Specifically, one has to address:

- Effective and optimal uses of natural resources, materials, products; reduction of waste;
- Promoting Public-Private Partnerships to enhance productivity and effective utilization of available resources;
- Maximizing the value of investment, reducing risks, and motivating innovation;
- Building human capital in the era of Industry 4.0; and
- Fostering leadership and coordination across all levels of government.

In closing, while there are several technical challenges that are posed by transformative technologies like those spurred on by Industry 4.0, the impact of these transformations transcends the industries where they took root. The digital transformation of enterprises can cause upheavals in the workforce, societies and nations. For these changes to bring lasting peace and sustained prosperity to everyone, it is necessary that we recognize these changes and learn to continuously adapt to them. We should also build the educational and societal infrastructure that will help people and organizations continuously adapt to these changes. Finally, we need to develop policies that promote ethical and socially responsible ways to harness Industry 4.0 in a sustainable manner while safeguarding the environment. We repeat Charles Darwin's prescient statement "It is not the strongest of the species that survive, nor the most intelligent but ones most responsive to change." Recognizing that there is a transformation coming as a result of Industry 4.0, let us ponder on how to build on these changes for a just, equitable, and prosperous world for everyone!

Glossary

AI Artificial Intelligence

CAD Computer-Aided Design

cDSP Compromise Decision Support Problem

Cloud Computing Cloud computing—virtualization of software and hardware resources allowing for networked resources to be accessed as a service in a ubiquitous way and on the basis of pay-as-you-go pricing Schaefer (2017)

Cloud-based Design and Manufacturing (CBDM) A cyber-physical integration and control of manufacturing machines with CAD, CAE, ERP, and MES systems across one enterprise as a vertical integration. CBDM is a precursor for extending IoT and IoS Schaefer (2017)

Computational Complexity Mathematical model of high complexity that requires extensive computational resources to instantiate

CRM Customer Relations Management

DFDM Design for Dynamic Management

Digital Platform Provides decision support for engineers/designers, collaboration between different users from different domains and trains them how to understand the impacts of design decisions in order to speed up the design process and facilitate the creation of quality cost-effective designs

Digital Thread Refers to the communication framework that allows a connected data flow and integrated view of the asset's data throughout its lifecycle across traditionally siloed functional perspectives

Digital Twin Refers to computerized companions of physical assets that can be used for various purposes. Digital twins use data from sensors installed on physical objects to represent their near real-time status, working condition or position

DOM Dynamic Operability Model

ERP Enterprise Resource Planning

FDIA Fault Detection, Identification, and Accommodation

Low-Level Controls Refer to control of machines and processes based on desired output and actual output. Low-Level Controls are characterized by real-time operation and take corrective actions based on real-time measurements (data). In contrast, High-Level Controls act on information and abstractions of data and deal with system level functionality as opposed to actual outputs

Health 4.0 A tactical deployment, and managerial model for healthcare inspired by the Industry 4.0

High-Level Decision Making Enterprise level decision making

Industry 4.0 Industry 4.0 is the subset of the fourth industrial revolution that concerns industry. Industry 4.0 describes the trend towards automation and data exchange in manufacturing technologies and processes that include cyber-physical systems (CPS), the internet of things (IoT), industrial internet of things (IIOT), cloud computing, cognitive computing and artificial intelligence (Wikipedia)

Inspection Stations Inspection Machines

Inspection System Control System
IoP Internet of People
IoS Internet of Services
IoT Internet of Things
MMP Multistage Manufacturing Process
NES Networked Engineered Systems
RIS Reconfiguration of the Inspection System
RMS Reconfigurable Manufacturing System
RMT Reconfiguration of Machine Tools
SF Smart Factory
Smart Manufacturing Digitized Manufacturing
SoV Stream of Variation model
SSOM Steady-State Operability Model.

References

Carlini, J., Grace, D., France, C., & Iacono, J. L. (2019). The corporate social responsibility (CSR) employer brand process: Integrative review and comprehensive model. *Journal of Marketing Management, 35*(1–2), 182–205. DOI: 10.1080/0267257X.2019.1569549.

Chandler, D. (2016). *Strategic corporate social responsibility: Sustainable value creation* (5th ed.) p. 480. Los Angeles, CA: SAGE Publications, ISBN 10: 1544351577.

Choi H.-J. (2005). A Robust Design Method for Model and Propagated Uncertainty, Georgia Institute of Technology, Ph.D. dissertation.

Conti, M., Pasarella, A., & Das, S. K. (2017). The Internet of People (IoP): A new wave in pervasive mobile computing. *Pervasive and Mobile Computing, 41*, 1–27.

Crane, A., Matten, D., Glozer, S., & Spence, L. (2019). *Business ethics: Managing corporate citizenship and sustainability in the age of globalization*. Oxford University Press, ISBN: 9780198810070.

Dassault Systems. (2019). Digitalization and continuity: Game changing approaches for manufacturers. In *How digitalization is revolutionizing the way manufacturers innovate, produce, sell and service their products*. https://discover.3ds.com/digitalization-and-continuity. Last accessed August 01, 2019.

Deutch Bank Research. (2015). *eHealth: Industry 4.0 can serve as the model for digital healthcare*. http://www.dbresearch.com/PROD/RPS_EN-PROD/PROD0000000000441850/eHealth%3A_Industry_4_0_can_serve_as_the_model_for_d.pdf. Last accessed August 01, 2019.

Ding, Y., Ceglarek, D., & Shi, J. (2000). Modeling and diagnosis of multistage manufacturing processes: Part I: state space model. In *Proceedings of the 2000 Japan/USA Symposium on Flexible Automation*.

Ding, Y., Shi, J., & Ceglarek, D. (2002). Diagnosability analysis of multi-station manufacturing processes. *Journal of Dynamic Systems, Measurement, and Control, 124*(1), 1–13.

Ernst and Young. (2017). *Life sciences 4.0: Transforming health care in India*. https://www.ey.com/Publication/vwLUAssets/ey-transforming-health-care-in-india/$FILE/ey-transforming-health-care-in-india.pdf. Last accessed August 01, 2019.

Fujitsu. (2019). https://www.fujitsu.com/global/solutions/industry/manufacturing/industry/. Last accessed August 01, 2019.

Georgakis, C., Uztürk, D., Subramanian, S., & Vinson, D. R. (2003). On the operability of continuous processes. *Control Engineering Practice, 11*(8), 859–869.

Guedim, Z. (2017). *Apple vs Google: Health Care in Industry 4.0. Edgy 2017*. https://edgy.app/google-vs-apple-healthcare-industry-4-0. Last accessed August 01, 2019.

IBM. (2019). *How smart is your factory?* https://www.ibm.com/industries/manufacturing. Last accessed August 01, 2019.

Intelligroup. (2019). *Industry 4.0: What's next?* https://us.itelligencegroup.com/whitepaper-sap-industry-4.0. Last accessed on August 01, 2019.

Leadbeater, C. (2000). *Living on thin air: The new economy*. p. 256, London: Penguinbooks. ISBN: 10: 0670876690.

Magic Software. (2019). *The executive guide to data-driven decision making in Industry 4.0*. https://lg.magicsoftware.com/the-executive-guide-to-data-driven-decision-making-in-industry-4.0. Last accessed August 01, 2019.

Milisavljevic, J. (2015). Accounting for uncertainty in the realization of multistage manufacturing processes. School of Aerospace and Mechanical Engineering, Norman, Oklahoma, The University of Oklahoma, MS thesis.

Milisavljevic, J., Commuri, S., Allen, J. K., & Mistree, F. (2017). Concurrent design exploration method (CDEM) of networked engineering systems. In *ASME International Design Engineering Technical Conferences*. Cleveland, Ohio: ASME.

Milisavljevic-Syed, J. (2018). *Architecting Networked Engineering Systems*. Doctoral Dissertation, The School of Aerospace and Mechanical Engineering, University of Oklahoma, Norman, Oklahoma.

Milisavljevic-Syed, J., Allen, J. K., Commuri, S., & Mistree, F. (2018). Steady-State operability in design for dynamic management in realization of networked engineering systems. In *ASME International Design Engineering Technical Conferences, Paper number DETC2018-85864*. Quebec City, Canada: ASME.

Milisavljevic-Syed, J., Commuri, S., Allen, J. K., & Mistree, F. (2019a). A method for the concurrent design and analysis of networked manufacturing systems. *Engineering Optimization, 51*(4), 699–717.

Milisavljevic-Syed, J., Commuri, S., Allen, J. K., & Mistree, F. (2019b). Concurrent design exploration method for realizing networked manufacturing systems for Industry 4.0. In *52nd CIRP Conference, PROC-D-18-00262*.

Ming, Z., Nellippallil, A. B., Yan, Y., Wang, G., Goh, C. H., Allen, J., & Mistree, F. (2018). PDSIDES—A knowledge-based platform for decision support in the design of engineering systems. *Journal of Computing and Information Science in Engineering, 18*(4), 041001–041014.

Mistree, F., Allen, J. K., & Mistree, D. F. A. (2019). *Non-technical, career-sustaining competencies for a workforce of a digitally transforming manufacturing enterprise, 2019*. Future Steel Forum, Budapest, Hungary, September 25–26, 2019.

Mistree, F., Smith, W., & Bras, B. (1993). *A decision-based approach to concurrent design. Concurrent engineering* (pp. 127–158). Berlin: Springer.

Monteiro, A. C. B., França, R. B., Estrela, V., Iano, Y., Khelassi, A., & Razmjooy, N. (2019). *Health 4.0 as an application of Industry 4.0 in healthcare services and management* (Vol. 4, pp 262–276). https://doi.org/10.26415/2572-004x-vol2iss1.

Rüßmann, M., Lorenz, M., Gerbert, P., Waldner, M., Justus, J., Engel, P., & Harnisch, M. (2015). Industry 4.0: The future of productivity and growth in manufacturing industries. http://www.inovasyon.org/pdf/bcg.perspectives_Industry.4.0_2015.pdf. Last accessed on August 01, 2019.

Schaefer, D. (2017). *Industry 4.0—A holistic perspective*. Future Steel Forum 2017, Warsaw, Poland, June 14–15, 2017.

Seepersad, C. C., Pedersen, K., Emblesvåg, J., Bailey, R. R., Allen, J. K., & Mistree, F. (2006). The validation square: How does one validate design methods? In W. Chen, K. Lewis & L. Schmidt (Eds.), *Decision-based design: Making effective decisions in product and systems design* (pp. 305–326). New York, NY: ASME Press.

Senge, P. M. (2006). *The fifth discipline: The art and practice of the learning organization*. p. 445. Broadway Business, New York, NY: Doubleday Publishing. ISBN 10: 9780385517256.

Shi, J., & Zhou, S. (2009). Quality control and improvement for multistage systems: A survey. *IIE Transactions, 41*(9), 744–753.

Smith, W. F., Milisavljevic, J., Sabeghi, M., Allen, J. K., Mistree, F. (2014). Accounting for uncertainty and complexity in the realization of engineered systems. In *International Conference on Complex Systems Design and Management*, Paris, France, November 12–14, 2014. Paper number 55.

Smith, W. F., Milisavljevic, J., Sabeghi, M., Allen, J. K., & Mistree, F. (2015). The realization of engineered systems with considerations of complexity. In *ASME 2015 International Design Engineering Technical Conferences and Computers and Information in Engineering Conference*, American Society of Mechanical Engineers.

Splunk. (2019). https://www.splunk.com/en_us/form/7-examples-of-the-incredible-value-of-industrial-data.html. Last accessed August 01, 2019.

Srivastava, A. K., Negi, G., Mishra, V., & Pandey, S. (2012). Corporate social responsibility: A case study of tata group. *IOSR Journal of Business and Management* (IOSRJBM). *3*(5), 17–27. ISSN: 2278-487X.

UNIDO. (2007). *Corporate social responsibility and public policy: The role of governments in facilitating the uptake of csr among smes in developing countries*. Expert Group Meeting, Vienna, Austria, November 2007. pp. 20–21. https://www.unido.org/sites/default/files/2008-06/Discussion_Paper_0.pdf

UNIDO. (2013). *Introduction to united nations industrial development organization – inclusive and sustainable industrial development*. https://www.unido.org/sites/default/files/files/2017-11/DG_Brochure_February_2015_Web_0_0.pdf

UNIDO. (2019). What is CSR? in Competitive Trade Capacities and Corporate Responsibility, United Nations Industrial Development Organization. https://www.unido.org/our-focus/advancing-economic-competitiveness/competitive-trade-capacities-and-corporate-responsibility/corporate-social-responsibility-market-integration/what-csr

VBCSD. (2018). *Corporate social responsibility—The TATA way*. http://en.vbcsd.vn/detail.asp?id=179. Last accessed September 16, 2019.

Wang, G., Shang, X., Yan, Y., Allen, J. K., & Mistree, F. (2018). A tree-based decision method for the configuration design of reconfigurable machine tools. *Journal of Manufacturing Systems*, *49*, 143–162.

Washington, L. (2018). *Transforming health in the fourth industrial revolution*. Diplomatic Courier, April 2018. https://www.diplomaticourier.com/posts/transforming-health-in-the-fourth-industrial-revolution. Last accessed August 01, 2019.

Index

J. Milisavljevic-Syed et al., *Architecting Networked Engineered Systems*,
https://doi.org/10.1007/978-3-030-38610-8

The manufacturer's authorised representative in the EU is Springer Nature Customer Service Centre GmbH, Europaplatz 3, 69115 Heidelberg, Germany. If you have any concerns regarding our products, please contact ProductSafety@springernature.com

Printed and bound by CPI Group (UK) Ltd, Croydon, CR0 4YY
22/09/2026
02221060-0014